PROJECT COST CONTROL
in Construction

Second Edition

Roy Pilcher

OXFORD
BLACKWELL SCIENTIFIC PUBLICATIONS
LONDON EDINBURGH BOSTON
MELBOURNE PARIS BERLIN VIENNA

© Roy Pilcher 1985, 1994

Blackwell Scientific Publications
Editorial Offices:
Osney Mead, Oxford OX2 0EL
25 John Street, London WC1N 2BL
23 Ainslie Place, Edinburgh EH3 6AJ
238 Main Street, Cambridge,
 Massachusetts 02142, USA
54 University Street, Carlton,
 Victoria 3053, Australia

Other Editorial Offices:
Librairie Arnette SA
1, rue de Lille
75007 Paris
France

Blackwell Wissenschafts-Verlag GmbH
Düsseldorfer Str. 38
D-10707 Berlin
Germany

Blackwell MZV
Feldgasse 13
A-1238 Wien
Austria

First Edition published by Collins
 Professional and Technical Books 1985
Second Edition published by Blackwell
 Scientific Publications 1994

Set by DP Photosetting, Aylesbury, Bucks
Printed and bound in Great Britain by
Hartnolls Ltd, Bodmin, Cornwall

DISTRIBUTORS

Marston Book Services Ltd
PO Box 87
Oxford OX2 0DT
(*Orders:* Tel: 0865 791155
     Fax: 0865 791927
     Telex: 837515)

USA
Blackwell Scientific Publications, Inc.
238 Main Street
Cambridge, MA 02142
(*Orders:* Tel: 800 759-6102
       617 876 7000)

Canada
Oxford University Press
70 Wynford Drive
Don Mills
Ontario M3C 1J9
(*Orders*: Tel: (416) 441-2941)

Australia
Blackwell Scientific Publications Pty Ltd
54 University Street
Carlton, Victoria 3053
(*Orders:* Tel: 03 347-5552)

British Library
Cataloguing in Publication Data

A catalogue record for this book is available
from the British Library

ISBN 0–632–03637–0

Library of Congress
Cataloging in Publication Data

Pilcher, Roy.
  Project cost control in construction/Roy
Pilcher.—2nd ed.
    p. cm.
  Includes bibliographical references and
index.
  ISBN 0–632–03637–0
  1. Construction industry—Cost control.
2. Building—Cost control.
3. Construction industry—Management.
I. Title.
TH435.P55 1994
690′.0681—dc20   93-46809
                                      CIP

Project Cost Control
in Construction

Contents

Preface

This is the second edition of a book which first appeared in 1985. The objectives of the text have not been reset. Its purpose is to assist students and practitioners of construction management – in the widest sense – to be made aware of and understand the concepts behind the many aspects of cost which impinge on the process of bringing a construction project to a successful conclusion from its first beginnings as an idea in someone's mind. Almost all the decisions which a construction manager makes will have an influence on some or other component of cost in a project. It is necessary, therefore, to understand the implications of such decisions on other, sometimes remote, aspects of the organization in which the project is intended to fit.

It is worth repeating from the previous preface that this is not intended to be a cost control manual in which systems are detailed for ready installation and operation. Systems need to be adapted and fitted to the particular needs of each organization and the people who operate in it. What the text sets out to do is to explain such principles as there are underlying cost control in the context of the wider influences of organizations so that systems design can be undertaken. This book will therefore be of use to undergraduate and post-graduate students, and practitioners alike.

No specific reference has been made in the text to the use of computers for the purposes of accounting and cost control. There is, of course, a wide variety of such programs available. However, it is considered that such reference, while being immediately useful, soon becomes out of date; the underlying principles are the most important aspect of the subject. Software companies need to produce updated versions of their products at regular intervals to take account of changes in both hardware and current practices and the technical press deals thoroughly with such changes as an on-going practice.

The major addition to the text since the previous edition is Chapter 2 which deals with financial control. At the end of the day, the financial state of companies is judged by their overall financial performance. To all except the privileged, the financial aspects of a company have to be judged by its published accounts. There is much information in these accounts that will enable an outsider to review the progress of a company and, of course, from the inside, the very preparation of the accounts throws up a myriad of management accounting information. Performance on a company's individual projects must inevitably be reflected in these accounts and hence on the financial performance of the company itself. It was therefore thought to be desirable to offer an introduction to these matters and Chapter 2 has been added as a result.

The general instructional sequence of the book starts with Chapter 1 which provides a brief overview of management activity in relation to finance and cost. This chapter has been realigned in a number of ways and some material has been added.

Chapter 2 is referred to above and is entirely new to the text. An introduction deals with some of the legal aspects of the formation of a limited liability company and how they affect its structure and reporting functions. The question of how finance is raised and what are the various sources is described and then two of the most important financial reports required of a company – the profit and loss account and the balance sheet – are described. This leads on to other resulting accounts such as the appropriation account and the funds flow statement. Finally, some measures of performance are discussed in describing the use of ratio analysis. As far as is possible this chapter is written in the context of construction. Some typical problems are set out at the end of the chapter and the suggested answers are set out at the end of the book.

Chapters 3 and 4 deal with the basic mechanics of investment appraisal. This chapter is little changed except for the updating of examples and the addition of some work on the calculation of a company's required rate of return for its investment projects. Additional work on the external rate of return method as a means of dealing with multiple yields is included in Chapter 4, as is additional work on the influence of tax on cash flows and hence the calculation of rates of return. The shorthand method of setting out the relevant equations and the pictorial display of cash flow diagrams have been included to bring the text into line with current teaching and presentation methods.

Chapter 5 deals with the questions of risk and uncertainty. This is an important chapter in the context of finance and cost. It remains largely as it did in the first edition though some of the examples in the text have been changed for the purpose of making the explanations more readily understandable. The main change in the subsequent chapter, *Chapter 6 – Management Information and Control Systems,* concerns some additional comments on data collection. This is an area of increasing importance and will become more so in the future. So much depends on the accurate, ready and cheap means of collecting information. Reference has been made to *barcoding* and *voice recognition* as two methods which may be developed for use in this field in the future.

A section has been added to Chapter 7 to illustrate the use of contract costing and some of the factors that have to be taken into account when deciding on the profit to be taken on a particular construction project. In addition, it illustrates the generation of some of the amounts that need to be transferred to the balance sheet in the reporting process. Chapters 8 and 9 take a similar form to those in the previous edition. The changes that have been made tend not to be major but rather refer to the examples that have been used. Some have been changed; some have been updated.

In the first edition, *Further Reading* was placed at the end of each chapter. In a number of cases this led to repetition where more than one chapter dealt with related topics. In this edition the reading list has been updated and collected under subject headings rather than chapters in the hope that this proves a more useful method of dealing with a necessary topic.

Finally to my acknowledgements. No person writes a book of this nature from entirely original material though frequently they bring to the work a fresh view or a modest contribution in interpretation and the like. Often the contribution is the bringing together of a collection of relevant material on a specialist subject. Everything from conversations to abstruse research papers tend to influence what one writes and, of course, many techniques are standard throughout professions. To all of those who have contributed to what is written here, whether wittingly or unwittingly, I wish to offer my gratitude and thanks.

Chapter 1
Background

1.1 Introduction

With the inevitable increase in size and complexity of construction projects, managers of projects, whether acting on behalf of an owner, a designer or a contractor, are faced with the need to have a greater understanding of all the relevant and associated economic aspects of their work. They must be able to implement a high degree of control, not only in respect of the day-to-day progress, cost and quality of the work, but also over the influences arising from the total external environment in which the project is being constructed. This book is directed principally towards the many aspects of analysing and controlling cost as part of a project manager's responsibilities.

At the conceptual and feasibility stages of a construction project, a project manager needs to understand, and to apply, the basic principles of investment and business finance in the context of the potential owner's current financial situation. (The term *project manager* is used in this book in the general sense of applying to the person who has responsibility for managing the whole or some major part of a total project. Where it is used more specifically, this will be noted at the time.) A project manager needs to understand the different sources and forms of project funding that are available and the implications of using each appropriate alternative. The important relationship between time and money must also be understood since, inevitably, there will be a delay between the investment of capital in a construction project and the anticipated returns or benefits which will flow back as a result of the utilization of the constructed facility. A project manager, therefore, at this initial stage, also needs to be able to gauge the risk of possible variability from the estimated costs for a project and the way in which the magnitude of the likely returns or benefits arising from it may vary from those on which the investment decision was based.

At the conceptual and detailed design stages a project manager needs to be familiar both with the design method and the construction processes that are proposed for use and to know how to make the best economic use of each of these technical aspects of the project. The technologies must be selected by taking into account total estimated capital costs for each alternative under consideration, not only during the design phase but also when determining

1

the construction method and process. In addition, they must be taken into account when considering the whole of the facility's future useful life. It is important at this stage, therefore, that the opportunity is taken to create a design that is likely to reduce the life costs associated with the facility whilst taking due cognizance of the owner's requirements.

In addition to the aspects of cost, time and quality, a project manager who will have a responsibility for ordering the design and construction of a project needs to be familiar with the variety of contractual relationships which can exist between an owner, the professional consultants that are engaged to work on the project in order to formulate, design, supervise and manage the creation of the facility, and the contractors/suppliers who may be engaged to supply and/or construct it. There needs to be a clear under-standing of the implicit apportioning of risks between the various parties to the contract which will almost certainly vary between the various forms of contractual relationship that may be adopted. It is important that the selected form of contractual relationship offers the parties the best oppor-tunity of completing a project within the required timescale, at not more than the cost which has been estimated in the feasibility studies, to the quality that will ensure adequate and proper serviceability of the facility during its life and with a distribution of risk in accordance with the requirement, under-standing and expertise of each of the parties.

It is necessary for a project manager to be able to monitor and to control the costs throughout a project. He must appreciate the consequences of varying the planned course of construction and the steps that need to be taken in order to bring a project back onto course in the event of such deviation. He needs to be aware of the benefits and penalties of accelerating a programme in terms of cost and other resources. A project manager needs to understand all the possible influences of such deviation on the well-being of each organization that is associated with the project.

When a project is completed and the facility commences its useful life, a manager concerned with its operation needs to have a general appreciation of the value of the asset for which he is responsible and the basis of the initial and subsequent financial investment. Thus, he will be aware of the level of efficiency, and thus profitability, with which the facility has to be operated in order to become a successful investment.

In addition to the above important responsibilities which will be under-taken by project managers during the construction of a facility are many other economic influences which may fall outside the immediate scope of the activities of many construction project managers. Such influences as infla-tion, taxation, legislation, etc., will all affect the economic outcome of a project both during its construction and when it commences production on completion. An awareness rather than a great depth of understanding of such matters is nevertheless important within the broad scope of project management.

1.2 Technological economics

Economic laws underlie the day-to-day functioning of successful business and commerce and must influence the determination of policies which will lead to the achievement of the objectives of a company in an economic fashion. All scientific and technological personnel should understand the general principles which are the basis of these laws if they are to play their part in the effective management of a commercial undertaking. In particular, many economic laws will affect the decisions that have to be made within a single construction project at a level where greater detail of performance must be considered than that of the overall company or group organization.

Within recent years it has become essential that scientists and technologists become involved in the economic evaluation of projects, many of which will involve the utilization of large amounts of resource in their creation. Previously there was a tendency for economics to be seen as a study based largely upon history and politics; science and technology was viewed as something quite different and apart. Engineering design and construction can no longer take place in an environment in which even the more abstract concepts of economic theory are entirely ignored. Technological economics now forms an essential expertise for engineers and managers and formal training in this area is now commonplace in technological and business courses.

Increasingly, for some time now, those managers who are concerned with decision-making have turned to mathematics, with its clarity of logic, in order to be able to evaluate alternative solutions to the particular problem at hand. Scientists and engineers have long since made decisions using mathematics as one of their tools and it is no longer uncommon for economists to turn to mathematics in order to explain, clarify and quantify some of their theories. It has often been said that there are no such problems as economic problems, engineering problems, mathematical problems or other subject-dominated problems, but rather that they are all just problems as seen through the eyes of economists, engineers, mathematicians, etc. The use of operation research is a process which encompasses the expertise of different disciplines in the arts, sciences and technology in order to assess problems and, through collaboration, to produce a solution(s) as a result of its multi-disciplinary approach.

The general principles which govern economic behaviour, as with social and technical behaviour, can be stated in the form of laws. An applied scientist, however, needs to be careful about the interpretation of such laws since economic behaviour, with its many influencing though inexact variables, is not always reflected by a precise calculation that is usually associated with the solution of many engineering problems. A technologist, with his analytical background and an affinity for numerical interpretation, is usually relatively satisfied with an answer so obtained and often accepts it without too much question as to the limitations of its validity. In contrast, the laws of economics tend to take the form of generalizations and cannot be used to predict future performance with absolute certainty – largely because of the unpredictability of human behaviour.

1.3 The management process

Before technological economics can be applied to the activities of a company, or of a lower level in the organizational hierarchy of an enterprise, it is necessary to look at the management process of decision-making as a whole in order to understand both the general details of the process and the sequence in which the elements of it will normally be undertaken. Figure 1.1 is a diagrammatic view of the management process. It shows, in broad outline, the general sequence of activities that need to take place within an enterprise and how they interrelate one with another. It can be elaborated in greater detail at each stage of the process. This process may be interpreted on a very wide basis for the whole of a company or in relation to a particular aspect of the company's operations. In this instance it is interpreted in the light of the application of technological economics to a company's operations.

Having determined the enterprise's objectives, it will be possible to produce a broad plan for the future, perhaps for a period of approximately five years, in which the details of the nature and extent of the company's operations will be laid down, the resources of labour, equipment and capital that will be made available to support the operations will be established and, as the third aspect of the process, there will be a schedule of the major capital

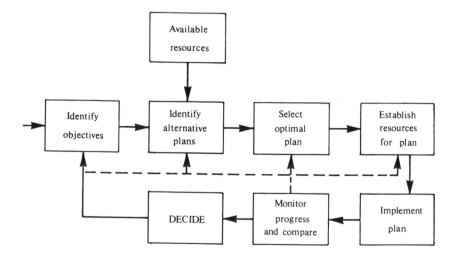

Fig. 1.1   The management process.

investments that the company anticipates making for this period of the plan. In some cases there may be a commitment to research and development work and other service aspects of the company's operations and, therefore, the investment in resources related to this work would also need to be included.

From the outline plan, the way in which the resources that are to be made available can be employed so as to achieve the objectives, can be established. Usually, there will be a number of different ways in which this can be done, each being feasible in itself and forming a plan that will become one of a series of mutually exclusive programmes. From this series of plans can be chosen the optimal solution, that, in view of the senior management of the company, will be the best arrangement in order to achieve the enterprise's objectives.

The chosen plan will form the basis for the control of the enterprise's operations. A budget is simply a statement, broken down into varying degrees of detail depending on the purpose for which it is to be used, of the estimated phasing of the resource use and requirement over the time period concerned. A budget is most frequently expressed in financial terms, but it is sometimes important that it can be seen in the terms of other resources such as labour and equipment. The budget statement must also necessarily include details of the expenditure that is anticipated under each head of the plan together with the income that is likely to accrue from it. This budget then allows the actual expenditure and income, as well as the use of other resources, to be compared with that which was anticipated at the time of the planning. A comparison of the actual and estimated expenditure and income will enable the variances between the two to be established and, if an appropriate method of control is used, the reasons for such variances can quickly be identified. They could, for example, be because of a low rate of production, for increased costs that were not foreseen, for major changes in the environment of the work and so on. It is important that the reasons are identified so that when it comes to the revision of a plan or to the establishment of another plan for a further period, they can then be taken into account. If at this stage it is seen that what is actually happening does not accord with the objectives that were set at the outset, then there will need to be some revision either to the original budgets or to the way in which they are being effected.

Whilst the management process is depicted by a diagram which appears to indicate possible cycles of activities, it is rarely, in practice, so clearly defined. The process is one of iteration and constant repetition of steps in the light of experience in order to make the necessary adjustments and the revisions. The period of time for which the plan is prepared will be a strong influence on the degree to which there will be iteration through the process and to which there will need to be revisions of the original budget. It is very much a matter of experience as to the choice of the length of period for the plan and, because of vastly differing conditions, it will almost certainly vary from one industry to another. In the construction industry, for example, it may be that a long-term budget, which might include such matters as major investments in buildings and plant, would be prepared for a period of five years ahead. The nature of the construction industry, and its frequent use as a regulator in national economies, means that five years can be a very long time. What might have been thought at one time to be good markets into which to venture might not exist five years after the preparation of the plan. Capital construction programmes, such as, for instance, those for motorways, have a habit of being slowed down or speeded up to suit current economic conditions and hence the future market for those projects tends to be uncertain. Competition tends to vary a great deal with newcomers seeking a project in the market and others falling out of it.

After establishing a long-term budget, it is then necessary to produce a budget for a shorter period, such as one year. In the yearly budget there will be the detail of an enterprise's production and turnover and the income that is expected from the resources that will also be detailed on a month-by-month basis. It will show the working capital that is necessary to finance the work and it will indicate the influence of major capital investments during that time period. One step further down the hierarchy of budgeting will be the weekly or the monthly budgets which will be used by managers in order to monitor and control the actual day-to-day operations of the enterprise at the operating level. The actual progress towards the achievement of the budget will be established and the reasons for the variances will be assigned. These will then result in adjustments to the other levels of planning if necessary.

1.4 Management control and decision-making

In the construction industry there is an increasing awareness of the need for projects of all sizes to be completed on time, within budget and to the required quality. Often, delays in placing an order for the work and for giving permission to proceed, do nothing except accentuate the urgency for completion of the work. Emphasis then needs to be placed on stricter control of progress, cost and quality. The control function tends to look at the past and the present rather than, as planning, the future.

The achievement of an enterprise's objectives and plans is all important and it is the managerial function of control that ensures, by the measurement and correction of performance of subordinates, that these objectives and plans are accomplished. There are, therefore, two elements of control that are of considerable importance. The first is that of measurement and the second is that of action which must be initiated if the measurements show some variance from those that were expected. It is clear that the measurements will only be meaningful if they can be compared with that which was expected.

Management by exception is a desirable aim for achievement. It results in a measure of freedom for subordinates to manage without close supervision so long as the performance of their particular responsibilities does not fall outside certain prescribed limits. Every manager does not need to be subjected to the same degree of control and even if, in some cases, that degree of control is strong, it must still allow certain freedom of action. It must, however, be recognized that the degree of control required for individual managers will inevitably vary from one manager to another and as between different levels of manager within an organization.

There are four basic principles of control:

(a) to ensure attainment of the enterprise objectives by the establishment of variances from the plan soon enough to enable corrective action to be taken;
(b) the cost of implementing a control system must certainly be less than the cost savings achieved by its implementation;
(c) managers, within their areas of responsibility, must be given the authority to put their plans into effect in order to achieve the objectives set for them;
(d) a system of control must allow direct action on the first indication (feedback) that an activity is off course from the predetermined plan so that it can be brought back onto the appropriate path.

There is little point in having a system of control which allows any facet of an organization's activities to end in failure or disaster before the control system allows corrective action to be taken.

The process of controlling can be divided into five phases:

(a) the preparation of a plan that will achieve the objectives of the work;
(b) the recording of the plan, preferably formally, in terms of the inputs to or the outputs from the system or both;
(c) the definition of the quantities and the organization of the resources that will be necessary for the conversion of inputs to outputs;
(d) the use of feedback in order to compare what is happening in practice with that which was planned;
(e) the evaluation of variances arising from the comparison in the previous phase leading to decisions as to whether corrective action is required and whether a change in resource allocation is necessary.

All of these five phases are essential in the management control function, though the test of effective management control takes place in phase (e). In construction project control there are a number of devices and techniques that can be used to assist in the control process. Probably the most important of these is the schedule – the plan of the sequence of activities that go to make up the project aligned to a timescale. The control of cost, for example, will inevitably need to be linked closely to the time schedule.

It is necessary not to be over-influenced by the concept that management control can be effected entirely through the control devices such as scheduling. It has long been accepted that there are many human aspects and personal traits of managers and the managed that need to be considered in the management information and control functions. Amongst them are the nature and design of the organization structure, the political, sociological and economic factors of the environment, the experience of the available manpower, the technology and the scale of the operations involved.

Decision-making and controlling form by far the largest and most important part of a manager's role. Before it can be decided what information is needed or required by each manager, it is necessary to analyse the types and structural levels of the decision that must be made. At the outset, these can probably be divided into two broad categories. The first consists of those decisions that are thought to be made by hunch, intuition, experience or flair. These are not logically structured, or necessarily made as a result of a logical problem-solving analysis, but rather rely on a distillation of experience, personal relationships and referring back to previous like-situations, whether consciously or subconsciously. In such instances the most important information supporting the decision may well not be quantifiable.

The second category contains the balance of the decisions – those for which the information required can be readily quantified leading to numerical and/or logical assessment of the available alternative courses of action and a ready comparison between them.

With these two broad categories that identify the nature of managerial decisions, a hierarchy structure applying to both can be postulated. At the lowest level will be operating decisions, following up through the hierarchy with tactical and then strategic decisions.

An operating decision can by typified by its simplicity, its relative lack of uncertainty, its almost immediate effects, its limited influence in its own right on tactical and strategic matters in an enterprise, the relative ease of definition of the decision processes and the data required to support it, and its immediate concern with the resources that will be affected by its implementation. Examples of operating decisions in construction works are the decision to supplement one excavator filling six trucks by another truck so as to improve the rate of excavation and removing the spoil thus avoiding the excavator continually waiting for a spoil truck; the decision to order another load of cement; a decision to make routine tests on samples of material delivered to a site for incorporation in the works.

A tactical decision is characterized by a moderate degree of complexity in that the situation to which it refers may need examination and alternative courses of action will need to be evaluated. There may be some uncertainty about the future course of events and, in general, the time horizon of the decision will be longer than for an operating decision; months rather than hours. Tactical decisions, as with operating decisions, will normally be made on the basis of data which are generated within the organization, though they will probably require a longer and more thorough investigation before a decision is made. Examples of tactical decisions in construction project management are the determination of a staff organization to manage a construction project; the determination of resource levels to complete an important section of a specific project within cost and time; the selection of one from a number of feasible alternative methods for the construction of an important facility.

Strategic decisions are quite distinctly of a different order from the other two. They are characterized by concerning a total enterprise, business or company unit and are about the management policy. Policy decisions at this level require data from outside the organization, are likely to have a longer time horizon than other types of decision – years rather than months, are more uncertain and hence more risky, are more complex and require an original and creative approach for their successful conclusion. It is unlikely that strategic decisions will be repetitions of previous decisions and due to their complexity and importance will almost certainly take an extended period for their formulation.

1.5 The investment decision

Though individuals, companies and countries tend to look after their affairs along the broad and general lines of economic laws, there are many situations in which specific techniques may be introduced to effect a quantitative assessment of problems. One such situation is that in which a decision to invest is to be made. This is almost certainly a situation where the economic problem necessitates a choice between alternative competing investments.

It should be borne in mind that managerial decision-making is not solely a process of applying a quantitative technique to the solution of a problem. It demands, in addition to the ability to use the technique expertly in a technical sense, the application of sound judgement concerning the qualitative aspects of the environment in which the decision is to be made. This judgement may be a blend of experience, commonsense, hunch (hopefully to a lesser extent) and, probably, a knowledge of human behaviour. Techniques, in this sense, are used as an aid to decision-making rather than a means by which a decision can be reached.

An investment may be defined as putting resource – usually money – into a scheme in the anticipation of getting a larger amount of resource in future returns. An investment decision is almost certainly the most important decision which is made within a company and it is usually made at the senior executive level. It is very rarely that the broad picture of the company's activities can be seen and that information from every aspect of the company's operations is available at any other organization level. This is not to say that demands for the investment of capital will not be initiated at other executive levels. There is usually keen competition between departmental managers for capital which is available for investment – so much so that there is clearly a need for an overall and unbiased supervision of such decisions.

An investment decision is important for a number of reasons. Firstly, it is important to the economic well-being of a company that investments chosen for implementation are profitable. The capital must be invested in projects that will result in a larger net return of income at some time in the future unless pursuing a loss-leader. If a company is profitable, then it will not find too much difficulty in raising capital by way of equity and/or borrowing, so long as its profitability is in some way comparable to other businesses having a similar risk to capital invested in them. Profitability will also provide a company with surplus funds to plough back in capital for investment, after having rewarded its owners who have supplied the capital (shareholders) and paid tax to the Government. Not only is it important that a company is profitable for its own sake, and for the sake of its owners, but it is important to the national economy as a whole. A Government of any country would find little pleasure in an excessively large number of unprofitable companies in that country. Not only is there the question of the tax, but in addition, for a sound economy scarce resources must be put to the best possible and the most efficient use. It is often necessary for a Government to adjust the tax system of a country to ensure that this is so.

Profitability comes to a company which is economically and technically efficient, rather than a company which is restricted to technical efficiency alone. The latter, though very desirable, can be no criterion of success unless supported by considerable research into the selection of the product, the market at which it is examined, the blend of resources which are required for production and hence the establishment of a minimum product cost to be compared with the price at which the product will sell. It is this pre-investment examination of the likely results of capital expenditures which is vital to the economic efficiency of a business. Poor selection of projects can only lead to poor profitability, or loss, in the long run. The management of a company is so frequently judged, in the main, against the yardstick of the profitability of that company.

Another reason why an investment decision is important is that investments are made on the basis of the long term. They are very often irreversible decisions to buy assets such as buildings, plant and equipment which, if not used for the purpose for which they are purchased, often have very little alternative ready use and low resale value. Selecting a project which will ultimately yield less than a required minimum profit may result in a drain on profitable areas within a company's operations for the whole life of the project. This will result in less funds being available for future investment and possibly the commencement of a downward spiral towards general unprofitability. Alternatively, a company might be forced to make the unpalatable decision to cut its losses completely and dispose of a project's assets before the end of their useful life. This is not an enviable position to be in and it is frequently a decision which requires considerable courage to make.

The efficiency of business management is so often judged on the basis of the profitability or economic efficiency of an enterprise. It can be appreciated, therefore, that if an increasing amount of capital is invested in a company, the profits must also increase at least proportionately to maintain the profitability when measured as a ratio of profit to capital invested. The management of a company, therefore, must protect its own position by ensuring that a thorough and critical appraisal is made of all investment proposals of any consequence. Active steps must be taken to ensure that by investing money in one particular proposal a company is not forced to forego the opportunity of investment in a better, more profitable one. It follows that the greatest care must be taken to ensure that every possible alternative is examined.

1.6 Why control construction cost?

The decision to invest capital in building and civil engineering structures and all types of plant and equipment is taken on the basis of an estimate of capital cost together with the need for the associated working capital (for raw materials, labour, work-in-progress, product stocks, etc.) and sundry other items such as future replacement, repair and maintenance costs. These investment costs are expected to be returned over the life of the asset plus an additional margin by way of return for the investment. It is important, therefore, that the capital cost of the facilities is controlled during design, construction and installation so that the basis of the investment appraisal can be sustained during this phase of the creation of capital assets. The estimated capital cost becomes a standard against which accumulating costs of design and construction are referenced during subsequent action. Once capital is invested in physical assets almost nothing can be done to reduce it and its effect on the economic appraisal of the investment. It is important that needless capital investment is therefore avoided and that control rather than monitoring of expenditure is undertaken during both design and construction.

Should the capital cost of a facility escalate during design and/or construction beyond that level where the asset being provided would cease to be financially viable, work can be stopped. At the design stage, with probably almost no direct financial commitment to physical construction, the relative penalty costs will not be too severe. However, if construction is underway, with a major part of the facility already constructed, the choices between alternatives at this stage are not attractive. On the one hand there is a part-finished facility probably incapable of performing any useful function and difficult to dispose of; on the other, the prospect of a final total capital investment in excess of that originally envisaged and a project that is no longer economically acceptable.

A construction contractor will have objectives for his own cost control which will differ from those of an owner's project management. Essentially, a contractor is controlling against his anticipated margin on the work for the well-being of his own organization. In doing so he needs to be in a position to pinpoint inefficient and uneconomic areas of his operations with a view to their investigation and correction. However, through these cost control procedures and the information flows between himself and an owner's project management, a contractor will become the first source of warning for possible variations in the total cost of the works which will affect the subsequent return on the investment project. It is important, therefore, to recognize that all the parties to an investment project involving construction activity have a role to play in controlling the construction costs.

1.7 Risk in construction

Construction is often considered to be a high risk industry. The risks not only stem from financial activities but also from physical activity in the field. Increasing attention is being paid to risk as an element in construction business and methods of assessing risk of all kinds exist and are being developed and extended as time passes and understanding improves. Many construction activities are very much concerned with the future. All estimates of cost at the design and tendering stages, for example, involve a degree of uncertainty because they almost always have to be prepared on incomplete information and the actual outcome cannot be predicted accurately.

At the outset of a construction project – before any design or construction has commenced – an owner generally has the opportunity to retain or distribute the responsibility and the risk associated with the project as is thought fit. This is carried out when making the choice of contractual arrangement between the parties to the contract. However, it must be borne in mind that whichever party is assigned the risk in this process, that party will require to be compensated for it. Alternatively, if the extent of the assigned risk, or its implications, are not clearly identified when the relevant party prepares an estimate of the costs involved in carrying out the work, problems may occur through the relevant party underperforming on the work, suffering financially and/or in the extreme becoming bankrupt. When this happens the owner invariably incurs greater cost than perhaps was foreseen.

Each type of contractual arrangement will involve a different apportionment of risk. For example, between a client and a contractor, a *cost reimbursable* contractual arrangement should result in the minimum apportionment of risk for a contractor and the maximum for an owner. A *design and build* contractual arrangement between a client and a contractor will represent the other end of the spectrum wherein a contractor is required to assume a wide variety of speculative risks, thus relieving the owner of much risk in this area.

Clearly risk analysis is becoming increasingly important as an aspect of construction to be investigated at all stages of a project. With new approaches and differing views becoming available, risk analysis will become, if it is not already, a standard acceptable aspect of project management's work. The 'what if' analysis can be applied in many relevant situations and gives managers an excellent insight into the possibilities and probabilities of a wide range of outcomes of investment and construction activity. However, it is far better to undertake such studies as early as possible in the investment and preferably before any of the parties to a project are committed to the expenditure of hard money.

Chapter 2
Financial Control

2.1 Introduction

The principal aim of this chapter is to introduce a reader to the subject of the financial control of a construction business. It is not intended to instruct readers in the preparation of company accounts since this is best left to accountants. It is intended to suggest ways in which a non-financial construction manager can put the question of the control of project cost into the context of a company's financial structure and how they can begin to interpret a company's principle financial statements for this purpose. The accounting requirements for different organizational structures and the fundamental accounting ideas and financial statements will be explained briefly. It may seem to be jumping to the end of the whole control process in examining financial control at this stage. However, it is necessary to view the objective of the control process before looking at the detail. There is considerable emphasis today on what is known as the *bottom line*. This expression summarizes all that has gone before in the financial process. It sums up the outcome over a period. The net result of the progress of a company or organization over a period, by which its endeavours can be judged, is stated in the bottom line.

The nature and management style of business activity of a particular industry, within the general framework of the statutory requirements, will influence the way of recording financial transactions and the preparation and presentation of financial statements for use in financial control. All sizes of business and many activities are undertaken within the construction industry, including building, civil engineering, property development, general contracting and mechanical and electrical services installations. Even taking this into account, financial control methods are likely to be broadly similar across the industry. Differences in the detail of control methods at the project level, however, may result from differences in operational activity between various sectors of the industry. How a company is administered is a factor that influences the recording and presentation of financial information. For example, some companies work through a centralized management system resulting in many administrative functions being undertaken at a centralized head office; others adopt a decentralized system with such functions being undertaken at dispersed regional offices or at single large work sites.

A business can take one of several organizational forms. The type of financial information that must be generated and to whom it must be made available is determined by the nature of its organizational form. The means by which an organization may raise capital and how it must conduct its affairs are also determined by its organizational form. These matters are frequently the subject of legislation. The most common forms of business organization are:

(a) *The sole trader*: a form of business which one person owns and operates for his/her sole benefit.
(b) *The partnership*: must consist of at least two and not more than twenty partners. (There may be more than twenty, however, if the professional rules of the partners in practice are required to adopt this organizational form alone.)
(c) *Limited liability companies*: can be either private or public. The ownership of the business may not necessarily be in the hands of its managers.
(d) *Public sector ownership organizations*: controlled by the Government and accountable to the public which has a bearing on the type of financial information that must be generated and to whom it must be made available. Examples are local authorities, regional health authorities, the British Broadcasting Corporation, British Rail and the Post Office.
(e) *Unincorporated associations*: usually social organizations, such as clubs, run for the benefit of their members who may or may not have the objective of making a profit.

Since the emphasis here is on the nature and principles of financial reporting rather than the legal issues involved in establishing and operating a construction company, the aim is to progress towards the main example of a limited liability company. The general principles can be applied readily for use in each of the organizational forms described above, though the statutory requirements for a specific organizational form may differ in detail.

2.2 Limited liability companies

A limited company may be formed by delivering to the *Registrar of Companies* at Companies House (an executive agency of the Department of Trade and Industry) the following documents with specific fees and stamp duties:

(a) The company's *Memorandum of Association* that defines the constitution of the company. It must set out the name of the proposed company, the country in which its registered office will be established and the amount of the company's share capital and the number of shares of a fixed amount into which it is divided, for example, £100 000 divided into 400 000 shares of 25 pence each. The Memorandum must also include the company's objectives, and a declaration that the members' (shareholders') liability is limited.

(b) The *Articles of Association* that govern the rights of the members and detail the regulations on how the company must conduct its affairs. Examples of the Articles are those that detail the provisions for general meetings, voting at meetings, the payment of dividends, and the increase and reduction of a company's capital.

(c) Statements as to the names and particulars of the directors and secretaries of the company in the first instance, the address of the company's registered office, the assessment of capital duty by the Inland Revenue and a statutory declaration of compliance with the registration requirements.

(d) A document, if applicable, that states the company meets the capital requirement of a public company, i.e. that the share capital stated in (a) above is not less than £50 000.

Once the limited company is formed, it becomes a legal entity and is seen to be separate from its owners (usually the holders of the ordinary shares). As such, the company can be sued in its own name. This may be contrasted with a sole trader who, while the business may bear a title other than that of the owner's name, e.g. *Fancy Pipe Fittings*, it is the owner who bears total responsibility for the business. Thus the organization and its owner are a single entity and would, if necessary, be sued in the name of the owner. It is similar with a partnership. If, for example, the partnership were unable to pay its debts, its partners would be sued individually for payment.

Once the limited company becomes unable to pay its debts it is possible that it must cease trading and go into liquidation. If this happens, the loss to the shareholders is limited to their investment in the company. Thus, the effect of forming a limited liability company is to reduce the risk of shareholders losing undefined amounts of money. Therefore several legal restrictions are placed on limited companies, some of which concern the accounting records of the business. These records are required to be in sufficient detail so that the company's transactions can be examined and the financial position of the company can be determined with reasonable accuracy at any time.

The legal requirements of a limited company have been established, and then amended and added to, in a succession of legislation first dating from the Joint Stock Companies Act 1844. Provisions have been made in this legislation that cover a wide spectrum of subjects including the financial reporting requirements related to limited companies. While these requirements are now complex and go beyond the mere presentation of information, the principal requirements in relation to this text are for a company's directors to produce, at the end of each financial year:

(a) a balance sheet;
(b) a profit and loss account;
(c) an auditor's report;
(d) a directors' report.

These annual accounts and reports must be laid before a company's Annual General Meeting. For this purpose, a copy of the final accounts must be sent to each shareholder of the company, each debenture holder and any other individuals who are entitled to attend the meeting, at least 21 days beforehand. Copies of the documents must also be deposited with the Registrar of Companies. In doing so, they become public documents, a copy of which may be purchased by any interested party. Of the above documents (a) and (b) are of prime interest in financial reporting. The auditor's report results from the examination of the company accounts, the balance sheet and the profit and loss account before they are put before the shareholders. The directors' report comments on the accounts, the activities of the company over the relevant accounting period, significant changes in the company and various other aspects of the company's activities such as policies for employment, training, career development, donations for political or charitable purposes, etc.

2.3 Sources and forms of finance

One essential aspect of planning in all forms of business organization is that of the preparation of a *business plan*. This plan represents a strategy by which a business should achieve its objectives over a given period. It describes the products, the technology, the potential markets and examines their sensitivity to changes in the environment. It goes on to link to the plan the necessary resources – time, money, labour, skills, plant and equipment – that are needed to fulfil it. A business plan is essential for all business organizations. Not only does a business plan provide the strategy for the future to ensure the effective and efficient operation of the business activities, but its formulation ensures that managers of a business think about all aspects of it. A business plan is fundamental to satisfactory financial planning and ultimately the plan will need to be formulated in financial terms. When approved and adopted, the business plan forms a basis for control and comparison. When it is set up, after approval and adoption, consideration of its financing must take place. This may be short-term, such as a bank overdraft, or long-term such as debt finance. A business plan must be designed so that it provides the answers to all the questions likely to be raised by prospective lenders.

The main sources of finance for a business are the following:

(a) banks;
(b) share issues;
(c) internal company finance;
(d) hire purchase and leasing;
(e) factoring.

Long-, medium- and short-term loans are provided by *banks*. Short-term loans are usually referred to as overdrafts and they are the simplest and relatively cheap form of loan. They suffer from the disadvantage that repayment can be required at very short notice. Long-term loans are those for periods greater than ten years. They are usually raised for the purchase of a major asset such as an office block or a large plant installation having a long life. These projects can provide security for a loan. Suitable security reduces the lender's risk of loss arising from a borrower defaulting on the repayments. The method of paying back the loan and the interest on it can vary. Medium-term loans span the time between overdrafts and long-term loans, that is between about three years and ten years. Such loans are usually made for a specific time period and are paid back with the interest in monthly, quarterly or half-yearly instalments to one of a number of acceptable patterns over time.

Sometimes a company will borrow money from people other than shareholders. It can do this by means of a *bond* called a *debenture*, which is a loan secured on the assets of the business. Debentures can be of two main types. Firstly, a *floating debenture* is secured on the circulating assets of a business, e.g. stock and cash which are constantly changing or circulating within the operating cycle of the business. Secondly, a *fixed charge (or mortgage) debenture* is secured on specific assets of the business, e.g. land or buildings. The business may not dispose of those assets without the prior agreement of the holders of the relevant debentures. Debenture holders are not *members* of the company, as are ordinary and preference shareholders, and they are entitled to interest on their loan whether the business is profitable or not. In the event of non-payment of interest, the debenture holders may seize the assets on which their loans are secured and sell them to recoup their money.

A limited liability company can issue shares, either by adding to its already issued equity or by setting up a new company to undertake a specific project. Shares for public purchase need to be attractive to a purchaser. This can be accomplished by offering to a shareholder a reasonable share of any profit made. Such a share of the profit is called a *dividend*. A good prospect of capital appreciation in a share's issue price is also likely to be an inducement for a subscriber. A sole trader or a partnership is unable to issue shares unless it becomes a limited company. Initially the cost of becoming a *public limited company* with the facility of issuing shares to the public is unlikely to be appealing. In this case a *private limited company* can be formed. Such a company cannot seek public investment in its shares though, on the other hand, it is unlikely to have the capital requirements of a public company. Often the shareholders will be members of the family running the business or their friends and acquaintances.

The amount of capital that a limited company may raise is prescribed in the company's 'Memorandum of Association'. It is known as the *nominal* or *authorized capital* because it is expressed in terms of the *par* or *nominal* value of the shares, e.g. 100 000 shares of 25 pence each. A company is required to show the amount of authorized share capital in its balance sheet, even though it may not all have been issued to its shareholders. When the shares are sold for cash or some other consideration, they become part of the *issued capital* of the company and the total cash or credit value received for them becomes the *paid up capital*. The issued capital may be made up of a variety of share types. That part of the issued share capital that gives its holders the right to receive dividends and capital distributions is known as the *equity share capital*.

The most common types of shares are *ordinary shares* and *preference shares*. Ordinary shares are the equity shares of a company and usually carry voting rights at company shareholder meetings. Preference shareholders are entitled to a fixed percentage rate of dividend based on the nominal amount of the share, e.g. the holder of a 7 per cent preference share issued at £100 will have a right of preference to a dividend of £7 per year.

While preference shares have a fixed rate of dividend (if one is authorized for payment) the dividend on an ordinary share may fluctuate from year to year. This is because the ordinary dividend is usually related to the extent of profit made by the company. Preference shares have a preferential right to a share of any profit and their dividend has a prior claim on profits as compared to the equity. In addition, preference shareholders often have a prior claim for the repayment of their capital over equity shareholders if a company fails. Preference shares may be *cumulative* when, in a year of insufficient profits to warrant a dividend, the dividend is carried forward for payment in the next year and so on. In the case of *non-cumulative* preference shares, arrears are not accumulated.

Another source of capital for a business is by using *internal company finance*. This is achieved by the generation of *capital* and/or *revenue reserves*. Capital reserves are profits made outside the normal operations of a business. For cxample, a new share issue may be made with the price asked for the shares being greater than the par value. A 50 pence share may be offered at 65 pence. A purchaser is therefore paying a premium of 15 pence on each share. This premium is placed in a *share premium account* and becomes part of the reserves of a business in the balance sheet. Other reserves can be generated by writing-up the valuations on fixed assets, popularly those of buildings in times when building values are appreciating significantly over time. *Revenue reserves* are created by the directors of a company deciding not to distribute all of the profits made in one year as a dividend. The surplus kept in the business is then available for financing its operations.

Hire purchase is used in many businesses since it allows the possession and use of the hired item in order to generate income in return for regular instalments of money. At the end of the agreed period of the hire there is an option for the hirer to purchase the item. *Leasing* is a similar arrangement for providing short- or medium-term finance. A financial lease, one that operates for the major part of the life of the asset, puts less risk on the *lessor* (the asset's owner). In this case the *lessee* (the user), in addition to paying the leasing fees, will usually be responsible for the service and maintenance of the asset. An operating lease is used when the period of lease is normally shorter than the asset's economic life. The asset in this case may be leased to several lessees over its life and the lessor is then usually responsible for the service and maintenance costs.

Another means by which finance can be raised by a business is *factoring*. It is possible to transfer the debts of a customer for services rendered into the hands of a factoring organization. Such an organization will then bill a customer for the debt and collect the outstanding money, pursuing any bad debts in the process. The business does not then have the need to undertake this process itself. A factoring company will charge the business for undertaking the work, usually up to about three per cent of the debts involved. A business will agree with the factoring organization a date by which the money arising from the debt is to be paid to it. The factoring organization will pass that amount to the business on the date whether it has been collected or not. Thus, the business can rely on the finance being paid by the agreed date and hence the risk of not getting it is reduced.

2.4 Definitions

The following definitions are used in the text:

Assets   The resources of value that can be expressed in monetary terms and are owned by a business. They are anticipated to be of future benefit in the business.

Authorized share capital   The amount of capital up to which a company is allowed to issue.

Balance sheet   A financial snapshot at an instant of time setting out the assets of a business, its liabilities and its capital.

Book value   The value of an asset as shown in the accounts and on the balance sheet of a business.

Budget   A statement of future activity, either intended or desirable, in either quantitative or financial terms, for a specified period of time.

Called-up capital   Where shares are issued only part of the amount that is payable for each may be demanded of the shareholder. This part is referred to as the called-up capital. The balance due at some future time is known as the uncalled capital.

Capital   The amount of money put in to finance a business at its start. When the business subsequently becomes established, capital is the excess of assets over liabilities.

Consolidated balance sheet   Shows the financial state of a parent or holding company together with its subsidiaries.

Creditors   Those persons, businesses, companies, etc. to whom the enterprise owes money.

Current assets   Resources provided by a business for consumption, use or to be sold almost immediately or certainly within approximately 12 months. Examples are cash, short-term investments, debtors (purchasers of goods for which they have not yet paid) and all forms of stock held by the company. Sometimes alternatively called *circulating assets*.

Debenture   A document which creates or acknowledges debt showing the details of the loan and the payment of interest.

Depreciation   The value of the extent to which an asset has worn out, has been consumed or deteriorated as a result of its use, technological advances or obsolescence.

Finance lease   A contract which substantially transfers the benefits of ownership of an asset to a lessee in return for paying instalments.

Fixed assets   Permanent or semi-permanent resources which are bought, not for resale by the business but for use in the business to produce revenue. Examples are buildings, plant and machinery.

Gearing   Also known as *leverage*. The ratio between debt and equity in the financial structure of a company.

Liabilities   What a business owes to others. Can be *long-term* where the business has more than one year to settle, or *current* where the period to settle is less than 12 months. Examples of long term liabilities are loans, mortgages and debentures. Examples of current liabilities include wages and salaries payable, declared dividends, accrued taxes payable, trade creditors for stock and VAT payable.

Liquidity   The ease with which a business can raise cash by selling its assets.

Par value   The face or nominal value of a share, printed on the face of the share certificate. It has no practical significance.

Profit and loss account   An account which summarizes the income and expenses of a business to show the net profit or loss for a given period.

Turnover (sales)   The value of work completed and invoiced sales. In construction this may include the value of construction work executed, the sales of property and housing (usually for which legal completion has taken place) and rents receivable.

2.5 The profit and loss account

Example 2.1

Wilson Probate is the sole owner of a small building company. In the year 19x3, the company had a turnover of £1 725 300 in constructing house extensions and modifications. During the year, Wilson Probate spent £475 000 on wages for labour, £837 600 on materials and £120 000 on sub-contracts.

The annual expenses of the company are as follows:

The company rents it main offices for £4000 a year and the rates amount to £2500. Heating the offices costs £950; electricity costs £750. The telephone bill for 19x3 amounted to £1800. Office supplies such as postage and stationery cost £2760. Contract cleaning for the offices amounted to £6000. Cost of insurance of offices, etc. equals £9000.

Legal, accountancy and audit fees amount to £7450; travel and entertainment expenses cost £1760. Depreciation of cars and the delivery van amount to £6350 and that for equipment to £2460. Salaries for office staff, gross pay plus insurance, etc., £50 000. Interest on loans and overdraft £12 000. Bad debts and provision for bad debts amount to £4600. Repairs and maintenance cost £1250.

Draw up a trading and profit and loss account for Wilson Probate for 19x3.

The profit and loss account for Wilson Probate is set out in Fig. 2.1. Strictly speaking the profit and loss account as shown has two parts. The first, down to and including the *gross profit* is known as the *trading account* and the remainder as the profit and loss account. It is customary to call both parts combined the profit and loss account. The trading account results in the calculation of the gross profit which is the sales of the business minus the *cost* of those sales. The profit and loss account results in a *net* or *operating profit*, which is the *gross profit* less *operating expenses*. In the case of Wilson Probate it is a clear profit before taxes are paid.

A profit and loss account may be compiled for any period of time such as a month, a quarter or a year. It is important that the period to which the account applies is clearly stated in its heading. In allocating costs to periods of time, the *matching convention* is used rather than by allocating the actual cash payments made during the period (see Example 2.2).

WILSON PROBATE

Profit and loss account for year ending 31 December 19x3

	£	£
Sales		1 725 300
Cost of sales:		
Labour	475 000	
Materials	837 600	
Sub-contracts	120 000	1 432 600
Gross profit		292 700
Operating expenses:		
Rent of offices	4 000	
Rates	2 500	
Heating	950	
Electricity	750	
Telephone	1 800	
Office supplies	2 760	
Contract cleaning	6 000	
Insurance	9 000	
Legal, accountancy and audit fees	7 450	
Travel and entertainment	1 760	
Depreciation – cars and van	6 350	
Depreciation – equipment	2 460	
Staff salaries	50 000	
Loan/overdraft interest	12 000	
Provision for bad debts	4 600	
Repairs and maintenance	1 250	113 630
Operating profit		179 070

Fig. 2.1 Wilson Probate profit and loss account for 19x3.

Example 2.2

Wilson Probate pays business rates on his office premises. The local authority's financial year runs from 1 April of one year to 31 March of the next. Rates are required to be paid in two instalments, one at the beginning of April and the other at the beginning of October. The rates for April 19x2 to March 19x3 amount to £2200 and those for April 19x3 to March 19x4 amount to £2600. Wilson Probate will therefore pay in cash in the year 19x3, £1300 on 1 April 19x3 and £1300 on 1 October 19x3, a total of £2600. However, 19x3 includes three months of the previous financial year. The expense must be matched to the actual cost rather than to the cash payments made. The actual cost $= (0.25 \times 2200) + (0.75 \times 2600) = £2500$. This sum will appear in the profit and loss in accordance with the matching convention, which must also be applied to all similar cases.

2.6 The balance sheet

The traditional form of balance sheet is divided vertically into two halves in what is known as the *horizontal format*. On one side of the sheet are set out the assets of the business; on the other side are the means by which the assets have been provided. The assets, such as land, buildings, equipment, motor vehicles and stock are the resources required to run the business. The means by which the assets are provided, such as capital and borrowings, are supplied by the company owner(s) and therefore are liabilities of the business for which the business is liable to the owner(s). The basic accounting equation of the balance sheet is:

ASSETS = CAPITAL + LIABILITIES

This equation must always hold true in the balance sheets for any point in time for all businesses. For example, if the assets are increased by purchasing further equipment, then either the capital and/or the liabilities must also increase or other assets of the business must decrease in value by a similar amount to keep in balance. Consequently, if any two of the components of the accounting equation are known, the third can be calculated.

Example 2.3

When Wilson Probate started his business on 1 January 19x3, he put in £100 000 capital. From this he purchased a motor truck for £20 000 and bought a small office for £55 000. He also purchased two second-hand concrete mixers for £2000 and spent £4000 on a stock of timber and steel reinforcement.

His balance sheet at the time is set out in Fig. 2.2

In constructing the balance sheet the *fixed assets* have been separated from the *current assets*. Buildings, trucks and machinery should have a production life of several or more years and are intended for long-term use; stock and cash are likely to be used over a much shorter period and will be expended on a day-to-day basis.

Checking by using the balance sheet equation above, it will be seen that the

WILSON PROBATE

Balance sheet as at 1 January 19x3

			£		£		£
Capital		100 000	*Fixed assets*				
			Buildings		55 000		
			Motor truck		20 000		
			Machinery		2 000		77 000
			Current assets				
			Stock		4 000		
			Bank		18 500		
			Cash		500		23 000
		100 000					100 000

Fig. 2.2 Wilson Probate balance sheet, 1 January 19x3.

assets equal £100 000, that the capital equals £100 000 and the liabilities are therefore zero.

Example 2.4

Two weeks after starting his business, on 15 January 19x3, Wilson Probate buys bricks to the value of £2000 and has this sum debited to his account at the local builders' merchant. He then proceeds to construct a brickwall for a client and uses bricks to the value of £1500 and other materials to the value of £650 which he pays for by cheque. On completion he receives £3000 in payment for the brickwall.

On receiving the payment of £3000 for the construction of the wall, Wilson Probate buys, for £500, a small mixer for mixing mortar and then withdraws £600 from the business for his own personal use. He also adds bricks to the value of £2000 to his stock which, as before, are purchased on credit from the builders' merchant.

On 1 February 19x3, Wilson Probate completes a wall for another client to whom he gives monthly terms. Payment for the second wall of £1500 is expected at the end of February 19x3. Stock used in the wall cost £750. Draw up a balance sheet for the sole trader, Wilson Probate, at 2 February 19x3.

The balance sheet for Wilson Probate at 2 February is set out in horizontal format in Fig. 2.3. An explanation of the detailed transition between the balance sheets shown in Figs. 2.2 and 2.3 follows.

Wilson Probate's first move is to buy, on credit, bricks worth £2000. The stock of his business, therefore, increases to £6000 and the business takes on a liability to pay the builders' merchant £2000. Since the liability is short-term,

WILSON PROBATE

Balance sheet as at 2 February 19x3

	£	£		£	£
Capital		100 000	Fixed assets		
Profit(1)	850		Buildings	55 000	
Profit(2)	750	1 600	Motor truck	20 000	
			Machinery	2 500	77 500
		101 600			
Less: Withdrawals		600			
		101 000			
Current liabilities			Current assets		
Trade creditors		4 000	Stock	5 750	
			Debtors	1 500	
			Bank	19 750	
			Cash	500	27 500
		105 000			105 000

Fig. 2.3 Wilson Probate balance sheet, 2 February 19x3.

probably only one month or so, the £2000 is shown under a general heading of *current liabilities* and since the accounting equation shows that any increase in assets must be accompanied by an increase in capital + liabilities, current liabilities are listed on the left-hand side of the balance sheet. The builders' merchant has become a *trade creditor* of the business.

The second transaction in the business is a small project to build a wall for which Wilson Probate is paid £3000. Materials costing £650 are paid for by cheque and hence the bank account is reduced by this amount. The adjusted balance is £18 500 – 650 = £17 850 and stock is increased by a similar amount to £6000 + 650 = £6650. The project is undertaken and stock is reduced by £1500 + 650 to £4500; the £3000 in payment for the work is deposited in the bank bringing the current balance up to £17 850 + 3000 = £20 850. Since the balance sheet is for a sole trader, the *profit* on the project, £3000 less the cost of the materials, £2150, belongs to the owner of the business. The business, in other words, has a liability to pay £850 to Wilson Probate and this sum is shown on the left-hand side of the balance sheet. Checking the balance sheet shows that, at this stage, the assets equal the capital and liabilities at £102 850.

Wilson Probate now buys machinery for £500 so that figure in the balance sheet becomes £2500 and the bank account is reduced to £20 350. This purchase simply amounts to a redistribution of assets of the business on the right-hand side of the balance sheet and no adjustment, therefore, is required to the left-hand side. Wilson takes £600 from the bank leaving a balance of £19 750 and the withdrawal must be recorded on the left-hand side of the balance sheet since it is, in effect, a withdrawal of capital. The purchase of further bricks adds £2000 to stock (£6500) and £2000 to trade creditors (£4000) under current liabilities.

Finally, Wilson Probate undertakes another project using £750 of stock which then stands at £5750. He grants a monthly account to his client so that the £1500 for the work may not be paid for a month or so. This sum is, therefore, owed to the business until paid and an entry under *debtors* on the assets side of the balance sheet is required. The profit made on the project is the payment of £1500 less the cost of the materials, £750, and thus amounts to £750, to be shown together with that for the previous project.

It will be noted that everytime a transaction takes place within the business, buying or selling and involving the use of resources, the balance sheet changes. It can only, therefore, be a snapshot at one particular point in time.

2.7 Vertical format balance sheets

Since the Companies Act of 1981, limited companies registered in the UK have been required to present their balance sheet in one of two closely specified forms and the layout of their profit and loss accounts in one of four alternative forms. The specified forms for balance sheets are of vertical format rather than the horizontal format which has been used in Figs. 2.2 and 2.3 to illustrate more clearly a number of points of explanation particularly in relation to the basic accounting equation. The balance sheet for Wilson Probate shown in Fig. 2.3 is transformed to vertical format in Fig. 2.4. The vertical format will be used in future examples in this text.

It should be noted that the vertical format presentation of a balance sheet provides more information than the horizontal type. The upper section of the balance sheet for a sole owner business shown in Fig. 2.4 highlights the *capital employed*, in this case £101 000. This is arrived at by deducting *current liabilities* from *total assets* – a calculation not made in the horizontal format balance sheet. The bottom half of the balance sheet shows how the capital was financed; the *owner's worth* being equal to the *capital employed*.

A further calculation is added to show the *net working capital* (or *net current assets*) at £23 500 being the *current assets* less the *current liabilities*. The figure for *working capital* is then added to the total for *fixed assets* in order to give the *capital employed*.

WILSON PROBATE

Balance sheet as at 2 February 19x3

	£	£	£
Fixed assets			
Buildings		55 000	
Motor truck		20 000	
Machinery		2 500	77 500
Current assets			
Stock	5 750		
Debtors	1 500		
Bank	19 750		
Cash	500	27 500	
Less: Current liabilities			
Trade creditors		4 000	
Net working capital			
(or net current assets)			23 500
Capital employed			101 000
The above capital was financed as follows:			
Capital			100 000
Profit(1)		850	
Profit(2)		750	1 600
			101 600
Less: Withdrawals			600
Owner's worth			101 000

Fig. 2.4 Wilson Probate balance sheet, 2 February 19x3, in vertical form.

2.8 The appropriation account

The majority of construction companies of significance operate under the protection of limited liability. As such, their accounts will differ in a number of respects from those of a sole owner previously examined.

Firstly, the capital of a sole owner's business is the sum of money put into his business at the outset. It is increased by the profit made and decreased by any withdrawals made by the owner for his personal use. The capital of a limited liability company is subscribed by either private persons or institutional investors such as banks, insurance companies, unit trusts and pension funds. The investors then become members or *shareholders* of the company. The liability of the investors is limited to the value of the shares that are held. If, for example, an investor has subscribed £10 000 to the share capital of a company, in the case of the company's liquidation only the £10 000 initially subscribed is lost. (If the shares are only partly paid at their purchase, the balance owing on the shares will also be lost.) The profit made by the company belongs to its shareholders who receive it as a *dividend*. Dividends are paid from that part of the profit that is available for distribution after the deduction of tax and also any retentions that may be made for future investment in the company. Dividends are recommended by the company's directors for approval by the shareholders who have the right to reduce the recommended dividend but not to increase it.

The ways in which the net profit of a limited liability company is distributed, such as to corporation tax, dividends and general reserve, is shown in an *appropriation account*. The exact liability for corporation tax is generally not established at the time of the preparation of the accounts and therefore an estimated amount is shown as being transferred to reserve for this purpose.

Dividends are sometimes declared more than once a year. Where this is the case, the usual practice is to declare an *interim dividend* at the mid-point of the year in addition to a final dividend at the year end. Dividend and corporation tax appropriation appear in a balance sheet under the heading of *current liabilities, creditors due within one year*.

The third part of net profit is a transfer to the *general reserve*. Such a transfer is made by the directors of a company to strengthen the company's financial position. Reserves may be used for the purpose of purchasing fixed assets, increasing the capital of the company or perhaps to equalize the level of dividend payments from year to year during periods of fluctuating profit. Reserves appear in a balance sheet under *capital and reserves*.

When the above appropriations have been made, since they are unlikely to total exactly the profit available for appropriation, there is likely to be a balance of profit shown on the profit and loss account which is as yet unappropriated. This is then shown on the balance sheet in the *capital and reserves* section.

	£million	£million	£million
Net profit (19x3)			125.0
Net profit brought forward (19x2)			0.5
Total available for appropriation			125.5
Corporation tax			42.0
			83.5
Appropriations:			
To general reserve		49.0	
Preference dividend	4.0		
Ordinary dividend, interim	10.0		
final	20.0	34.0	83.0
Balance carried forward			0.5

Fig. 2.5 Typical appropriation occount.

Reserves can be generated in several different ways other than appropriation from the profits of a company. One class of reserve is that of a *capital reserve*. Capital reserves are profits/losses which do not come through the profit and loss account in the usual way. For example, shareholders may pay a premium when buying stock in a company. If stock with a nominal share value of £1 is issued by a company at £1.75 per share, for each share sold, the 75 pence over its nominal value is placed in a share premium account. This account appears in the balance sheet under the general heading of *reserves*. Another example of capital reserve is that generated by the revaluation of fixed assets belonging to the company. If the revaluation of properties owned by the company, for example, produces a greater total value than is shown in the previous balance sheet, the increase in value is shown as a *capital reserve*. The reverse is also true.

A typical appropriation account is shown in Fig. 2.5.

2.9 Balance sheet standards

Figure 2.6 illustrates typical balance sheets for two consecutive years of a small/medium construction company. It is usually the case that such balance sheets, when published and circulated to the shareholders and other interested parties, are condensed to the main headings. The main headings are then indexed to a series of notes on the accounts, each of which enlarges the detail under the relevant heading.

It should be noted that, in addition to the Companies Acts, an important controlling document for the construction industry is the Statement of Standard Accounting Practice (SSAP) Number 9 (revised), *Stocks and Work in Progress*. This Statement is one of a series which is prepared by the Accounting Standards Committee of the Consultative Committee of Accountancy Bodies and is issued by the professional institutions governing

TACKS CONSTRUCTION PLC

Balance sheet as at:		31 December 19x3			31 December 19x4	
	£000	£000	£000	£000	£000	£000
Assets employed						
Fixed assets						
Land and buildings		740.0			800.0	
less depreciation		120.0	620.0		129.7	670.3
Vehicles, plant and machinery		425.0			600.0	
less depreciation		114.5	310.5		161.6	438.4
Fixtures and fittings		205.0			155.0	
less depreciation		52.9	152.1		41.2	113.8
Investments			26.4			26.4
Total fixed assets			1109.0			1248.9
Current assets						
Land and developments	1207.4			882.2		
Raw materials and consumables	22.4			91.0		
Work in progress	332.3			251.3		
Debtors	942.1			1302.1		
Short-term investments	20.0			10.0		
Cash at bank and in-hand	521.7			240.7		
Total current assets		3045.9			2777.3	
Less current liabilities						
Creditors, amounts falling due within one year:						
Bank loans and overdrafts	423.1			501.4		
Obligations under financial leases	54.3			24.1		
Payments received on account	120.3			20.1		
Trade creditors	750.2			807.4		
Taxation on profits	94.0			62.3		
Proposed dividends	100.0			50.0		
Other creditors	52.6			16.2		
Total current liabilities		1594.5			1481.5	
Net current assets (working capital)			1451.4			1295.8
Total assets less current liabilities (capital employed)			2560.4			2544.7
Less **creditors, amounts falling due after more than one year:**						
12% bank loan repayable June 1995		75.0			75.0	
10% unsecured loan stock 1992/97		100.0			100.0	
Obligations under financial leases		37.2	212.2		2.0	177.0
Net worth of company			2348.2			2367.7
The above net worth of the company was financed by:						
Called-up share capital						
7% cumulative preference shares of £1 each		200.0			200.0	
1 750 000 ordinary shares at £1 each		1750.0	1950.0		1750.0	1950.0
Reserves						
Share premium account		188.1			188.1	
Revaluation reserve		(70.0)			(109.3)	
Other reserves		27.0			66.3	
Profit and loss account 1 January 19x3/4		253.1	398.2		272.6	417.7
			2348.2			2367.7

Fig. 2.6　Tacks Construction PLC balance sheets for 19x3 and 19x4.

accountancy practice. The Statements are not enforceable by law but their use is expected of the members of the accounting bodies that subscribe to their development. The intention of the Statements is to promulgate rules which will ensure, as far as is possible, that there is consistency and comparability between the financial accounts of different companies. Each Statement issued by the Committee is numbered and deals with a particular subject or topic. Another similar and relevant series of publications is the Statements of Recommended Practice (SORPs) which have a similar standing to the SSAPs.

SSAP 9 sets out the rules the establishment of the two items *Raw materials and consumables* and *Work in progress* shown under *Current assets* in the balance sheets set out in Fig. 2.6. These two items, and particularly the latter, have an important influence on the profits declared through the balance sheet for a construction company in any particular year. The value of work in progress can also be a key item in valuing a company for take-over purposes. Work in progress arises because a construction company will have projects, at any one time, in all phases of progress – at their start, at completion and at any point in between. It is also accepted practice for companies to retain projects within work in progress until their final account is approved and paid, thus, in some cases, extending the work in progress period by several years after completion of the work. At the point of final payment for the site work, the financial outcome must be included in the current year accounts. If some profit has been taken in previous years, this amount is then subtracted from the final figure and the balance is shown in the current year accounts. The reverse is true where losses are incurred and/or provision has been made in the accounts for future losses where these have been anticipated.

Prior to the introduction of SSAP 9, it was not uncommon for a conservative approach to be adopted with regard to bringing profit into the company accounts. Because contract completions tend to be haphazard in their occurrence, such a policy often led to the accounts showing other than a consistent, even progress of a company over several years. SSAP 9 requires that profit from long-term contracts should be taken as it arises, as long as it is reasonably certain and the profit should be established reflecting a comparison of turnover with the contract costs incurred as the work proceeds. Where variations occur to the scope and value of the work, allowances for these should only be included when accepted by the client. In addition, provision should be made for all losses in the year when they are first foreseen.

Other points to note about the balance sheets shown in Fig. 2.6 are as follows:

(a) For each category of fixed assets (except *Investments*) the costs of the assets at the last balance sheet are shown together with the amount by which the asset has been depreciated since then. The depreciation is subtracted from the cost and the resulting *net book value* or *written down value* is shown.

(b) Note that the term *current liabilities* has been used as well as *Creditors, amounts falling due within one year*. The latter is the modern expression and the former was used until the recent past but now may be excluded.

(c) Short-term *bank loans and overdrafts* are shown under *current liabilities* even though loans and overdrafts of this nature are a form of borrowing which becomes long-term by renewal. However, they are normally included here because, technically, this type of loan can be repayable on demand.

(d) The company's debt is shown under *Creditors, amounts falling due after more than one year*. The date of redemption of each loan and its interest rate, where applicable, are usually shown.

(e) The *called-up share capital* is detailed as a source of company financing. The values of the shares are shown as the *nominal* or *par* values. The *share premium account* shows the total amounts that shareholders have paid for shares where the price at the time of offer was in excess of the nominal value.

(f) The *revaluation reserve* contains any excess (or reduction) that has occurred during the year due to revaluing the fixed assets. In this case the total value is reduced by £70 000 and £109 300.

(g) Other figures under reserves are transferred from the profit and loss account.

2.10 Ratio analysis

The two successive balance sheets of Tacks Construction shown in Fig. 2.6 can be compared by inspection. The following points are noted:

(a) The fixed assets of *land and buildings* have been increased by £60 000 and *vehicles, plant and machinery* by £175 000.

(b) The fixed assets of *fixtures and fittings* have been decreased by £50 000. There may have been a sale of some of these assets with a credit to reserves but this cannot be confirmed from the statement. *Investments* have remained the same as the previous year.

(c) *Land and developments* show a marked decrease of £325 200 and *work in progress, short-term investments* and *cash at bank and in hand* are reduced by £81 000, £10 000 and £281 000, respectively.

(d) *Debtors* and *raw materials* have increased by £360 000 and £68 600, respectively.

(e) *Bank loans and overdrafts* have increased by £78 300, *trade creditors* by

£57 200 whereas *payments received on account, taxation and dividends* have all fallen.
(f) The balance sheet total has gone up by £19 500 from one year to the next. The financial side of the balance sheet incorporates an increase in reserves to effect the balance.

Another method of examining the financial statements of the profit and loss account and the balance sheet is by the use of *ratio analysis*. However, it must be noted that these two financial statements are not intended to provide a detailed analysis of a company's performance and great care needs to be exercised in interpreting such an analysis. A ratio analysis may be most useful in indicating the direction of further enquiries and seeking further information when the performance or the liquidity of a company is under review.

A ratio analysis can be used for each year over a period of years to determine possible trends in performance – whether the company is improving or declining in the particular area in which the examination is being made. The information so derived may also be used to make comparisons with industry averages which have been established through inter-firm comparisons and the like, or with similar firms that adopt similar accounting conventions. In some cases, ratios so established can be compared with company budgets if these are prepared with such an exercise in mind.

Ratio analysis normally has three main functions:

(a) to enable an assessment of the *performance* of a company to be made;
(b) to enable an assessment of the *liquidity* of a company to be made;
(c) to enable an assessment of the *stock exchange investment potential* of a company to be made.

The following ratios will be discussed:

(a) Profitability ratios: Gross profit
                                   Return on capital employed
(b) Liquidity ratios: Current
                                   Acid test
(c) Activity ratios: Stock turnover
                                   Debtor turnover
                                   Asset utilization
(d) Financial structure ratio: Gearing.

The balance sheets of Tacks Construction PLC (Fig. 2.6) and the associated profit and loss accounts (Fig. 2.7) will be used to illustrate the calculation of the ratios.

TACKS CONSTRUCTION PLC

Trading and profit and loss accounts
for years ending 31 December 19x3 and 19x4

	19x3	19x4
	£000	£000
Turnover	13 265.6	11 427.4
Cost of sales	10 673.2	9 075.2
Gross profit	2 592.4	2 352.2
Other operating expenses:		
Employee costs	1 064.2	972.3
Administrative costs	632.5	723.1
Directors' salaries	92.5	105.3
Depreciation	287.4	332.5
Operating profit before interest	515.8	219.0
Interest	92.4	87.2
Operating profit after interest	423.4	131.8
Corporation tax	94.0	62.3
Net profit after tax	329.4	69.5
Dividends	100.0	50.0
Retained profit for the year	229.4	19.5

Fig. 2.7 Tacks Construction PLC trading and profit and loss accounts for 19x3 and 19x4.

Profitability ratios

Gross profit ratio
The gross profit ratio measures the profitability of the company.

$$Gross\ profit\ ratio = \frac{\text{gross profit}}{\text{sales}} \times 100$$

For 19x3 the gross profit ratio is $2592.4/13\ 265.6 \times 100 = 19.54\%$
For 19x4 the gross profit ratio is $2352.2/11\ 427.4 \times 100 = 20.58\%$

This important ratio shows that, for 19x3, for every £1 of turnover (sales) that was achieved, 19.54 pence were available to pay for overheads and profit. The ratio for 19x4 improved slightly and probably resulted from increasing price rates for work carried out, reduced labour costs for unit output (improved productivity), lower labour rates, or lower materials costs. There may be other reasons peculiar to the situation under consideration. The increase of approximately 1 per cent in the ratio is a beneficial trend.

Return on capital employed ratio
In order to measure the efficiency with which capital is used within a company, the *return on capital employed* ratio is used.

$$Return\ on\ capital\ employed = \frac{operating\ profit}{capital\ employed} \times 100$$

Care is needed in using this ratio in that capital employed can be defined in a number of ways. However, here (*total assets – current liabilities*) will be used and it should be noted that (*share capital + reserves + long-term loans*) or (*fixed assets + working capital*) will give the same figure. In the following the figures of the balance sheet for 19x3 are used as an example.

(a) Capital employed = total assets – current liabilities
    = 1109.0 + 3045.9 – 1594.5 = 2560.4
(b) Capital employed = share capital + reserves + long-term loans
    = 1950.0 + 398.2 + 212.2 = 2560.4
(c) Capital employed = fixed assets + working capital
    = 1109.0 + 1451.4 = 2560.4

For 19x3 the return on capital employed is $515.8/2560.4 \times 100 = 20.15\%$
For 19x4 the return on capital employed is $219.0/2544.7 \times 100 = 8.61\%$

This ratio is used to indicate the effectiveness of the utilization of assets. It shows the return that company management is getting from the capital invested in assets. In the case of Tacks Construction it will be seen that the return has declined markedly between 19x3 and 19x4.

Liquidity ratios

These ratios are concerned with the *liquidity* of the business rather than its profitability. In essence they are a measure of a company's ability to raise cash in the short-term to meet short-term obligations.

Current ratio
This is a simple measure given by:

$$Current\ ratio = \frac{curent\ assets}{creditors:\ amounts\ due\ within\ one\ year} \times 100$$

For 19x3 the current ratio is $3045.9/1594.5 = 1.91$
For 19x4 the current ratio is $2777.3/1481.5 = 1.87$

Current assets are, in theory, those which, if they are not already liquid, should become liquid within one year. The implication of this is that they will be available at relatively short notice to pay creditors if required. On paper, a ratio of 1.00 means that there will be sufficient of the one to meet the other. However, it is usual to look for at least a ratio of 1.50 in most businesses. If the ratio falls below 1.00, the business is almost certain to be suffering a shortage of working capital even though it may only be a temporary situation.

Acid test ratio

An improved measure for liquidity is the *acid test* or *quick ratio*. In this ratio the numerator becomes the liquid assets, *debtors, short-term investments* and *cash at bank and in hand*. Stock is excluded from current assets since it may take some time to convert this to sales or turnover.

$$Acid\ test\ ratio = \frac{\text{current assets} - \text{stock}}{\text{creditors: amounts due within one year}} \times 100$$

For 19x3 the acid test ratio is 3045.9 − 1562.1/1594.5 = 0.93
For 19x4 the acid test ratio is 2777.3 − 1224.5/1481.5 = 1.05

Clearly, the liquidity of Tacks Construction is just adequate. Debtors are a high proportion of current assets in both years and pressing for quicker payment will increase the margin, if the policy is successful. Good managers will be constantly aware of the current liquidity of a business by using properly prepared cash flow forecasts. The liquidity ratios are therefore unlikely to be a great deal of use internally in businesses where this is the case but rather to external analysts wishing to obtain a measure of the stability of the company.

Activity ratios

Stock turnover ratio

Stock turnover is a ratio that measures the amount of stock, in cost terms, in relation to its throughput. It is an essential ratio in retailing businesses where large stocks may suffer from obsolescence, deterioration and high storage and financing costs.

$$Stock\ turnover = \frac{\text{stock}}{\text{cost of goods sold}} \times 365$$

The ratio gives an indication of the average time an item is in stock. It suffers from difficulty of interpretation because average stock is difficult to establish and production needs to be relatively constant for the ratio to be of value. There are few useful applications for this ratio in mainstream construction activity since large quantities of finished stock are not normally held, the units of sale are usually relatively large and hence their overall stock control is not complex. It may have applications in building supply businesses, for example, and the manufacture and production of the many components that go into constructing a facility, such as precast concrete products, etc.

Debtor turnover ratio

The *debtor turnover* ratio measures the average length of time that debtors are outstanding and, therefore, the effectiveness with which the business is collecting debts from its customers.

$$Debtor \; turnover = \frac{debtors}{sales} \times 365$$

For 19x3 debtor turnover is $942.1/13\,625 \times 365 = 25.23$ days
For 19x4 debtor turnover is $1302.1/11\,427.4 \times 365 = 41.59$ days

These figures are very good for construction though there may be many hidden factors which are not obvious which will influence the figures. Examples in contracting are payments may be made in advance of work commencing, establishment charges may be paid, or favourable final settlements from previous contracts may have been received or a company may have a very wide range of products from large to small with different credit terms applying to each. Such payments would need to be sorted and ordered before a realistic ratio can be calculated.

Asset utilization ratio

The *asset utilization ratio* is a measure of the operating performance of the company showing the extent to which the assets of the business have been used to generate sales.

$$Asset \; utilization \; ratio = \frac{sales}{operating \; assets} \times 100$$

Operating assets are commonly defined as *total assets minus current liabilities.*

For 19x3 the asset utilization ratio is $13\,265.6/(1109.0 +$
$3045.9 - 1594.5) = 5.18$
For 19x4 the asset utilization ratio is $11\,427.4/(1248.9 +$
$2777.3 - 1481.5) = 4.49$

Asset utilization for Tacks Construction is good in 19x3 but reduces slightly in 19x4. Total assets have increased over this period and sales have fallen by nearly 14 per cent. Management need to increase turnover and look to current assets for economies. It must be borne in mind, however, that the value of fixed assets in the balance sheet is very dependent on the past depreciation policies that have been pursued and also that, in many instances, they will be *historic* rather than at *current* cost.

Financial structure ratio

Gearing ratio

The *gearing ratio* compares the extent to which a company uses long-term debt to finance its activities as opposed to equity. There are four or five different versions of this ratio. One is as follows:

$$Gearing\ ratio = \frac{\text{preference shares} + \text{long-term loans}}{\text{ordinary shares}} \times 100$$

For 19x3 the gearing ratio is $((200.0 + 212.2)/1750.0) \times 100 = 23.55\%$
For 19x4 the gearing ratio is $((200.0 + 177.0)/1750.0) \times 100 = 21.54\%$

An alternative name for *gearing* is *leverage*. By most standards this is a *low-geared company* and the trend is only very slightly down from 19x3 to 19x4. If more external capital is required in the company then clearly one source to be considered is that of *long-term debt*. However, it must be remembered that the interest on debt, and eventually the debt itself, has to be paid back whatever the state of trade and profits, whereas dividends can always be passed over in the hard times of recession for the company.

In times of recession or reduced trade, an overburden of debt, and hence interest payments, can prove to be disastrous to a business that cannot service it.

There are a number of investment ratios that are of interest to analysts, financial managers and investors rather than operating management. These are not discussed here.

2.11 The funds flow statement

In order to have a clear picture of how the company's asset/liability state has changed between the dates of any two balance sheets, a *funds flow statement* or a *source and application of funds statement* can be prepared. Such a statement will now be prepared for the period between the two balance sheets shown in Fig. 2.6. It will illustrate how the company's net wealth has changed between 31 December 19x3 and the same date in 19x4. In addition, it will show the detail of the areas in which the changes have taken place and in which direction.

A funds flow statement is prepared in two stages; firstly, by itemizing the changes under *liabilities* and then *assets*; secondly, by summarizing the changes into two sets, one to show the sources of the funds and the second to show how they have been applied. These two stages are shown in Figs. 2.8 and 2.9, respectively.

In Fig. 2.8 there are two adjustments to be made to the figures shown in the financial statements for the purpose of determining funds flow. Firstly, depreciation is not an actual cash flow but a value flow for adjusting the assets values in the accounts. It appears in the profit and loss account as a deduction from gross profit and therefore needs to be added back into the increase in reserves. For example, in 19x4, total depreciation is £332 500 from the profit and loss account and the balance sheet. The increase in reserves is, therefore, £19 500 + 332 500 = £352 000. Secondly, the expenditure on fixed assets must be adjusted by adding to it the depreciation:

TACKS CONSTRUCTION PLC

Balance sheet changes for the years 19x3 and 19x4

	19x3	19x4	Differences	
			Positive	Negative
Liabilities				
Ordinary shares	1750.0	1750.0	—	—
Preference shares	200.0	200.0	—	—
Reserves	398.2	417.7	19.5	—
Loan capital	175.0	175.0	—	—
Long-term financial leases	37.2	2.0	—	35.2
Bank loans and overdrafts	423.1	501.4	78.3	—
Short-term leases	54.3	24.1	—	30.2
Payments received on account	120.3	20.1	—	100.2
Trade creditors	750.2	807.4	57.2	—
Taxation	94.0	62.3	—	31.7
Dividends	100.0	50.0	—	50.0
Other creditors	52.6	16.2	—	36.4
	4154.9	4026.2	155.0	283.7
Assets				
Land and buildings	620.0	670.3	50.3	—
Plant and machinery	310.5	438.4	127.9	—
Fixtures and fittings	152.1	113.8	—	38.3
Investments	26.4	26.4	—	—
Land and developments	1207.4	882.2	—	325.2
Raw materials	22.4	91.0	68.6	—
Work in progress	332.3	251.3	—	81.0
Debtors	942.1	1302.1	360.0	—
Short-term investments	20.0	10.0	—	10.0
Cash at bank and in hand	521.7	240.7	—	281.0
	4154.9	4026.2	606.8	735.5

Fig 2.8 Tacks Construction PLC balance sheet changes for 19x3 and 19x4.

Expenditure on fixed assets $= 50\,300 + 127\,900 - 38\,300 +$
$332\,500 = £472\,400$

There is a wide variety of ways of displaying a funds flow statement such as that shown in Fig. 2.9. The amount of detail shown will also vary widely. It must be remembered that this example is relatively simple but in practice such statements may be much more complicated. The key points to be noted from the *sources of funds* concern the total of *cash flow from trading*, which is usually the major source of cash entering a business, the large reduction in *stocks – land and developments* and *work in progress –* at £406 200, and the large decrease in *cash at bank and in hand*.

On the application side of the statement, the *purchase of fixed assets* will normally be the largest cause of expenditure. In this example, however, there is an alarming increase in the *debtors* cash flow, only partly offset by a decrease in *creditors*. Some attention would need to be given to this. Also there is an increase in stocks of *raw materials* which should be investigated.

TACKS CONSTRUCTION PLC

Funds flow statement for period ending 31 December 19x4

Sources of funds

Cash flow from trading	352.0
Bank loans and overdrafts	78.3
Increases in trade creditors	57.2
Decrease in land and developments	325.2
Decrease in work in progress	81.0
Decrease in short-term investments	10.0
Decrease in cash at bank and in hand	281.0
	£1184.7

Applications of funds

Purchase of fixed assets	472.4
Tax paid	31.7
Dividends paid	50.0
Increase in raw materials	68.6
Increase in debtors	360.0
Decrease in creditors	36.4
Decrease in financial leases	65.4
Decrease in payments in advance	100.2
	£1184.7

Fig. 2.9   Tacks Construction PLC funds flow statement for 19x4.

Problems

2.1 When Proconbuild Limited was established as a limited company, it obtained the right to issue 1 000 000 ordinary shares of £1 each. Currently 870 000 of the ordinary shares have been issued though none of these has as yet been fully paid. The company has since called for 75 pence per share but has only received this amount on 800 000 of the shares.

What is the authorized share capital, the issued share capital, the called-up capital, the calls in arrears and the paid-up capital?

2.2 A company makes a net profit of £100 000 for the current year and brought forward from the profit and loss for the previous year the sum of £17 500. Corporation tax for the current year is £33 000.

The directors of the company decide to pay out of the profit a final dividend of 3.0p per share, in addition to the interim dividend of 1.5p, then transferring most of the balance of the profit to the general reserve. This leaves the sum of £2 500 to be carried forward to the profit and loss account for next year.

If the company has issued 1 500 000 ordinary shares that will qualify for the dividends, draw up an appropriation account for the current year.

2.3 Cheery Cabins PLC has a turnover in 19x4 of £4 300 000 and a cost of sales of £2 350 000. During the year the administration expenses amount to £1 400 750 and taxation of £95 000 is anticipated. If dividends amounting to £50 000 are to be paid, compile the 19x4 profit and loss account for the company and determine the retained profit for the year.

2.4 Partners Construction had a turnover of £345 000 in 19x5. Labour costs amounted to £83 400, for materials £163 000 was paid and sub-contracts amounted to £47 000. The following costs were also incurred:

	£
Administrative expenses	21 700
Depreciation on owned assets	2 600
Auditors' remuneration	760
Directors' emoluments	990
Taxation	7 000
Dividends	2 500
Office expenses	22 000
Insurance	5 500
Legal and accountancy fees	11 000
Repairs and maintenance	7 900

Prepare a trading and profit and loss account for Partners Construction.

2.5 The following information is available from the books of Jasmin Construction PLC for the year ending 31 December 19x5:

	£000	£000
Turnover		1500
Cost of labour	250	
Cost of materials	320	
Sub-contracts	210	
Administrative expenses	250	
Office expenses	82	
Buildings	300	
Cumulative provision for depreciation on building		108
Plant and equipment	175	
Cumulative provision for depreciation on plant and equipment		66
Motor vehicles	110	
Cumulative provision for depreciation on motor vehicles		32
Debtors	320	
Cash at bank	125	
Creditors		90
Share capital:		
Ordinary shares issued: 330 000 at £1 each		330
General reserve b/f at 1/1/19x5		12
Profit and loss account b/f 1/1/19x5		4
	2142	2142

In conjunction with the above, the following information is provided:

(a) Depreciation for the year should be taken as 5%, 20% and 20%, respectively, for buildings, plant and equipment and motor vehicles;
(b) dividends are proposed to be 50 pence per ordinary share;
(c) transfer to general reserve is to be £100 000;
(d) corporation tax should be provided for at £34 000;
(e) stock (work in progress plus raw materials, etc.) is £134 000.

Prepare a trading and profit and loss account, appropriation account and a balance sheet for the year ending 31 December 19x5 for Jasmin Construction PLC.

2.6 A small company, Turbotruss Limited, sells timber products to the building industry. The need arises to raise some extra capital and the Chief Executive decides to review the financial accounts for the past two years with a view to looking at the company's recent performance. The accounts are as follows:

TURBOTRUSS LIMITED

Trading and profit and loss accounts for the years ending
31 December 19x5 and 19x6

	19x5 £000	19x6 £000
Turnover	1250	1340
Cost of sales	821	963
Gross profit	429	377
Office expenses	23	32
Wages and salaries	109	132
Administrative costs	27	41
Directors' emoluments	29	38
Depreciation	32	32
	220	275
Net profit before interest	209	102
Interest	40	57
Net profit after interest	169	45
Corporation tax	53	21
Net profit after tax	116	24
Dividend	75	20
Profit retained for year	41	4

	19x5	19x6
	£000	£000
Balance sheet		
Fixed assets		
Buildings	210	202
Plant and equipment	195	206
Motor vehicles	65	60
	470	468
Current assets		
Stock	121	117
Debtors	111	119
	232	236
Creditors: amounts falling due within one year		
Creditors	75	131
Bank loan	25	45
Corporation tax	53	21
Dividend	75	20
	228	217
Net current assets	4	19
Total assets less current liabilities	474	487
Creditors: amounts falling due after more than one year		
12% bank loan	220	229
	254	258
Financed by		
Called up share capital	150	150
Profit and loss account	104	108
	254	258

Chapter 3
Introduction to Investments

3.1 Types of investment problems

Investment problems will usually fall into one of two broad classifications:

(a) investments to reduce cost;
(b) investments to expand income.

Some situations will have an element of both categories either by implication or consequence.

In the first group, a project is generally one which is concerned with reducing the cost of production. For example, the introduction of mechanical plant to undertake tasks which consequently reduces the total labour costs, is a typical and common cost-reduction proposal if it is carried out with that in view. The cost-reduction so achieved is then not put into effect solely with a view to increasing the total quantity of production and hence a firm's income, although clearly the difference between the cost and the income, the gross profit, may be increased. The cost of a modernization scheme will be considered as an initial capital investment and the saving in the cost of production is considered as the return from that investment.

It may be necessary to choose between a number of different ways of effecting the cost-reduction. Hence the problem to which an answer is required may not only be concerned with ensuring an adequate return for the outlay on the new equipment, but also with deciding which scheme, of all those alternatives that are available, will produce the best return, taking into account, where necessary, the risk involved with each.

In the second category of decision group, an investment is made with a view to increasing the total amount of income arising from invested capital. This may be effected through buying additional plant, providing more working capital to increase turnover or by increasing the labour force, for example. However, there is little point in expanding income if the cost of expansion does not bring with it a satisfactory net profit. Not only is it necessary that the cost of expansion will ultimately be recovered from the increase in income, but it is also necessary that it is covered to the extent that the net profit anticipated must at least meet a predetermined standard.

It is possible to visualize some situations wherein investment decisions clearly have an element or characteristics of both classifications. The decisions may be made with a view, not only to increasing income, but also to reducing cost. For example, an analysis of production methods may yield an improvement in the utility or life of the product at the same time as reducing the cost of production. This in turn may enhance the value of the product to a customer and result in a legitimate and acceptable price increase. Income will thereby be increased on the basis that there is no drastic change in the level of sales resulting from a change in product price.

Within the two broad classifications, a number of other common types of problem can be placed. One such group is that of the replacement problem. The need for replacement may arise for many reasons; old age and hence possibly low production, frequent breakdowns, increasingly expensive maintenance, production of an inaccurate product, technological obsolescence and a host of other reasons, many of which are interrelated and contributory one to the other.

Income-expansion projects themselves can often be broken down into two groups, those which arise because of the increased demand for a product and those which arise from the development of new products. In the latter case, the market is as yet unlikely to be established and must be the subject of market research. The elements of risk and uncertainty have considerable influence on the outcomes of each of these two types of problem.

Research and development problems may fall into either of the two broad classifications but to which one may not be obvious at the outset. In fact, the outcome of the research and development itself may be unsuccessful to the extent that it never reaches a stage where classification is even possible or necessary.

Whatever the nature of an investment problem, it will always have one particular characteristic. It will always be such that it necessitates making a choice between alternatives. It may be, with some frequency, that the choice rests between investing or not investing, to carry on as before or to continue doing nothing may be very valid alternatives. However, so long as they are recognized as being so and examined on that basis, the outcome will have a very much greater chance of being successful.

3.2 The stages in an economic appraisal

As well as the various techniques which are available to make the calculations of the rate of return that may be expected from a given investment, it is essential to know of the other necessary steps for a complete statement of the problem so as to give a better understanding of how the use of the techniques fits into the broader picture of investment.

Three such steps which can be clearly identified are:

(a) estimation;
(b) calculation;
(c) evaluation.

Each of these three steps may involve several minor stages in itself and, moreover, not all the constituent stages of each step may be required in every instance.

The first, *estimation*, is concerned with the establishment of all those data concerned with the project. One of the most important estimates is that of the capital construction costs that are required to establish a project. In addition, such estimates as the probable size of the market for the product, the cost of raw materials for production, the project life and the varying rate of production due to obsolescence or depletion will need to be compiled. It will be necessary to estimate the likely national and/or international economic climate during the life of the project, that is both during its construction and production periods. It will be necessary to estimate the possible competition during this period since this will obviously affect the proportion of the available market that may be obtained.

Where possible, estimates should be made in, or converted to, financial terms, since this facilitates the calculations and particularly so where a choice between alternatives is being made. Not only should these estimates be in terms of currency, where possible, but the timing of their movement in or out is important. It is necessary, therefore, to prepare an accurate statement of when a particular expenditure is estimated to be made or when a particular receipt may be expected.

The second stage, that of *calculation*, is predominantly concerned with the establishment of a yield or rate of return for a project or, where the evaluation is concerned with choosing between mutually exclusive alternatives, the most attractive one. At this stage, the data that have previously been obtained from established sources, or those which have been established during the course of the study, are marshalled and summarized. An appraisal technique is then used to obtain a rating of their financial and investment merit.

The use of an appraisal technique may consist of two or more steps. In the first place, the process of calculation may reveal projects which are obviously not viable and they can then be discarded. As an alternative, relatively simple and crude techniques can be used to act as a coarse screen, sifting out projects which are clearly not going to approach the required rate of return. Those that do pass this test can then be subjected to a closer, more accurate, and probably a more sophisticated analysis.

Thirdly, it is necessary to *evaluate* a project in terms other than purely numerical ones. Judgement must be exercised with respect to the intangible aspects of the project. Judgement must also be exercised with regard to the risk involved and with regard to the degree of confidence with which the project may be undertaken successfully. The more detail that can be confirmed in the first two stages above, the easier and less complex will be the judgement decision of the third stage. Resulting from this analysis will also be a measure of the degree of control which will be required in order to ensure that the project proceeds, as anticipated in the estimation stage, in all its phases.

No matter how well and how thoroughly an investment situation is investigated, if will never be possible to express every aspect of the problem in numerical, quantitative and/or financial terms. There are always likely to be some factors which must be resolved out of the experience and with the aid of judgement of the decision-maker. There may be questions of conflict with the policy or corporate strategy of the business; there may be conflict with what a company believes to be its moral obligations as an employer; the conflict may come on humanitarian grounds or on the grounds of conflict with the firm's social obligations. In such situations there is still a need to be aware of the precise nature and the extent of the problem and to investigate closely the overall effect of a final decision on the problem as a whole. In reporting on a particular project, the report-writer must be careful to include recognition of the points which may also require the exercise of judgement in the preparation of a solution. The decision-maker's attention must necessarily be drawn to such factors.

An example of a *judgement factor* or *intangible* occurs when an exercise to reduce cost is undertaken. Almost invariably in a major scheme, there will be involved a possible and probable redundancy of labour resulting from improvements in the mechanization of a production process. This situation might conflict with the company's moral and social obligations arising from its considered policy in this respect. The problem may well be accentuated by the fact that the company is the sole employer of specialized labour in their locality and redundancy may have very far-reaching effects on the surrounding community. This is a situation in which the value of a technique as such is tempered in the preparation of a solution. In many cases, the nature of the situation obviously precludes the decision being made entirely as a matter of course, having only reference to the numerical answer which has been obtained by the application of those techniques to the established facts.

3.3 Risk and uncertainty in investment decisions

When attempting to establish a plan and a schedule for all varieties of activity the associated uncertainty increases the further the predictions range into the future. A planner will often be reluctant to look too far ahead because of this uncertainty. As a result there tends to be a reluctance to plan and, when planning is carried out, a scepticism about its value. It is known from our personal experiences that, whilst we can probably forecast with some accuracy just where we will be next week there is much more difficulty in forecasting our whereabouts next month and to do the same for next year is, as a rule, virtually unthinkable. The reluctance associated with uncertainty similarly applies to an investment decision, since this is little more than a plan, a programme, or a prediction for the future – very often for tens of years ahead. It is commonly true that preoccupation with the events of today tends to exclude thoughts about the future. Both of these deficiencies militate against thorough investment appraisal and the subsequent decision-making.

Very few investments of capital are made under conditions of certainty, that is, in circumstances where the future outcome of an investment is known with a reasonably high degree of precision. If a decision is made in circumstances wherein it is possible to predict from historic data the probability or chance that a particular outcome will occur, no matter how great or small that chance, then it is said that the decision is made *under risk*. Alternatively, many major decisions in investment problems have to be made under conditions of *uncertainty*. These are problems different from those where decisions are made under risk in so much as they lack the background of experience and past records on which the likely outcomes of a specific situation can be assessed in numerical terms. The condition of uncertainty is therefore a much more difficult one with which to deal than those of risk or certainty.

It might be argued that the use of precise and logical mathematical techniques for the evaluation of project investments is, therefore, a waste of time in situations of uncertainty, and little better for risk situations and that because of the uncertainty surrounding a situation, sophistication in analysis makes little or no contribution to getting a correct answer. Such argument is false and tends to perpetuate a situation in which decisions are made without the support of the full facts or a rational evaluation of the information available. One important point concerning all methodical, analytical appraisal is that it lays the facts of the situation open to thorough examination and creates awareness about the detail of the situation under examination. In this way, it is possible to gain a good idea of the critical areas in which there is likely to be more risk or uncertainty than others and to highlight what further information or investigation may be justified in order to eliminate as many doubts as possible. It separates those areas which are factual and therefore those which will have a predictable, or near-predictable, outcome from those which are at risk or are uncertain.

A technique for dealing with the situation of risk or uncertainty must, therefore, always be combined with a manager's judgement in order to arrive at a proper and optimal decision; it will provide a manager with some measurement of how risky or uncertain a situation is likely to be. From this assessment it is then possible to judge what reserve or latitude there is in different parts of the problem situation between the proposal becoming one that is seen to be undesirable from one that will probably be successful.

3.4 The cost of capital

All investments are made in the expectation of getting rather more by way of return than the extent of the original investment. They are made in anticipation of something favourable happening in the future. It is necessary, however, to provide the original stake in the first place.

Once a business is established and in progress, funds should be flowing into the firm to provide a source of cash for its immediate day-to-day requirements. It is, however, a characteristic of business that cash flows inwards are most unlikely to coincide, either in timing or magnitude, with the outward flow necessary to pay for the resources which are being used. This disparity of cash flows inevitably necessitates borrowing money to keep the business going until the next inflow of sufficient money to cover any large outgoing payments. Alternatively it may mean saving from income over a period, against the times when the payments will have to be made. The provision of money in a business, when it is required in this way, is known as *financing*.

It has been shown in Chapter 2 that the finance for a business can be obtained in a large number of different ways – from issuing preference shares, from equity stock, from debentures, from bank borrowing and other loans, from trade credit, from retained earnings and so on. Some of these sources of finance may be described as *short-term*, for example bank borrowing, and some are *long-term*, for example equity funding. Generally, a company will employ a number of these sources of finance and it is desirable that a company has a policy setting down the broad proportions of each type of funds that it will employ.

In order to attract funds into a company it is necessary to offer a promise to repay their providers a larger amount of money at some time in the future. When the finance is in the form of a loan, then the addition to the original sum is known as *interest*. When in the form of equity stock then it is known as a *dividend*. This incentive is required for providing finance for a business in order to compensate for the loss of use of the money for other things and as compensation for the risk that is being taken. As a rule, the compensation is directly proportional to the uncertainty and risk which surround the investment. In some circumstances, however, other factors, such as the likelihood of capital appreciation through business growth, may play a significant part in the decision to lend capital.

In setting a standard against which the extent of this compensation will be measured, it is necessary for a firm to compare favourably in its performance with other businesses of like risk. If it does not, then it may well be faced with a withdrawal of funds. A company can determine its cost of capital by reviewing the amounts it is drawing from various types of funding, the cost of raising them and the compensation it is paying the providers thereof. It is possible to arrive at a weighted average cost which reflects the proportions of the company's funds which are raised in each specific manner. Knowing this average cost of capital, it would be pointless for a firm to invest in a project which did not earn at least enough surplus to allow the cost of capital to be repaid and then, in addition, something for the required profit and to reinvest in the business. This is putting the return required at its minimum level, since there will usually be a need to make sufficient return in order to finance further projects within the company and for which there will be little or no measurable return in monetary terms. Many socially desirable welfare projects fall into this category.

It is often argued, though rarely convincingly, that new projects should be required to produce a return in excess of the company's *marginal cost* of finance. The simplest case of marginal finance is where the funds are raised for the purpose of financing a specific project. In times of inflation and/or high interest rates, the cost of marginal finance may be considerably greater than the finance previously raised. Where funds are raised specifically for one project, there is no doubt as to the source and the applicable cost of the funds. This situation may be acceptable in certain circumstances where there is a particularly expensive source of money or where the uncertainty and risk involved in a particular project is exceptional and outside the general run of the firm's commitments. Such an example is a construction project overseas in comparatively untried and unknown circumstances. On the whole, however, the minimum level of return required should be that of the weighted average of the company's cost of the whole of its capital, since moneys for projects are allocated, as a rule, from a central fund wherein the money from each source is not readily identifiable. In addition, since in practice the investment in different projects is staggered over a period of time, returns accrue in the company reserves as a result of earlier investments made on original costs of capital. One particular project, therefore, will be financed from a mixture of funds.

A company must obtain new funds in the light of its previous financial performance and by careful monitoring of its likely performance in the future. Appropriate proportions of debt and equity finance must be maintained so that one or the other, as required, can be used for the company's next expansion. An example of the need for balanced finance and the risks involved in having a biased emphasis can be provided by a study of the *capital gearing* of a company. Capital gearing is a term which describes the balance sheet proportions of debt and equity capital. If there is a high ratio of debt to equity capital, the company is highly-geared and with the converse, is low-geared.

In a low-geared company, the providers of the fixed interest debt finance are reasonably well-protected against fluctuations in profits made by the company, though it may be difficult to raise debt finance if the fluctuations are too great and extend, on occasions, to a loss. This is so, in part, because the providers of fixed interest debt capital are not likely to be attracted to a situation wherein they are exposed to the risk of no return in bad years and a return limited by the agreed interest rate in highly profitably years. On the other hand, a company achieving steady performance can raise the yield on its ordinary shares by becoming highly-geared and paying out a high proportion of its profits for fixed interest debt.

Hence, the extent to which a company can use various forms of finance and also the times at which it can call on these additional sources will, to a certain degree, depend on that particular company's financial performance. It would be wrong, therefore, to use debt finance, say at a cost of 8 per cent, to finance a series of projects where that rate of return is used as the accept/ reject criterion, and subsequently find that the next series of projects will be financed with equity capital at 16 per cent. A good project accepted in one period may then be rejected in another.

The cost of capital is therefore, properly calculated as a weighted average of the cost of finance already raised as demonstrated in the following example.

Example 3.1

Dick Shonry Limited has ordinary shares issued which have a current market price per share of £27.60. This values the total issue at £6 900 000. Next year's dividend is anticipated to be £2.10 and the company confidently expects the dividend to grow at an annual rate of 5 per cent.

In addition, the company has issued preferred shares to the total current value of £250 000. These shares bear an annual dividend of £6.00 and they each have a current market value of £42.00.

Dick Shonry Limited has also £750 000 worth of debt in the form of debenture bonds with a par value of £100.00, bearing an interest rate of 7.25 per cent and which matures in 12 years' time. The market price of each bond is £110.00.

Calculate the weighted average cost of capital for Dick Shonry Limited if tax is paid at 30 per cent.

The rate of return on ordinary shares can be calculated by the use of a number of different *valuation models* based upon one or both of the two main expectations of a purchaser of such shares. Firstly, a holder of ordinary shares will look for a dividend yield not less than that which can be obtained from other equivalent investments bearing approximately the same risk; secondly, a capital appreciation in the market price of the share will be expected. One such valuation model is the *dividend growth valuation model*:

$$P_0 = \frac{D_1}{r - g} \qquad\qquad (3.1)$$

where P_0 = present worth of the stream of future dividends or the current price per share;

r = required rate of return;
g = expected annual growth in dividends;
D_1 = annual dividend for next year.

This formula may be transposed to give:

$$r = \frac{D_1}{P_0} + g \tag{3.2}$$

The dividend growth valuation model can only be valid if r is greater than g.

For the example, the cost to the company of capital raised by the ordinary shares will be:

$$r = (2.10/27.6 + 0.05)100 = \underline{12.6\%}$$

The dividend on preference shares is paid out of after-tax profits and, therefore, the cost of the preferred stock can be established more precisely without the need to estimate future dividend/interest rates, from:

$$K_p = \frac{D}{P} \tag{3.3}$$

where K_p = cost of preferred stock;
D = annual dividends per share;
P = market price per share.

The cost of the preferred stock is therefore:

$$K_p = 6/42 \times 100 = \underline{14.29\%}$$

As far as the debt or debenture bonds are concerned, this is long-term debt. It bears a fixed rate of interest until a specified date in the future when the debt will be repaid. If the debenture bonds are both issued and then redeemed on maturity at their face or par value (that is the value that appears on the bond itself – often but not necessarily £100 – quite apart from the market price since issue) then the cost to service the bond is the same as the fixed rate of interest. If the bond is issued and/or redeemed at a discount to its par value (less than par) or at a premium to par (greater than par) the effective cost of capital so raised will not be the same as the fixed rate of interest and will need to be calculated by the methods set out in Chapter 4. Alternatively an approximate formula may be used:

$$Y_m = \frac{I + (P - P_m)/n}{(P + P_m)/2} \tag{3.4}$$

where Y_m = yield to maturity for existing loan;
I = annual interest paid, £;
P = par value of loan stock, £;
P_m = current market value of stock, £;
n = number of years to maturity.

Therefore, in the case of Dick Shonry Limited the yield to maturity will be:

$$Y_m = [7.25 + (100 - 110)/12]/[(100 + 110)/2] = \underline{6.11\%}$$

The interest paid on debt capital can be deducted from profit before tax, therefore the rate of tax to be paid may be deducted from the gross cost of the capital. The effective cost of the long-term debt of Dick Shonry Limited, so long as the company pays tax, is therefore 6.11 (1 − 0.30) = 4.28%.

The cost of capital to the company taking all the different forms of capital used is calculated as set out in Table 3.1 and amounts to 11.81%.

Table 3.1

Source	Market value (£)	Cost after tax	Capital proportions	Weighted cost
Ordinary shares	6 900 000	0.1260	0.873	0.1100
Preferred stock	250 000	0.1429	0.032	0.0046
Debt	750 000	0.0428	0.095	0.0041
	7 900 000		1.000	0.1181

A company should invest only in projects which have a *rate of return* in excess of its cost of capital. The expression *rate of return*, in this sense, is synonymous with *yield, profit, interest* or *gain* and often these words are used in an interchangeable sense. *Interest* is more usually selected with reference to the gain which arises from an investment about which there is considerable certainty as to the return of the capital more or less on demand. For example, *interest* is paid on a deposit in a bank account because there is a great deal of certainty that the capital in the account can be withdrawn at a few days' notice.

3.5 The time value of money

Consider £100 invested today in a bank account bearing an interest rate of 12 per cent per year. In one year's time the interest will be added to the capital sum and the original £100 will have grown to £112. If the capital plus interest is then left in the account for a further year the total at the end of the second year will amount to £125.44 and so on. The calculation over three years can be set out as follows:

	£
Initial capital	100.00
Interest @ 12 per cent – Year 1	12.00
Total at end of Year 1	112.00
Interest @ 12 per cent – Year 2	13.44
Total at end of Year 2	125.44
Interest @ 12 per cent – Year 3	15.05
Total at end of Year 3	£140.49

The above process of calculating interest on accumulated interest as well as on capital is known as *compounding* and the total compound interest paid on £100 at 12 per cent for three years is thus £40.49.

Given a rate of interest as 12 per cent, then it can be said that £100 at today's date is *equivalent* to £125.44 at the end of two years or £140.49 at the end of three years.

Compounding can be expressed mathematically by the following formula:

$$F = P\,(1 + i)^n \qquad\qquad (3.5)$$

where F = compound total;
$\quad\ P$ = original sum or the principal;
$\quad\ i$ = rate of interest per period;
$\quad\ n$ = number of periods.

The reverse process of compounding is known as *discounting*:

$$P = \frac{F}{(1 + i)^n} \qquad\qquad (3.6)$$

£100 is said to be the *present worth* or *present value* of £140.49 at 12 per cent per year for three years.

3.6 Cash flows

One of the most difficult facets of problem-solving is to define the problem to be solved exactly. Once this is done, the technique to aid solution can be established and then the solution itself is often not difficult. In all investment problems there will be capital flowing out from an investor's resources at some stages of a project in the expectation that, sooner or later, more cash will flow back into them. Movements of cash, either into or out of the project account, are referred to as *cash flows*. Income or receipts are known as *positive* cash flows and expenditure or payments are known as *negative* cash flows. Capital outlays fall into the latter category.

The resultant cash flow, or net cash flow, for a year of a project's life may well be the result of summing, with due regard to sign, a large number of separate cash flows that arise from a wide variety of causes – some receipts, some payments. The establishment of the best and most accurate net result for one period of time may in itself be a complicated task involving forecasting, estimation, a knowledge of taxation, economic trends and the likely timing of events. At the stage, therefore, when it is possible to arrive at the net cash flows for a project throughout that project's life, the problem has been reduced to comparatively simple terms. The calculation of its merit on a financial basis then becomes relatively straightforward. To facilitate further the final calculation, a *cash flow diagram*, built up on a time base, is of considerable use in visualizing the problem. Figure 3.1 illustrates one form of such a diagram. The upper part of the diagram represents, from an investor's point of view, the deposit of £100 in a bank at an interest rate of 6 per cent, the resulting sum being withdrawn after three years. In the lower part of Fig. 3.1 are represented the same cash flows from the banker's point of view. In the cash flow diagram arrows are used to depict the actual cash flows. Arrows pointing downwards conventionally represent negative cash flows or payments; arrows pointing upwards represent positive cash flows or receipts. The cash flows shown on such a diagram may be referred to as *net cash flows* when they are the result of adding positive and negative cash flows which occur at one point in time.

It will be noted from these diagrams that although interest accrues on an annual basis, since there is no actual physical movement of cash between lender and borrower, nothing is recorded on the diagram at the intermediate stages. There are only two cash flows, the first when the original sum is

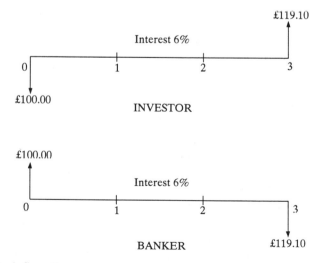

Fig. 3.1 Cash flow diagrams.

deposited and the second when the original sum, plus accumulated interest, is removed from the bank.

When appraising investment projects, it is usual to adopt a *year-end* convention for the purpose of establishing the timing of cash flows. For most projects a period of one year is the shortest time that can conveniently or realistically be considered. It is assumed that all of the individual cash flows which occur during one particular period of a year are concentrated in a single cash flow at the end of that year. Whilst it is known, for example, that wages are paid weekly, salaries usually monthly, material accounts and the like very often several months after the materials are delivered and incorporated in the work and so on, these and other cash flows for that year are all added together and indicated as an accumulation at the end of the period of one year. Any more detailed calculation using shorter time periods is often not warranted in the circumstances.

3.7 Equivalence for comparison and selection

It has been demonstrated that, given an interest rate, a sum of money has equivalent, though different, numerical values at all other points in time. If the interest rate for any given condition is known, or alternatively, if it is known what rate of return is required from each investment, it will now be apparent that there is a means of comparing different patterns of cash flows as between one investment project and another. Which project should be accepted as between *mutually exclusive* projects, and which from a collection of proposals should receive investment attention, can thus be determined. In some methods of appraisal it is necessary to establish a rate of interest or a required rate of return *before* carrying out the calculation. In others, the estimated rate of return earned by an investment can be calculated and compared with a required standard or, in special circumstances, with that of competing investments.

The following example illustrates the effect of taking the time-value of money into account in making a choice between mutually exclusive projects.

Example 3.2

It is found necessary to purchase an additional compressor and a company decides that it will be one of two models, *both of which will provide adequate and acceptable performance.* As is common with an investment in mechanical equipment, the choice is to spend more initially in return for the expectation of having lower subsequent maintenance costs and less interrupted service. This is reflected in the cash flow diagrams for the compressors shown in Fig. 3.2. The useful life of each compressor is estimated to be five years. The

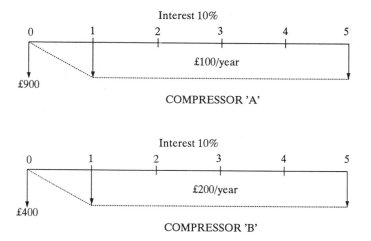

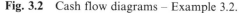

Fig. 3.2 Cash flow diagrams – Example 3.2.

company has a cost of capital of 10 per cent. Which compressor should it purchase?

If the figures in each case are added up as they are displayed, both compressors will cost a total of £1400. This sum recognizes that an initial commitment to buy either one of the models is also a commitment to incur the further associated expenses throughout its useful life. Such an addition, however, does not take account of the phasing and timing of the payments and hence the time-value of money. The equivalent values for each compressor, on a comparable basis, will be established if each cash flow to Year 0 is discounted at the company's cost of capital of 10 per cent. The present worth or present value of the series of cash flows will thus have been calculated.

In order to discount each cash flow back to Year 0 it is necessary to calculate the *discount factor*, $1/(1 + i)^n$ for a value of $i = 10$ per cent and values of n from 1 to 5. This can be a somewhat tedious process unless discount tables or a computer program are used for the purpose. (Some typical discount tables are provided at the end of this book.)

The cash flow diagrams for each compressor can now be restated to illustrate this calculation and these are shown in Table 3.2. Discounting all flows to Year 0 gives a total of £1279 in the case of Compressor 'A' and £1158 in the case of Compressor 'B'. (It should be noted that in this case Year 0 has been used as the point at which the equivalent sums have been evaluated. It makes no difference to the ultimate result if any other year, either during the life of the compressors or outside it, is selected, so long as both the equivalent values which are calculated are related to the same base year.)

The significance of the equivalent sums is as follows. In the case of Compressor 'A', £1279 at Year 0 buys the compressor, the remainder is

Table 3.2

Compressor 'A'

+	PERIOD	−	Discount factor 10%	Discounted cash flows
(£)	(year)	(£)		
	0	900	1·000	900·00
	1	100	0·909	90·90
	2	100	0·826	82·60
	3	100	0·751	75·10
	4	100	0·683	68·30
	5	100	0·621	62·10
	Total	1400		1279·00

Compressor 'B'

+	PERIOD	−	Discount factor 10%	Discounted cash flows
(£)	(year)	(£)		
	0	400	1·000	400·00
	1	200	0·909	181·80
	2	200	0·826	165·20
	3	200	0·751	150·20
	4	200	0·683	136·60
	5	200	0·621	124·20
	Total	1400		1158·00

conceptually invested at 10 per cent from which £100 is taken at the end of each year to pay for maintenance. On this basis, the choice must rest with Compressor 'B', since apparently this can be financed with a lower initial capital sum than can Compressor 'A'. The equivalent cost at Year 0 is lower. A general point arises – the sooner income is received and the later payments can be delayed, the greater will be the financial advantage.

There may be intangibles to be considered in this case. The company may possibly have a considerable number of type 'A' compressors and spares and maintenance policy based on it. There may be a reluctance to dispense with them or to set up two stocks of facilities. Such factors should be recorded with the calculation and be taken into account in making the final evaluation and decision.

3.8 Effect of rate of return

The rate of return, or interest, features prominently in the above calculation. From a glance at discount tables it can be seen that, for a given number of periods of time, the discount factor reduces for an increase in the percentage discount rate. Since there is this variation of discount factor, then clearly, at different rates of interest, the present worth of cash flows will vary and so too may the decision as to which is the most favourable series.

Example 3.3

Which is the most favourable of the following series of cash flows at rates of return of 5 per cent; 10 per cent; 20 per cent?
Series (a) – £1000 in the first year followed by £200 each year for the next four years.
Series (b) – £1000 in the first year followed by £200 at the end of Year 2 and £686 at the end of Year 4.

From the calculations in Tables 3.3 and 3.4 it appears that with a 5 per cent rate of return, series (b) is the most favourable, at 10 per cent both are equally favourable and at 20 per cent, series (a) is most favourable. Hence it can be seen that, where a method of comparison is used in which a rate of return of interest needs to be established before the calculation is made, care must be taken to see that as accurate and realistic a rate as possible is selected. The choice between alternatives may well hinge on the selection of an appropriate interest rate.

This example also illustrates the desirability of having further discount factors beyond those used in the tables. In series (a) above, the size of cash flow is constant for Years 1 to 4. Hence if the four relevant discount factors

Table 3.3

Series (a)

+	PERIOD	–	D/F	DCF	D/F	DCF	D/F	DCF
(£)	(year)	(£)	5%	(£)	10%	(£)	20%	(£)
1000	0		1·000	1000	1·000	1000	1·000	1000
200	1		0·952	190	0·909	182	0·833	167
200	2		0·907	181	0·826	165	0·694	139
200	3		0·864	173	0·751	150	0·579	116
200	4		0·823	165	0·683	137	0·482	96
			Totals	1709		1634		1518

Table 3.4

Series (b)

+	PERIOD	—	D/F	DCF	D/F	˙DCF	D/F	DCF
(£)	(year)	(£)	5%	(£)	10%	(£)	20%	(£)
1000	0		1·000	1000	1·000	1000	1·000	1000
	1							
200	2		0·907	181	0·826	165	0·694	139
	3							
686	4		0·823	565	0·683	469	0·482	331
			Totals	1746		1634		1470

are added up and multiplied by 200 (in this case), the same answer is obtained as by making each of the four calculations separately. Discount tables are drawn up to give the discount factors for the present worth of 1 per period; that is a uniform series of cash flows at the end of each period. In this case, for an interest rate of 5 per cent and for a period of four years, the factor is 3.546 – the same as that obtained by adding the four 5 per cent discount factors listed in the appropriate column above. Some discount tables for '1 per period' are also given at the end of this book.

3.9 The calculation of achieved rate of return

In making the calculation above it has been necessary to select a rate of return, or an interest rate, before commencing the appraisal. With a subsequent change in the cost of capital, or the required rate of return, the calculation must be made afresh. If, however, the actual rate of return is established, the calculation can stand, so long as the basic data remain substantially unaltered. Comparison of the calculated rate of return with that required then gives guidance as to acceptability or not.

Example 3.4

What is the rate of return achieved if £1000 is invested to bring in a net profit of £100 by the end of the first year after investment, and this profit then increases by £100 per year for each of the next four years?

A trial and error approach is needed in order to establish the rate of return. This can be achieved by looking at each year's cash flow in turn and applying the appropriate discount factor. A first guess of interest rate is made from inspection and experience and the discounting of each cash flow is then carried out. If this results in a positive answer, the discount factors are too large and another trial must be made at a higher interest rate and vice versa. The aim is to get two trial interest rates close together but either side of the required rate of return. An interpolation can then be carried out.

By interpolation (see Table 3.5), the interest rate at which the cash flows will be discounted to £1000 at Year 0 will be almost exactly halfway between the two trial rates chosen, that is 12 per cent. A linear interpolation between the two trial rates is not strictly exact but, providing the two rates are close enough together, the discrepancy is of no practical significance. If the cash flows are uniform throughout the study period, discount factors can be used in place of considering each year separately.

Table 3.5

+ PERIOD −			Discount factor 11%	Discounted cash flow		Discount factor 13%	Discounted cash flow	
(£)	(year)	(£)		+	−		+	−
	0	1000	1·000		1000	1·000		1000
100	1		0·901	90·1		0·885	88·5	
200	2		0·812	162·4		0·783	156·6	
300	3		0·731	219·3		0·693	207·9	
400	4		0·659	263·6		0·613	245·2	
500	5		0·593	296·5		0·543	271·5	
		Totals		1031·9	1000		969·7	1000
		Net Totals		31·9				30·3

3.10 Capital rationing

The cost of borrowing capital, or raising funds from other sources, clearly plays a major role in the determination of the cost of capital. Having established the cost of capital, those projects which are expected to give rise to a greater return are acceptable and vice versa. However, it is frequently the case that not all investments meeting the set parameters can be accepted, because a company finds that it has more opportunities in which to invest than funds to meet the total requirements. Either the funds may not be available to the company or it may have chosen to restrict its capital investment over a given period, perhaps because it is not eager to go to outside sources of supply for more capital. Where there is a limit to the funds available, a system of rationing must be imposed and projects must be ranked in such an order that it represents a schedule of priorities.

Table 3.6

No.	Project investment (£)	Rate of return (%)	Cumulative demand (£)
1	1 000 500	47	1 000 500
2	250 000	40	1 250 500
3	735 000	35	1 985 500
4	1 724 000	32	3 709 500
5	675 000	24	4 384 500
6	300 000	21	4 684 500
7	400 000	15	5 084 500
8	240 000	8	5 324 500
9	408 000	4	5 732 500

Where a rationing situation exists, the principle to be adopted in selecting the best projects is to list the demands in descending order of rate of return as in Table 3.6.

If a company now has £4.5m at a cost of 10 per cent in its current budget for investment, the line will be drawn under project number 5 in this list. The minimum acceptable rate of return now becomes 24 per cent even though it may have been assumed to be considerably less than that before conducting the exercise. The dangers, therefore, in such a situation of investing in projects indiscriminately until the available capital is used up are apparent. Even though the established minimum acceptable rate of return for a company may be 10 per cent, it will be seen from Table 3.6, given that situation, that projects arising in the current situation with a rate of return of less than 24 per cent, cannot compete with those appearing above the broken line.

The above situation can be displayed graphically as in Fig. 3.3. The horizontal line represents the ration of capital which is available, illustrating

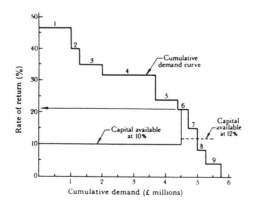

Fig. 3.3 Capital rationing curve.

that the cut-off rate is greater than the minimum acceptable rate of return. In practice, the supply of capital is usually subject to the laws of supply and demand, and additional capital beyond that allocated in the rationing position can be made to be available at an increased cost. This extension of the availability of capital is illustrated by the dotted line in Fig. 3.3 and shows the adjustment of the cut-off rate which results.

Chapter 4
Establishing Feasibility by Project Appraisal

4.1 Investment appraisal methods

The simpler methods of investment appraisal, of which there are two principal variations, are often now referred to as *rule-of-thumb* methods. In their favour it can be said that they are simple to use, but there is no doubt that their use can result in poor and inappropriate investment decisions being made.

So often in a capital budgeting situation, there is a surplus of apparently profitable projects competing for capital funds which are in short supply. In this case it is not only required to know whether the project in itself is a viable undertaking (in so much as it will give an acceptable rate of return on the investment) but there is also a need to rank these competing projects in an order of merit. Capital which is in restricted supply can then be allocated in the most efficient way. In view of the great importance of the decision to invest funds in projects, largely because in doing so the long-term future is being shaped, there is a need to use methods of appraisal which make a realistic and accurate contribution to the ultimate decision.

4.2 Payback

Payback is alternatively known as *payout* or *pay-off* and it is a simple rule-of-thumb method. It is a method which has received considerable attention wherever an attempt has been made to appraise projects. It is still widely used. The basis of the method is to determine the number of years which it will take to repay the capital invested, taking into account the cash flows which arise once tax has been deducted from them, if necessary. The shorter the payback period so calculated, the greater the favour with which the project can be viewed. Calculations are sometimes made using the cash flows before deduction of tax. Such a convention is quite unrealistic. When using payback as a means of determining acceptability, a standard is frequently set as a measure of the number of years to payback for specified types of investment projects (commonly three to five years) and any projects within given categories offering lower paybacks are considered acceptable.

Table 4.1 Project X.

Year	Cash flow (£)
0	(10 000)
1	2 000
2	5 000
3	4 000
4	4 000

Payback = 2¾ years

A simple example of a calculation for payback is set out in Table 4.1. It will be seen that this method has the advantage of simplicity. Where single investments, obviously showing a very high rate of return and a very short payback are concerned, probably little more sophistication is required in order to assist with an investment decision. Such a situation might well arise in an industry where technological advancement is being made at an extremely rapid rate. The length of the payback calculated in such situations is simply a crude measure of the time that the investment is under risk either from obsolescence or competition. Such situations frequently occur in the chemical, pharmaceutical, petrochemical and oil industries, for example. There is, however, little justification for using such a method where the magnitude of the investment is large and the payback as calculated is spread over a longer period.

It is often assumed that the payback method gives a conservative evaluation of an investment. In situations where only short payback periods are favourably considered, the overall policy may well be described as being conservative. It may also mean, however, that many perfectly suitable and adequate opportunities to invest may be lost.

Another of the disadvantages of using the payback method arises from its neglect of the time-value of money. Because of this it cannot sensibly be used in order to rank projects in their order of profitability. As demonstrated in Table 4.2, three projects, obviously having cash flows of widely differing merits, can be rated by this method as being of equal virtue. It is a major disadvantage that it does not take account of the relative magnitude of cash flows during the overall life of a project. It penalizes a project which is slow to produce large cash flows in its early stages as against a project which has high initial flows, perhaps quickly dying away after the first few years. (This in itself may not be contrary to the general principles arising from taking into account the time-value of money, but when combined with the restricted view taken of the life of the project it does not enhance the merit of the method.) Table 4.2 has already demonstrated three projects which would all be ranked of equal merit by the payback method, but which clearly have different virtues when the whole life of a project is taken into account.

Table 4.2

Year	Cash flows (£) for projects			
	S	T	U	
0	(10 000)	(10 000)	(10 000)	
1	1 000	6 000	1 000	
2	1 000	2 000	6 000	
3	4 000	1 000	2 000	Payback
4	4 000	1 000	1 000	↓
5	6 000	0	1 500	
6	8 000	0	500	
Payback (years)	4	4	4	

Another disadvantage of the payback method, though of less importance, is that it does not allow the introduction into the calculation of the possible residual value of assets which accrue after the payback period. In certain projects, particularly those involving buildings, the residual values may be considerable.

Payback probably has its optimum use as a means of making a preliminary screening of all those projects which are available for consideration. In this way, the more detailed and sophisticated analysis can be reserved for those cases which have passed this first test of suitability. Another suitable use for this particular method might be in a situation where an investment is constantly under risk from political or like influences, and a measure of the period of risk is desirable. Such influence might be very strong in, for example, the more unstable overseas territories where large assets can virtually be lost overnight because of political instability and/or nationalization.

4.3 Rate of return

Like the previous rule-of-thumb method, *rate of return* goes under a number of alternative names, *accountants' rate of return* and *book rate of return* being two of the alternatives. Not only are there alternative names for the method but also there is a variety of ways of making the calculation.

Rate of return is a ratio of profit to capital invested. Clearly, profit can be variously interpreted. It can, for example, be gross or net of tax. It is certainly more meaningful to use a figure net of tax, since this represents money that can be put to use. The profit figure used in calculating the ratio is normally after allowance for depreciation of the assets involved has been made. Further, it is necessary to make a decision as to which year's figure for the surplus will be used. To use the initial profit may be misleading because few projects give rise to immediate surpluses which are typical of their subsequent performance. In choosing any other specific year's profits it may be difficult to find a representative figure. This leads to using an average figure or at least consideration of a project's returns over a significant period in order to arrive at a figure more likely to represent true performance.

Having defined the figure to be used for profit, consideration must be given to that used for the capital invested. Capital invested is normally taken to be the initial capital for purchasing plant and other fixed assets together with the associated working capital. Working capital in this context is that finance which is required to provide resources in order to get a project into production during its initial stages. This may be necessary, for example, before any revenue is forthcoming from the operational use of the assets. There are generally two variations in the way that the capital sum is incorporated in calculating the ratio. More generally the total figure is used, but an alternative is to take an average of the initial amount plus the residual value at the end of the useful life of an investment.

It is often claimed that the percentage rate of return which is obtained as the result of this calculation is meaningful and can realistically be compared with a cost of capital. This, however, is not generally true except for those projects which have a very long life and fairly uniform cash flows throughout, together with projects which have but a one-year life. In other instances, wherein the figures used are on a different basis, it will be found that the rate of return tends to underestimate the true yield of a project. Thus, it might well lead to incorrect decisions, an investment in certain proposals not being made because of apparently insufficient yield. The example set out in Table 4.3 demonstrates the difference between a yield as calculated on the rate of return method and that as calculated on a discounted cash flow basis. It will be seen that the discounted cash flow yield is in excess of the calculated rate of return. As an example, when money can be borrowed at, say, 6 per cent, there would be no incentive to borrow and invest on this basis if one uses the rate of return method of calculation.

The rate of return method takes no account of the incidence of cash flows throughout the life of a project. On the basis that £1 now is worth more than £1 in the future, this is a serious deficiency. Table 4.4 illustrates the cash flows for three projects 'P', 'Q', and 'R'. For each, a rate of return over a five year period has been established by dividing the average profit by the total of the capital invested. In each case the rate of return is found to be $33\frac{1}{3}$ per cent. Each project will be seen to have quite different merits from the investment point of view. Project 'P' has an increasing rate of cash flow throughout its life; Project 'Q' has a more or less decreasing cash flow; the life of Project 'R' as a money-earning project is extremely short with high initial cash flows.

Table 4.3 Project A

Year	Cash flow (£)
0	(10 000)
1	2 505
2	2 505
3	2 505
4	2 505
5	2 505

Yield on discounted cash flow basis $=8$ per cent

$$\text{Rate of return} = \frac{\text{Average profit}}{\text{Capital invested}}$$

$$= \frac{(5 \times 2505 - 10\,000)}{10\,000 \times 5}$$

$$= 5.05 \text{ per cent}$$

If money to finance the project is borrowed at 8 per cent per annum interest, then the profit on a discounted cash flow basis or any other realistic basis should just balance the interest as follows:

Year	Capital outstanding at beginning of year (£)	Interest for year at 8% (£)	Total funds outstanding at year end (£)	Income for year (£)	Capital outstanding at end of year (£)
1	10 000	800	10 800	2505	8295
2	8 295	664	8 959	2505	6454
3	6 454	517	6 971	2505	4466
4	4 466	358	4 824	2505	2319
5	2 319	186	2 505	2505	—

Table 4.4

Year	Project cash flows (£)		
	P	Q	R
0	(15 000)	(15 000)	(15 000)
1	1 000	10 000	13 000
2	1 000	5 000	12 000
3	6 000	5 000	0
4	8 000	2 000	0
5	9 000	3 000	0
Rate of return $= \dfrac{\text{Average profit}}{\text{Capital invested}}$	$33\frac{1}{3}\%$	$33\frac{1}{3}\%$	$33\frac{1}{3}\%$

Applying the basic premise concerning the time-value of money to these projects, there is no doubt that Project 'R' would be accepted as being the best of the three in which to invest £15 000. The rate of return method does not differentiate between these projects. It is worth noting, however, that payback would, in this instance, have supported a recommendation for 'R'. Table 4.4 shows that rate of return does not generally take account of the earning life of a project and it discriminates *against* cases where there are large cash flows in the early stages of a project's life. This particular situation is highlighted frequently in the construction of works where there is often a considerable delay between the initial investment of capital and the first returns from the production operations. Finance that has been invested in this way is at considerable cost and risk to the investors for at least two to three years.

4.4 Depreciation

One of the more important aspects of engineering economy is that of *capital recovery*. It is of paramount importance for all project investments or purchases of assets that a deduction is made from the gross income arising from an investment to take care of its depreciation. In the simplest case, depreciation may arise solely from wear and tear of the asset. Every time a motor car engine is turned over, some wear takes place and the engine is then just that much nearer to the end of its working life. Ultimately, the time will come when an asset needs replacing and at that time a sum of money will have to be found to purchase the replacement. Replacement, however, need not necessarily arise solely from wear and tear. It may well be necessitated by obsolescence due to technical developments either in its function or its product; by marketing developments, perhaps caused by either the market contracting or expanding; because of depletion, for example, of a ballast pit or of a mine.

It is desirable, therefore, to distribute the initial cost of an asset over its predicted life and to make allowances for it in all investment appraisals. In making this distribution, due regard must be paid to any residual value of the asset at the end of its useful life, even if this is but a scrap value. It will be appreciated that it is essential to have an accurate estimation of an investment's working life so as to distribute the recovery for depreciation as realistically as possible.

There are various ways of determining the financial depreciation of an asset, some of which are purely empirical and have differing purposes. One aspect of an accountant's work is frequently described as *stewardship*. An accountant has a responsibility, in financial accounting, to prepare profit and loss accounts and balance sheets for a company. Such statements are necessary in law and they should reflect a fairness to the shareholders in the company who are the owners of the business. Depreciation in the accounting sense is concerned with reducing the value of an asset, for example a tractor, in the appropriate account and transferring that reduction of value to an account which shows it as an expense of the business. This expense can then be shown with other like provisions, in the profit and loss account. Their prime purpose is to ensure that a restriction, to the extent of the depreciation, is placed on the funds which are allowed to flow out of the business by way of dividends and so on. The depreciation provisions, therefore, reduce the profit that is available for distribution and will normally be shown as an entry in a profit and loss account before the *profit before taxation* is calculated.

Depreciation in an accounting sense can differ appreciably from that as used in the sense of an economist. In an accounting system, assets may be depreciated at very rapid rates, especially if the finance invested in those assets is at considerable risk. The rate of the depreciation may bear little or no relationship to the actual life of the asset, a situation which is often true of a construction contractor's mechanical plant which is frequently depreciated in full over a two to four year period. Its working life may be ten years or even longer. An economist thinks in terms of depreciating equipment over its true life; in company accounts depreciation rates are set, out of different considerations such as the implications of tax, in accordance with a policy set by a Board of Directors.

Example 4.1

A simple example, ignoring the time-value of money, concerns the purchase of a motor car. A motor car is purchased for £5000 and it is estimated to have a working life of five years when it will be sold for £1000. The amount, in money terms, of the asset which will be used over the period of five years is, therefore, £4000, or £800 per year.

If the owner travels 10 000 miles per year in the car, pays £120 per year for tax, £300 per year for insurance, £350 per year for repairs, oils and other consumables and finds that the car travels 30 miles to a gallon of petrol which, on average, costs £2.50; what is the average cost of travelling 100 miles in the car?

Total cost per year:

		£
Tax		120.00
Insurance		300.00
Repairs, etc.		350.00
Petrol $\dfrac{10\,000}{30} \times 2.50$		833.33
Depreciation		800.00
Total		£2403.33

Therefore, average cost of 100 miles $= \dfrac{2403}{100}$

$\qquad\qquad\qquad\qquad\qquad\qquad\qquad\quad = \underline{£24.03}$

This example illustrates the need to take account of the wasting value of the asset in the running cost. For every 100 miles which are covered, £8.00 must be taken into account (and in theory must be put on one side) against the day when the car is finally sold and must be replaced. If depreciation is ignored and the running costs for 100 miles are taken as £16.03, then at the end of five years the car owner has neither the asset left (except for its resale value) nor the money with which to replace it. In the case of a company, if adequate allowance is not made in the accounts for depreciation, then in the case of a profitable operation, there may be tendency to be distributing the company's capital to the shareholders in the form of dividends.

This calculation does not take into account the time-value of money. The £4000 depreciation on the motor car is effectively locked up on the purchase of the vehicle and is only released at a rate proportional to that at which mileage is accumulated. The £1000 residual value is effectively locked up in the vehicle from its purchase until its sale five years later. This money is denied to the car owner during the car's life, and cannot take on some other interest or profit-earning role. This must be recognized in the calculation.

4.5 Algebraic relationships between interest, time and capital

Let, as before, P = the initial capital invested in a project

$\qquad\qquad\quad F$ = a compounded amount

$\qquad\qquad\quad n$ = the number of time periods

$\qquad\qquad\quad i$ = interest rate per period expressed as a decimal

$\qquad\qquad\quad A$ = a uniform series of end-of-period payments or receipts which extends for n periods of time

$\qquad\qquad\quad S$ = the salvage or resale value at the end of n periods of time

If a sum, A, is invested at the end of each of n periods, then the first instalment to be invested will earn interest for $(n - 1)$ periods and after n periods it will have compounded to a sum $A(1 + i)^{n-1}$ ... from Equation (3.5).

In the case of the sum, A, invested at the end of the second period it will amount to $A(1 + i)^{n-2}$ by the end of n periods. Finally the sum, A, invested at the end of n periods will accumulate no interest.

The total sum accumulated as a result of these investments of A amounts to:

$$F = A(1 + i)^{n-1} + A(1 + i)^{n-2} + \ldots + A$$

$$= A[1 + (1 + i) + (1 + i)^2 + \ldots + (1 + i)^{n-1}] \qquad \text{(A)}$$

Multiplying both sides of this equation by $(1 + i)$,

$$(1 + i)F = A[(1 + i) + (1 + i)^2 + \ldots (1 + i)^n] \qquad \text{(B)}$$

Subtract Equation (a) from Equation (b)

$$iF = A[(1 + i)^n - 1]$$

and $$F = A \left[\frac{(1 + i)^n - 1}{i} \right] \qquad (4.1)$$

The expression in the square brackets of Equation (4.1) is known as the *uniform series compound amount factor* since when applied to a uniform series of investments, A, at the end of each n period of time, at interest rate, i, they will accumulate by compounding to a sum, F.

Equation (4.1) can be alternatively stated as,

$$A = F \left[\frac{i}{(1 + i)^n - 1} \right] \qquad (4.2)$$

In this case the factor inside the square bracket is known as the *uniform series sinking fund deposit factor*.

Equation (3.5), $F = P(1 + i)^n$, can now be substituted in Equation (4.2) to eliminate F and give,

$$A = P(1 + i)^n \left[\frac{i}{(1 + i)^n - 1} \right]$$

therefore,

$$A = P \left[\frac{i(1 + i)^n}{(1 + i)^n - 1} \right] \text{ or } P \left[\frac{i}{(1 + i)^n - 1} + i \right] \qquad (4.3)$$

Either of the square bracketed factors in Equation (4.3) is known as the *uniform series capital recovery factor* and by rearranging (4.3) we get

$$P = A \left[\frac{(1 + i)^n - 1}{i(1 + i)^n} \right] \qquad (4.4)$$

giving $P = A \times$ *uniform series present worth factor*.

The above equations (including Equations (3.5) and (3.6)) can be expressed in a symbolic form which is much more convenient when formulating problems involving their use. Table 4.5 summarizes the various abbreviations. These symbolic forms will be used in future problem formulations. The symbolic form can, for example, in the case of the *uniform series recovery factor* be read as 'factor to find A given P at an interest rate of i over a period of n years'.

Table 4.5

Equation	Factor	Symbolic form
3.5	Single payment compound factor	$(F/P,i\%,n)$
3.6	Single payment present worth factor	$(P/F,i\%,n)$
4.1	Uniform series compound amount factor	$(F/A,i\%,n)$
4.2	Uniform series sinking fund deposit factor	$(A/F,i\%,n)$
4.3	Uniform series capital recovery factor	$(A/P,i\%,n)$
4.4	Uniform series present worth factor	$(P/A,i\%,n)$

In order to avoid considerable and tedious mathematical work each time one of the above formulae is applied, interest tables are used which list the values of the factors in Equations (3.5), (3.6), (4.1), (4.2), (4.3) and (4.4), for various combinations of i and n. Such tables are given at the end of this book. Alternatively many pocket calculators are programmed to provide ready answers to problems involving these formulae.

4.6 Total equivalent annual cost

The first of three methods of investment appraisal to be described, which uses discounting principles, is that of *total equivalent annual cost* to be referred to as *annual cost*. Alternative names are *annual capital charge* and the *annuity method*.

The annual cost method is one which has had decreasing use in recent times. The basis of the method is to convert all payments and receipts concerned with a proposed investment of capital into uniform series cash flows so that a comparison with alternative proposals is a relatively simple operation. This is more acceptable where estimated cash flows are already reasonably constant from one period to another, because only the capital costs involved need to be converted to equivalent annual cash flows in order to effect an overall comparison.

In principle, if a capital investment is associated with a series of annual outgoings which are constant or near constant, then, having converted the capital and the cash flows to a uniform series, using the required interest rate or rate of return, the latter can be compared with the series of receipts arising from the investment. If the equivalent annual cost of outgoings is less than the equivalent annual cost of incomings at the given rate of return used in the calculation (and assuming no untoward risks) the investment is acceptable. In similar fashion, for mutually exclusive projects it is relatively easy to convert and compare two or more series of annual costs on the basis of competing investments in order to aid a decision as to which should be chosen.

The annual cost method of analysis has been used in public undertakings and similar large organizations where cash flows result from long-term budgets and reasonably readily predictable sales of utilities such as gas, electricity and water.

The annual cost method takes into account two elements of cost in the use of capital for investment purposes. Firstly, it recognizes that where capital is invested in one project or asset it is either to be borrowed in the first place or, if it is already available for investment, denied to another possible profitable project. There is, as a result of this, an interest cost on the one hand or a 'loss of profit' cost on the other. The second element of cost is depreciation or recovery of the capital invested, less residual values. An investment can only be profitable, in the financial sense, if the net annual income arising from it is sufficient to cover both of these elements.

The conversion of an initial capital expenditure into an equivalent annual cost can be achieved by multiplying that capital sum by the uniform series capital recovery factor given in Equation (4.3). The principle of the calculation is illustrated by referring to the lower tabulation in Table 4.3. The capital recovery factor for an interest rate of 8 per cent over five periods of time is 0.2505. Hence, if it is required to convert a capital sum to uniform equivalent annual cost, the magnitude of the annual costs can be established from £10 000 × 0.2505 = £2505. It can be appreciated, therefore, that capital recovery does allow for the two elements of cost referred to above, and is based on a concept of money being borrowed, at the firm's cost of capital or profit rate requirement, to finance an investment. If the sum of annual payments arising from the investment plus the annual cost derived from distributing the capital cost over the project's life is in excess of the net income resulting from the investment, the investment should not be made, since it will not achieve the desired rate of return.

A simple adjustment needs to be made to the calculation where an asset to be purchased has a residual value. If an investment of £10 000 is to be made in a machine having a five-year life and it is estimated that the machine will have a residual value of £1000 at the end of that life, then the investment can be considered as two clearly identified parts. Firstly, there is £9000 which is invested in the machine and, since it will not be realized at the end of the machine's life, must be depreciated, in the economic sense, on an annual basis. The other part of the debt is the £1000 residual value, which is, in effect, only loaned to the investment for the length of its life. This residual value need not be depreciated since the 'debt' will be paid back when the machine is finished. There is thus only a need to charge interest at an annual rate in this example for five years. If AC is the equivalent annual cost of such an investment it may be expressed as

$$AC = (P - S) \left[\frac{i(1 + i)^n}{(1 + i)^n - 1} \right] + Si$$

where S is the residual value.

Example 4.2

An investment in a microcomputer amounts to £2500 and it is estimated to have a residual value at the end of a three-year life of £400. A monthly maintenance agreement to keep the machine in good working order is signed to the value of £8 per month. This is paid as a lump sum at the end of each year. It is expected that annual savings of £900 will be achieved as a result of using the machine. If capital costs the company 8 per cent per year, is the machine a worthwhile investment? See also Fig. 4.1.

$$\text{Annual cost} = (2500 - 400)(A/P,\ 8\%,\ 3) + 400 \times 0.08 + 96$$
$$= 2100 \times 0.3.3880 + 128$$
$$= 815 + 128 = £943$$

The machine is not, therefore, economically viable.

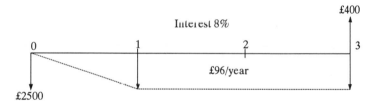

£400

Interest 8%

0   1   2   3

£96/year

£2500

Fig. 4.1 Cash flow diagram – Example 4.2.

Example 4.3

An examination is made of the lifting requirements for the construction of a large residential development consisting of tower blocks and three-storey blocks of flats. The contract for the whole works is awarded on the basis of a staged development providing continuous work for a period of eight years. The contractor, in setting out his work programme and method statement, makes a decision to purchase outright the cranes he requires and then to write-off their initial cost on the one contract. Three schemes are prepared for evaluation:

Scheme A Three Type A tower cranes with an initial cost of £240 000 each and having a residual value of £15 000. Maintenance and repair costs amount to £1800 per year each, with a major overhaul at the end of Year 3, costing £6000. Each crane driver is to be charged at £160 per week.

Scheme B One Type A tower crane as above, plus four Type B tower cranes, each having an initial cost of £116 000 and no residual value. Type B tower cranes are believed to have maintenance and repair costs of £1200 per year each and require a major overhaul at the end of Year 4 costing £5000 for each one. Crane driver costs are the same for both types of crane.

Scheme C Use seven Type B tower cranes having costs as above.

It is assumed that the service given by each scheme is satisfactory and the work can be carried out in a profitable manner. Which scheme should be adopted if capital costs the company 14 per cent?

Since many of the costs are already in constant annual sums, this problems lends itself to solution by the total equivalent annual cost method. Conversion of the capital costs to annual costs over eight years in each case will make for relatively simple calculations. See also Fig. 4.2.

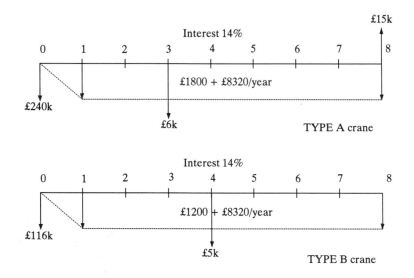

Fig. 4.2 Cash flow diagrams – Example 4.3.

Scheme A

Annual cost of investment $= [3(240\,000) - 3(5000)]\,(A|P, 14\%, 8)$
$$+ 3(5000)0.14$$
$$= (705\,000)0.2156 + (15000)0.14$$
$$= \underline{£154\,098}$$

Annual cost of major overhaul $= 3(6000)\,(P|F, 14\%, 3)(A|P, 14\%, 8)$
$$= (18\,000)(0.6750)(0.2156)$$
$$= £2620$$

Annual cost of maintenance $= 3(1800) = £5400$

Annual cost of wages $= 3(160)52 = £24\,960$

Equivalent annual cost $= £154\,098 + 2620 + 5400 + 24\,960$
$$= £187\,078$$

Scheme B

Annual cost of investment $= 51\,366 + 4(116\,000)(A|P, 14\%, 8)$
$$= 51\,366 + (464\,000)0.2156$$
$$= £151\,404$$

Annual cost of major overhaul $= 873 + 4(5000)(P|F, 14\%, 4)$
$$(A|P, 14\%, 8)$$
$$= 873 + 20\,000(0.5921)(0.2156)$$
$$= £3426$$

Annual cost of maintenance $= 1800 + 4(1200) = £6600$

Annual cost of wages $= 5(160)52 = £41\,600$

Equivalent annual cost $= 151\,404 + 3426 + 6600 + 41\,600$
$$= £203\,030$$

Scheme C

Annual cost of investment $= 7(116\,000)(A|P, 14\%, 8)$
$$= 7(116\,000)0.2156$$
$$= £175\,067$$

Annual cost of major overhaul $= 7(5000)(P|F, 14\%, 4)(A|P, 14\%, 8)$
$$= 140\,000(0.5921)(0.2156)$$
$$= £4468$$

Annual cost of maintenance $= 7(1200) = £8400$

Annual cost of wages $= 7(160)52 = £58\,240$

Equivalent annual cost $= 175\,067 + 4468 + 8400 + 58\,240$
$$= £246\,175$$

Therefore, Scheme A is the most economic.

4.7 Present worth

The present worth of a future sum of money can be found by discounting it to
the present point in time or year 0, if that is different, at the required rate of
interest. It is a feature of the discounting process, as with the calculations to

establish equivalent annual cost, that the selection of an interest rate or the cost of capital must be made *before* the arithmetical processes are undertaken. Such processes presume that a figure for interest or cost of capital can be established. The accuracy of all such calculations from the appraisal point of view, therefore, depends on accurate knowledge of the interest rate or the rate of return required.

If A_1 is the cash flow which arises in Year 1 (and it is assumed to arise specifically at the end of Year 1 by convention) then the present worth of that cash flow, $P = A_1/(1 + i)$ from Equation (3.6). If A_2 arises at the end of Year 2, A_3 at the end of Year 3, etc., then the present worth of all the cash flows associated with a project

$$P = \frac{A_1}{(1+i)} + \frac{A_2}{(1+i)^2} + \frac{A_3}{(1+i)^3} + \ldots + \frac{A_n}{(1+i)^n}$$

$$= \sum_{k=1}^{n} \frac{A_k}{(1+i)^k} \tag{4.5}$$

Referring back to Example 4.2, the present worth of the series of annual costs, i.e. £815 for each of the three years can be found:

$$P = \frac{815}{(1+0.08)} + \frac{815}{(1+0.08)^2} + \frac{815}{(1+0.08)^3}$$

$$= 815 \ (P/A, 8\%, 3) = 815 \times 2.577$$
$$= £2100$$

In calculating the present worth of a series of cash flows which arise out of a project, each of the cash flows, A above, will normally be *net* cash flows (capital outlays and payments of all kinds or cash flows *out* of the project).

The present worth method of investment appraisal is more conveniently used in its *net present worth* form by subtracting the initial capital outlay, at time 0, from the total present worth of the future net cash flows. The *initial* capital outlay is important because all other cash flows, whether positive or negative, must be taken care of (with due attention being paid to arithmetic sign) in calculating the present worth.

If C represents the initial capital outlay, then Equation (4.5) can be modified to give the net present value, NPV as follows:

$$\text{NPV} = \sum_{k=1}^{n} \frac{A_k}{(1+i)^k} - C \tag{4.6}$$

It will be seen from this formula, that if i is the minimum acceptable rate of return at which the future cash flows are discounted, then so long as NPV is positive, the investment will be acceptable. A negative value for NPV indicates future cash flows which, when discounted at the minimum acceptable rate of return, fail to exceed the capital investment. This, therefore, leads to the conclusion that the investment is not worthwhile because the minimum rate of return will not be achieved.

Example 4.4

If it is required to earn a rate of return of at least 12 per cent on invested capital, is it worth investing £10 000 to produce returns of £5000 in Year 1, £4000 in Year 2 and £2000 in Year 3?

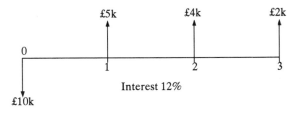

Fig. 4.3 Cash flow diagram – Example 4.4

$$\begin{aligned}
\text{NPV} &= 5000\,(P/F, 12\%, 1) + 4000(P/F, 12\%, 2) + 2000(P/F, 12\%, 3) \\
&\quad - 10\,000 \\
&= (5000 \times 0.893) + (4000 \times 0.797) + (2000 \times 0.712) - 10\,000 \\
&= -\pounds 923
\end{aligned}$$

The investment is not, therefore, worth making.

The NPV method, as demonstrated above, only establishes whether or not a single investment is likely to be profitable at a given interest rate. The answer is given in absolute terms, and there is no way of ranking projects so investigated in their order of profitability.

However, the *profitability index* or *benefit/cost ratio* is a means of doing so. Using the same symbols as before

$$\text{Profitability index} \;=\; \frac{\displaystyle\sum_{k=1}^{n} \dfrac{A_k}{(1+i)^k}}{\text{COST}} \tag{4.7}$$

COST in Equation (4.7) is interpreted as the initial amount invested or, where the capital outlay is made over more than one year, then the present value of all the capital outlays is used.

Example 4.5

If two investment projects are under consideration with a view to accepting only one of them and the cash flows are as below, which should be chosen if profitability is the sole criterion on which the choice will be made and a minimum acceptable rate of return is 10 per cent?

Project A				Project B		
+ £	**Year**	**−** £		**+** £	**Year**	**−** £
	0	10 000			0	20 000
	1	5 000		15 000	1	
12 000	2			15 000	2	
12 000	3			15 000	3	
12 000	4				4	
36 000	**Totals**	**15 000**		**45 000**	**Totals**	**20 000**

$$\text{Profitability Index (A)} = \frac{12\,000\,(P/A, 10\%, 3)(P/F, 10\%, 1)}{10\,000 + 5000\,(P/F, 10\%, 1)}$$

$$= \frac{12\,000 \times 2.487 \times 0.909}{10\,000 + (5000 \times 0.909)}$$

$$= \frac{27\,128}{14\,545} = \underline{1.864}$$

$$\text{Profitability Index (B)} = \frac{15\,000(P/A, 10\%, 3)}{20\,000}$$

$$= \frac{15\,000 \times 2.487}{20\,000}$$

$$= \underline{1.865}$$

Therefore, Project B is marginally more profitable.

4.8 Capitalized costs

A specialised variant of present worth analysis is called the *capitalized cost* method. The capitalized cost of a project is the present worth of providing a project in perpetuity and has been commonly used where assets of a civil engineering nature are concerned.

When facilities are constructed, it is often the intention of the owner that they should remain useful in perpetuity. This is particularly true of public bodies who provide facilities for the use of tax- and rate-payers. The assumption, in such a case, is that the asset will be maintained over its life and replaced at the appropriate time with one providing a similar service. Often, with such structures as embankments and rock tunnels, the question of replacement of the asset rarely arises. As far as engineering economy is concerned, perpetual life cannot be significantly differentiated from a life of, say, 100 years, as an examination of the uniform series capital recovery factor will show.

This factor at an interest rate of 12 per cent for a 50-year period is 0.12041; in other words, rather less than one-twentieth of one per cent more than the subject interest rate. For a 75-year period, the capital recovery factor is 0.12002 and for 100 years, five figures beyond the decimal point are not sufficient to make any distinction between the factor and the interest rate expressed as a decimal. If one requires, therefore, to convert a present sum to its equivalent uniform series over perpetual life, it can be effected by multiplying it by the appropriate interest rate. Conversely, if the long-term uniform annual cost associated with an asset is known, then the present worth of its future cash flows can be established by dividing it by the relevant interest rate.

Example 4.6

Calculate the total present sum of money, given an interest rate of 12 per cent, that is required to finance and maintain a public water supply scheme having the following details:

	£000
Initial cost of pumphouses	15 000
Annual maintenance cost for pumphouses	2 000
Initial cost of pumps	40 000
Annual maintenance cost of pumps	5 000
Initial cost of pipelines	150 000
Annual maintenance cost of pipelines	5 000
Estimated life of pumphouses	25 years
Estimated life of pumps	12 years
Estimated life of pipelines	30 years

The first step is to convert the replacement of the pumphouses, pumps and pipelines to an equivalent uniform annual cost. Since the replacement cost for the pumphouses, for example, must be accumulated prior to when it will be spent, the implication is that a form of *sinking fund* is required. Such a fund is used in order to provide the facility to invest a constant annual sum of money over the estimated life of an asset. In the fund, the moneys accumulate and earn interest until they reach the replacement cost of the asset over the life of the asset.

Present worth of series of renewals in perpetuity of pumphouses, pumps and pipelines:

$$= [15\,000\,(A/F, 12\%, 25) + 40\,000(A/F, 12\%, 12)$$
$$+ 150\,000\,(A/F, 12\%, 30)] \div 0.12$$
$$= [15\,000 \times 0.0.0075 + 40\,000 \times 0.0414 + 150\,000 \times 0.0041] \div 0.12$$
$$= [113 + 1656 + 615] \div 0.12 = £19\,867\,000$$

(*Note*: There are alternative percentage rates which can be used in such a calculation. As a general rule, the return on an investment is directly proportional to the amount of risk which is undertaken. A sinking fund, where used, needs to be a safe source of investment and hence will usually bear a relatively low rate of interest, perhaps of the order of 3 to 7 per cent. In a commercial undertaking the funds set aside for the replacement of assets are an important source of working capital and, depending on the nature of a firm's business, may therefore give rise to returns of 15 per cent or more if the firm is operating on a profitable basis.)

The second step is to convert the already stated uniform annual costs to present worth (all sums except answer are in £000):

Present worth of annual costs in perpetuity

$$= \frac{(2000 + 5000 + 5000)}{0.12} = £100\,000$$

$$\text{Initial costs} = 15\,000 + 40\,000 + 150\,000$$
$$= £205\,000$$

$$\text{Capitalized cost} = 19\,867 + 100\,000 + 205\,000$$
$$= £324\,867\,000$$

Example 4.7

It is intended to build a small timber toll bridge for pedestrian traffic at a cost of £1 500 000 on the basis that it will need to be widened, at a cost of £2 000 000 eight years after opening. It is estimated that maintenance costs during the first eight years will amount to £6000 per year.

If the foundations for the bridge widening are installed at the time that the first phase is carried out, then the initial cost of the work will be £1 900 000 and the maintenance costs during the first eight years will increase to £8000 per year. This alternative will, however, reduce the cost of the widening after eight years to £1 850 000. After widening, by whichever alternative, the maintenance costs are expected to be £15 000 per year in perpetuity. If the minimum required rate of return is 8 per cent, which plan should be adopted?

If the average annual traffic flow for the first eight years is expected to be 350 000 pedestrians per year, what uniform charge must be made per person during this period in order to pay for the initial construction of the cheapest alternative as established and the maintenance, before commencing construction of the second phase?

Present worth of £15 000 per year in perpetuity (at Year 9)

$$= \frac{15\,000 \times 100}{8}$$

$$= £187\,500$$

Present worth of £187 500 at end of Year 8
$$= 187\,500\,(P/F, 8\%, 8)$$
$$= 187\,500 \times 0.540$$
$$= £101\,250$$

Cash flows for method (a)			*Cash flows for method (b)*		
−ive cash flows (£)	Year	+ive cash flows (£)	−ive cash flows (£)	Year	+ive cash flows (£)
1 500 000	0		1 900 000	0	
6 000	1		8 000	1	
6 000	2		8 000	2	
6 000	3		8 000	3	
6 000	4		8 000	4	
6 000	5		8 000	5	
6 000	6		8 000	6	
6 000	7		8 000	7	
6 000	8		8 000	8	
15 000 + 2 000 000	9		15 000 + 1 850 000	9	
15 000	10		15 000	10	
15 000	etc.		15 000	etc.	

Present worth of £6000 per year for 8 years
$$= 6000 \times 5.747$$
$$= £34\,482$$

Present worth of £2 000 000 at Year 9
$$= 2\,000\,000 \times 0.500$$
$$= £1\,000\,000$$

Therefore, total present worth of Method (a)
$$= £101\,250 + 34\,482 + 1\,000\,000 + 1\,500\,000$$
$$= £2\,635\,732$$

Present worth of £8000 per year for 8 years
$$= 8000 \times 5.747$$
$$= £45\,976$$

Present worth of £1 850 000 at Year 9
$$= 1\,850\,000 \times 0.500$$
$$= \underline{£925\,000}$$

Therefore, total present worth of Method (b)
$$= £101\,250 + 45\,976 + 925\,000 + 1\,900\,000$$
$$= \underline{£2\,972\,226}$$

Method (a) is therefore the cheaper.
Annual cost over 8 years
$$= 1\,500\,000 \times (A/P, 8\%, 8) + 6000$$
$$= 1\,500\,000 \times 0.174 + 6000$$
$$= \underline{£267\,000}$$

Therefore, realistic toll charge

$$= \frac{267\,000}{350\,000} = \underline{£0.76 \text{ per person}}$$

4.9 The calculation of yield

In both the annual cost and net present value (NPV) methods it is necessary to know the appropriate interest rate before the calculations can be made. The decision to invest may change with interest rate (see Example 3.3). If the acceptable or required rate of return changes, then all supporting calculations need to be reviewed in order to establish the current viability of each project under the different condition.

The third method of appraisal which makes use of discounting techniques is commonly referred to as DCF (Discounted Cash Flow). It also bears a series of different titles, some of which are *internal rate of return, marginal efficiency of capital, yield, actuarial return* and *investor's method*. In this book the method generally will be called the *yield method*, since in effect, and in contrast to the previously discussed methods, it is a technique which allows the calculation of the actual yield, or rate of return, for a series of cash flows rather than applying a set rate of return and interpreting the result then obtained. The yield method, therefore, allows direct comparison of a calculated percentage, with a minimum return that is required.

If a series of four annual cash flows of £631 are discounted at an interest rate of 10 per cent to give a present value of £2000, it can be interpreted conversely that if £2000 is invested to bring a net return of £631 per year for each of four successive years, then a rate of return of 10 per cent is being made on the investment. The NPV of the proposal, at 10 per cent, is therefore nil. At different percentages for the required rate of return, the NPV will vary and its relationship with the discount rate in this example is illustrated by the curve of Fig. 4.4. The curve illustrates that for any given interest rate used in

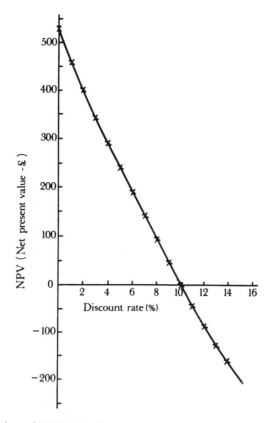

Fig. 4.4 Variation of NPV with discount rate.

a discounting operation, a positive NPV indicates a better return than that and a negative NPV indicates that the rate of return is not achieved.

Study of the implications of the curve of Fig. 4.4 leads to a clearer understanding of the yield method and in particular the definition of yield rate which is *that percentage at which the future cash flows arising from an investment, when discounted, will equate to the initial investment.* This has the same effect as seeking the interest rate at which an investment has a zero NPV.

Equation (4.6) may then be adjusted to read

$$C = \sum_{k=1}^{n} \frac{A_k}{(1+i)^k} \tag{4.8}$$

where i is the yield rate.

Moreover, it should be noted that C is the initial capital outlay only, and even if an investment takes place over more than one period, all but the first capital outlay must be incorporated as negative cash flows in the right-hand side of the above equation.

The yield for a situation in which the cash flows are irregular can only be found by trial and error and once again the graph of Fig. 4.4 indicates the means. Firstly, an interest rate must be arbitrarily selected, preferably (although at the time without the assistance of a graph of the form of Fig. 4.4) close to the point where the curve of NPV against discount rate crosses the X-axis. Having selected an interest rate, the NPV is calculated. If it is positive, a larger interest rate is selected and the calculation is repeated. This process is continued, if necessary, until a negative NPV is obtained. The required interest rate at which the NPV is zero can then be obtained by interpolation between trial rates which are closest together – one giving a positive NPV and the other a negative NPV. The following example illustrates the method further.

Example 4.8

What is yield for a project where £10 000 is invested to produce cash flows of £5000, £4000 and £3000 during Years 1, 2 and 3 respectively?

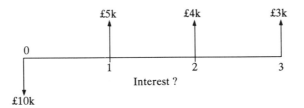

Fig. 4.5 Cash flow diagram – Example 4.8.

Firstly, it is necessary to estimate the likely rate of return. The absolute return to be made over the three years is £12 000. This represents an average annual profit of approximately £4000 and the uniform series capital recovery factor is therefore 4000/10 000 = 0.40.

By reading interest tables given three time periods and a uniform series capital recovery factor of 0.40, it will be found that an interest rate of 10 per cent is likely to be of the right order. However, the positive cash flows of the investment are rather larger in the early years of the project and the yield is therefore likely to be higher than if they had been constant; 12 per cent is therefore chosen as a suitable starting point.

This calculation, detailed in Table 4.6, shows that with a 12 per cent interest rate the NPV is negative and therefore a lower rate must be chosen for the second trial if a positive NPV is the object of the calculation. Ten per cent is chosen in Table 4.7. Ten per cent interest rate results in a positive NPV

Table 4.6

+	Period	−	Discount factor (12%)	Discounted cash flow	
(£)	(Year	(£)		+	−
5000 4000 3000	0 1 2 3	10 000	1·000 0·893 0·797 0·712	4465 3188 2136	10 000
			Totals	9789	10 000
			Net total		211

Table 4.7

+	Period	−	Discount factor (10%)	Discounted cash flow	
(£)	(Year)	(£)		+	−
5000 4000 3000	0 1 2 3	10 000	1·000 0·909 0·826 0·751	4545 3304 2253	10 000
			Totals	10 102	10 000
			Net totals	102	

and therefore by interpolation, as illustrated in Fig. 4.6, a yield rate of 10.65 per cent is calculated. It should be noted that linear interpolation is not strictly mathematically correct, but there is no practical difficulty arising from this assumption.

4.10 The use of yield in economic choice

In normal circumstances, the evaluation of a single project by either of the two discounting methods, NPV or Yield, will lead to the same conclusion. That is, if a minimum acceptable rate of return can be established and a straightforward yes/no, accept/reject decision is to be made, the supporting evidence provided by each method of evaluation should lead to similar judgements. The use of the total equivalent annual cost method and the NPV method will result in similar choices being made between a number of competing, mutually exclusive investments and similar rankings being established for projects competing for limited finance. These situations will be discussed later. For the time being it should be borne in mind that, under certain conditions, and given certain specified characteristics of cash flow, NPV and Yield can support different conclusions based on the result of the analysis carried out.

So far, in the explanations given of the various techniques of appraisal, there has been no significant discussion of the effects of the life of a project and, particularly, of the comparison of projects having different lives. One of the important facets of the discounting methods is that they take into account the timing and pattern of the cash flows. Hence the life of a project is

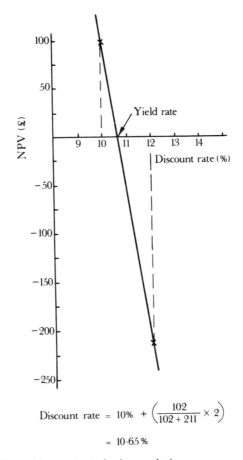

$$\text{Discount rate} = 10\% + \left(\frac{102}{102 + 211} \times 2\right)$$

$$= 10 \cdot 65\%$$

Fig. 4.6 Calculation of interest rate by interpolation.

important, more particularly so when the interest rate, or minimum rate of return required, is low. For example, if interest rates of the order of 25 per cent are considered, £100 in 10 years' time has a present value of £10.70, and in 20 years' time the present value is only £1.10. Conversely, if interest rates of 8 per cent are considered to be appropriate, a similar sum discounted over similar periods of time will have present values of £46.30 and £21.50, respectively. With the higher rate the estimation of project life is not one of the most significant factors in an overall calculation since most emphasis is placed on cash flows in the early years of the project.

The common case is that of selection from two or more mutually exclusive projects using the Yield method. If one of the projects is selected it automatically signals the rejection of the other. For example, if a factory is to be built there will be a large number of possible alternative designs and layouts which will fulfil the required function. Only one can be accepted and the acceptance of one design must therefore automatically mean the rejection of the others.

The Yield method must be used with care when it is required to rank competing or mutually exclusive projects. In some, though not all instances, the NPV and Yield methods may indicate different projects as giving the better return. Example 4.9 illustrates the method.

Example 4.9

One investment opportunity must be chosen from two alternatives. The initial investment in both cases is £10 000. Project A (Table 4.8) shows returns of £6000, £4000, £2000 and £1000 at the end of Years 1, 2, 3 and 4, respectively. Project B (Table 4.9) shows returns of £2000, £2000, £4000 and £6000 at the end of Years 1, 2, 3 and 4, respectively. Which investment should be chosen if its minimum acceptable rate of return is 6 per cent?

First, using the NPV method:

Table 4.8

Project A

Cash flows (£)	Discount factors 6%	Discounted cash flows (£)
(10 000)	1·000	(10 000)
6 000	0·943	5 658
4 000	0·890	3 560
2 000	0·840	1 680
1 000	0·792	792
Net cash flow 3 000	NPV = +1 690	

Table 4.9

Project B

Cash flows (£)	Discount factors 6%	Discounted cash flows (£)
(10 000)	1·000	(10 000)
2 000	0·943	1 886
2 000	0·890	1 780
4 000	0·840	3 360
6 000	0·792	4 752
Net cash flow 4 000	NPV = + 1 778	

Therefore, using the NPV method, Project B appears to be the better investment, since it will result in an increase in the capital resources by a larger amount than if Project A is accepted.

Secondly, using the yield for comparative purposes, it will be found that the yield for Project A is 15.9 per cent and that for Project B is 12.2 per cent. This would indicate that Project A is the better proposition and an explanation is needed as to why there should be this difference in outcome resulting from the two analyses. It is particularly necessary to establish a means of ensuring that the results of such analyses are not misinterpreted.

The explanation of how the above situation arises can best be explained by illustrating the problem graphically. Figure 4.7 shows the curve of NPV against discount rate for each of the Projects A and B. At zero NPV, the curve for Project A cuts the *X*-axis at 15.9 per cent and that for Project B at 12 per cent, to confirm the above calculation. However, at zero discount rate, Project B is showing a greater NPV (£4000) than Project A (£3000) and does so for all discount rates below approximately 6.9 per cent. If NPV is the criterion by which the Project will be selected, it will support Project B for all costs of capital less than 6.9 per cent and Project A for costs of capital in excess of this. If the yield rate is to be used as the criterion, then Project B will always be the better project because a single yield rate only can be calculated; that is when the NPV is zero.

This changeover arises from the fact that Project B has larger cash flows arising in its later years and the effect of the *single payment present worth factor* on the present worth of the project as a whole is more marked. The NPV of Project B will therefore fall at a greater rate as the discount rate rises. In general, it may be expected that there will be some conflict between the results of using the NPV and Yield methods as a means of choosing between two mutually exclusive projects, where the cash flows in one tend to increase in magnitude, and those of the other tend to reduce in magnitude, with

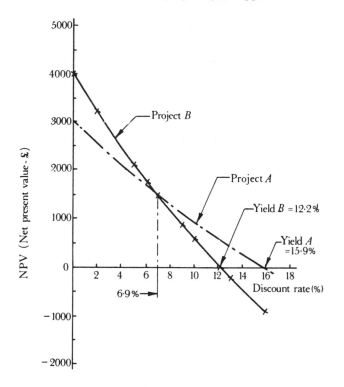

Fig. 4.7   Comparisons of NPV and yield.

project time. In addition, a similar result may well arise where projects are expected to have a significant difference between their lives and/or there is a large difference between the initial costs of the projects.

When considering two or more mutually exclusive projects it is necessary to choose between them on the basis of *incremental* cash flow or the differences between them. Example 4.10 demonstrates the principle of selection between three projects on an incremental basis.

Example 4.10

A choice is to be made between three methods for carrying out a construction process. The process is expected to be in use for a period of five years. The first alternative, A, involves the use of labour only and has total annual costs of £10 000. Alternative B involves the initial investment of £12 000 in capital equipment and is then estimated to have a total labour cost of £7000 per year. The third alternative, C, involves the initial capital investment of £20 000 which should further reduce the annual cost to £5300. If the firm's minimum acceptable rate of return is 5 per cent in this instance, which alternative should be selected?

Table 4.10

Year	Scheme A	Scheme B	Scheme C
0		(12 000)	(20 000)
1	(10 000)	(7 000)	(5 300)
2	(10 000)	(7 000)	(5 300)
3	(10 000)	(7 000)	(5 300)
4	(10 000)	(7 000)	(5 300)
5	(10 000)	(7 000)	(5 300)

The first step of the appraisal, as a matter of convenience, is to display the cash flows for each available alternative in tabular form, listing that project with the minimum capital investment on the left-hand side. From left to right the listings should be in increasing magnitude of capital investment as in Table 4.10.

The alternative having the lowest capital investment, that is, the one on the left-hand side of the table (Scheme A in this case) is then set as the standard against which the next one (Scheme B) is measured. The cash flows of Scheme B are subtracted from those of Scheme A to give the series set out in column 1 of Table 4.11.

In effect the subtraction of these cash flows asks the question, in arithmetic terms, whether it is worth investing £12 000 in order to save £3000 per year for five years? The yield of this saving on this capital investment is found to be 8 per cent. It is therefore worth making the investment of £12 000 from the point of view of a minimum acceptable rate of return of 5 per cent. Hence Scheme B now becomes the standard against which any further scheme must be measured. By subtracting the cash flows of Scheme C from those of

Table 4.11

Year	Column 1 Compare B with A	Column 2 Compare C with B	Column 3 Compare C with A
0	12 000	8 000	20 000
1	(3 000)	(1 700)	(4 700)
2	(3 000)	(1 700)	(4 700)
3	(3 000)	(1 700)	(4 700)
4	(3 000)	(1 700)	(4 700)
5	(3 000)	(1 700)	(4 700)
Yield	8%	2%	5·6%

Scheme B, the cash flows of column 2, Table 4.11 are obtained. Saving £1700 per year for each of five years for an investment of £8000 produces a yield of only 2 per cent and therefore is not acceptable by the required standard. Scheme C is therefore rejected. It is rejected because the £20 000 of capital investment required in Scheme C contains £8000 which only produces a return of 2 per cent. If Scheme C is now compared with Scheme A (Column 4, Table 4.11) it will be seen that the yield amounts to 5.6 per cent, which is above the minimum return required. It might well be asked if this is not sufficient reason to accept Scheme C. If Scheme B had not existed, C would have been acceptable. However, the fact that B exists, precludes the recommendation of C with its substandard element of £8000.

Had there been schemes D, E, etc., with increasing capital investment, each one would have been compared with the standard B, until an alternative with at least the acceptable rate of return on the incremental cash flow was found. Such an alternative would then become the standard against which all others with larger capital investment would have to be measured.

4.11 Multiple yields

One of the criticisms frequently levelled at the Yield method as a means of evaluating project proposals, is the situation in which it is found that there is more than one value of i offering a solution to Equation (4.8). Such situations are not commonly found in practice and there is a ready means of identifying them.

In general, the situation which gives rise to multiple yields is one in which there is more than one change of arithmetic sign in the net cash flows of a project throughout its life. A typical situation is one in which there are large cash outflows, or negative cash flows, towards the end of the project life. Such a situation can arise, for example, in an opencast coal-mining project or other open quarrying venture. Considerable expenditure is required on these projects at the end of the commercial extraction, in order to restore the site to an acceptable condition. A further example is that of a project for the removal and sale of timber from woodland. The stripped land is replanted with young trees at the end of deforestation.

In a situation where one, or a series, of initial capital investments is succeeded by a series of positive cash flows, the flows can be equated by discounting with a single interest rate. Where there are two capital investments or negative cash flows separated in time by a series of positive cash flows, then two solutions for i will sometimes satisfy the requirements of Equation (4.8).

In general, more than one rate of return may result from a calculation of the yield where the cumulative cash flow at any point in the overall duration of the proposed investment switches from positive to negative, and vice versa, on more than one occasion through the series. The cumulative cash flow at any point in time is calculated by starting at zero time and summing all negative and positive cash flows, taking due regard of sign, until that point in time is reached. In cases where more than one yield satisfies Equation (4.8), it becomes difficult to rank the subject rate of return with the required standard.

The difficulty of dealing with multiple yields, should they arise, can be overcome by alternatively using the *external rate of return method*. It is assumed in this method that all positive cash flows in the project are invested at the minimum acceptable rate of return to be used in the project or at some other acceptable return to suit the situation. The external rate of return can be calculated by equating the future worth of all the positive cash flows compounded at i, the interest rate for reinvested funds, to the future worth of negative cash flows compounded at i', the external rate of return (ERR). In this way the positive and negative cash flows are discounted to zero NPV and allow the ready calculation of the ERR, i', in the process. There is only one solution of i' in the above equation and where it is greater than i, with i taken to be the minimum acceptable rate of return, the investment is assumed to be satisfactory.

Example 4.11

The cash flows in Table 4.12 result in changes in cumulative cash flows as illustrated.

Table 4.12

Year	Cash flows £000	Cumulative cash flows £000
0	(400)	(400)
1	320	(80)
2	250	170
3	270	440
4	120	560
5	(620)	(60)

The cumulative cash flows begin by being negative, then change to positive and back to negative again – two reversals of sign. Applying the ERR method where the minimum acceptable rate of return is 12%:

$$320(F/P, 12\%, 4) + 250(F/P, 12\%, 3) +$$
$$270(F/P, 12\%, 2) + 120(F/P, 12\%, 1) = 400(F/P, i', 5) + 620$$
$$320(1.5735) + 250(1.4049) + 270(1.2544) + 120(1.12)$$
$$= 400(F/P, i'\%, 5) + 620$$

Therefore,

$$(F/P, i'\%, 5) = 1.7696 \text{ and } i'\% = 12\%$$

The investment cash flows of Table 4.12 are therefore acceptable since they should achieve the minimum acceptable rate of return.

Perhaps a simpler and more obvious check of a situation in which multiple yields are likely to arise can be made graphically, following the principles of the curves of Figs. 4.4 and 4.6. In plotting NPV against discount rate for a project having dual yields, a general curve of the shape shown in Fig. 4.8 will result.

Where multiple yields are suspected then NPV could be a more reliable means of appraisal.

4.12 Investment life

Not all alternatives, either when considering mutually exclusive investments, for example when considering alternative forms of construction for a building, or competing investments, will have similar lives. When taking account of a less permanent form of construction, a lower cost will probably be established, but the building will usually then be suitable for use over a shorter period of time. To make a reasonable comparison between the alternative structures it is necessary to take into account the relative lives of these alternatives. The interpretation of investment life will, to a certain extent, depend upon the method of appraisal used in the evaluation.

The following example illustrates the analysis of a problem involving two mutually exclusive projects and affords a basis for comment on the

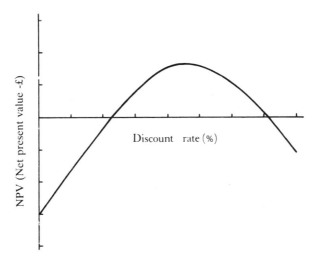

Fig. 4.8   Multiple yield situation.

equivalent annual cost method as applied to the comparison of projects having different lives.

Example 4.12

Alternative designs for a building are prepared, one in structural steelwork clad with asbestos cement sheets and the other with a reinforced concrete frame and brickwork cladding. Each alternative is considered to provide adequately for the requirements of the building. The initial capital cost of the former is £600 000 and of the latter is £1 600 000. The life of the concrete building is estimated to be 60 years, and although it is anticipated that there will be no maintenance costs incurred during the first 10 years of its life, there is likely to be a cost of £10 000 per year incurred thereafter. The steel-framed building is estimated to have a life of 20 years with a maintenance cost of £11 050 for each of those years. At the end of its useful life the concrete building is expected to fetch £400 000 and the steel-framed building £60 000. Which building offers the better economic proposition if the future building owner has a cost of capital of 10 per cent?

Reinforced concrete building:

$$\text{Capital recovery cost} = (1\,600\,000 - 400\,000)(A/P, 10\%, 60)$$
$$+ 400\,000\ (0.10)$$
$$= 1\,200\,000(0.10032) + 40\,000$$
$$= 120\,384 + 40\,000$$
$$= £160\,384 \text{ per year}$$

Present value of £10 000 per year from years 11 to 60 inclusive
$$= 10\,000\ (P/A, 10\%, 50)(P/F, 10\%, 10)$$
$$= 10\,000\ (9.915)(0.3855) = £38\,222$$

Equivalent annual cost of delayed maintenance over life of building
$$= 38\,222\ (A/P, 10\%, 60)$$
$$= 38\,222(0.10032) = £3\,834 \text{ per year}$$

$$\text{Total equivalent annual cost} = 160\,384 + 3\,834$$
$$= £164\,218$$

Steel-framed building:

$$\text{Capital recovery cost} = (600\,000 - 60\,000)(A/P, 10\%, 30)$$
$$+ 60\,000(0.10)$$
$$= 540\,000(0.11745) + 6\,000$$
$$= £69\,423 \text{ per year}$$

$$\text{Total equivalent annual cost} = 69\,423 + 11\,500$$
$$= £80\,923$$

On this basis the steel-framed building is the better economic choice.

The above calculation strongly supports the choice of a building with a steel frame. The lives of the two alternatives, however, are vastly different. Part of the final decision must be whether either building life will suit the purpose of the user or owner better than the other. In many respects this will be a matter of judgement and prediction as to what the future holds. There is an advantage in investing in the building that has the shorter life in so much as, at the time that replacement becomes due, it is possible to reconsider the type of building in relation to its function. It makes it possible to take advantage of new developments in construction materials and techniques and to take advantage of advances in technique with regard to the use to which the building is being put.

On the other hand, investing in the building with the longer life can have advantages in times of high inflation, since replacement costs are ever likely to increase. At the same time it must be borne in mind that as interest rates get higher, alternatives with longer lives become less competitive. Similarly, it is never worthwhile to increase the initial investment in a project in order to obtain a proportionate increase in the project life.

In using the total equivalent annual cost method of investment appraisal where the life of one project is a multiple of the other, as in this example, the costs can be tabulated on the basis that the investment with the shorter life will be repeated at current costs so as to give the same period of service. In the case of project lives without this relationship, the lowest common multiple of the lives can be used. It is usual to assume that replacement costs will be similar to initial capital costs at the investment stage. Whilst it is known that this is a situation which is extremely unlikely, the effect on the analyses of doing so is unlikely to be as severe as might at first be thought.

When using the NPV method of appraisal it is essential that comparisons are made over the same period of time. Unless this is done, the NPVs so obtained are not comparable. Procedures available for doing this, amongst others, are similar to those outlined above in so much as the least multiple of the various project lives can be established in most instances. Where this is difficult, because it gives rise to an excessive period over which consideration must be made, then the capitalized cost method can be used.

Alternatively, if for example two projects are being considered which have different lives, the longer life can be reduced to that of the other. This means that the cash flows in the latter years of the longer life of the one are ignored, though some compensation must be added in the form of an enhanced salvage or resale value of the shortened project. In other words, it must be assumed that the project with the longer life will be curtailed at an earlier date than was originally anticipated and, in effect, due allowance must be made for the enhanced residual value of the asset.

In using the yield method to compare alternatives, both projects should be considered over similar periods of time using one or other of the above methods of equating the project lives. Instances where projects with widely differing lives are compared on an incremental basis, without due allowance being made, can result in a solution having multiple yields.

Example 4.13

After successfully completing a course in construction management, Harold Hoof decided to set up his own company for the purpose of hiring out earthmoving equipment to construction contractors. An over-indulgent father promised to give Harold the necessary capital to buy the initial equipment in view of Harold's insistence that he could make a return on capital invested of at least 10 per cent per year.

Harold subsequently bought three loaders (A, B and C in Table 4.13) at a total cost of £71 000, all of which was given to him outright by his father.

Table 4.13

Loader	A	B	C	D
Total initial cost (£)	25 000	22 000	24 000	30 000
Useful life (years)	4	5	4	4
Total owning and operating costs excluding depreciation (£ per working hour)	5	4·5	5	6
Salvage value at end of useful life (£)	3 000	2 000	6 000	4 000
Average annual hours of working	2 000	1 800	2 000	2 200

After operating for exactly three years Harold purchased, out of the business returns, a fourth loader (D in Table 4.13) for the sum of £30 000. Subsequent to this purchase, however, due to a hardening of the economic situation Harold did not replace any of the original loaders as they came to the end of their useful life.

At the end of the sixth year of the company's operations, Harold's father sensed a decline in the tempo of trading by the company. He asked an accountant friend to check the actual return on the original gift to his son. In order to do so the accountant had to assume that the one remaining operating loader now had a resale value of £7000.

The gross annual receipts by the company are listed in Table 4.14. Had the performance of the company over its six years of trading justified Harold's original optimism?

Table 4.14

Year	Income (£)
1	52 100
2	53 100
3	59 000
4	50 300
5	45 000
6	20 000

There is a need in this problem to set out the cash flows in an orderly fashion before carrying out the calculation. This is done in Table 4.15.

Loader A is bought for £25 000 at the commencement of the project and costs an average of £5 × 2000 = £10 000 per year to operate. This cost of operation is alleviated by the resale value at the end of the fourth year. The cash flows for each scraper can be set out on a similar basis. The income can then be set against the negative cash flows as is done in the last column of Table 4.15. If the cash flows are now discounted at 10 per cent, it will be found that the NPV is positive and therefore the return is in excess of 10 per cent.

Table 4.15

Year	Loader A (£)	Loader B (£)	Loader C (£)	Loader D (£)	Income (£)	Net cash flows (£)
0	(25 000)	(22 000)	(24 000)		—	(71 000)
1	(10 000)	(8 100)	(10 000)		52 100	24 000
2	(10 000)	(8 100)	(10 000)		53 100	25 000
3	(10 000)	(8 100)	(10 000)	(30 000)	59 000	900
4	(7 000)	(8 100)	(4 000)	(13 200)	50 300	18 000
5		(6 100)		(13 200)	45 000	25 700
6				(6 200)	20 000	13 800

Example 4.14

A project involves the initial capital investment of £100 000. It is expected that following this investment the company will benefit from it by obtaining net cash receipts of £40 000, £40 000, £30 000 and £20 000 in each of the first four years of the project's life, respectively. Assuming that these values are in money terms as opposed to real terms which take account of the rate of inflation, what is the NPV for the project if the firm's cost of capital is 12 per cent? Calculate the yield on the investment. Examine the results of your calculation if inflation is at the rate of 3 per cent per year.

Tabulated cash flows

−ive cash flow		+ive cash flow
£100 000	0	—
	1	£40 000
	2	40 000
	3	30 000
	4	20 000
£100 000		£130 000

The following two calculations for NPV and yield are carried out in money terms making no allowance for an inflationary trend. Following the calculation of yield, the cash flows are expressed in real terms, assuming a rate of inflation of 3 per cent per year. The calculations are then repeated.

$$
\begin{aligned}
\text{NPV} &= 40\,000 \times 0.893 + 40\,000 \times 0.797 + 30\,000 \times 0.712 \\
&\quad + 20\,000 \times 0.636 - 100\,000 \times 1.000 \\
&= 35\,720 + 31\,880 + 21\,360 + 12\,720 - 100\,000 \\
&= 101\,680 - 100\,000 \\
&= \text{£}1680
\end{aligned}
$$

−ive cash flow		+ive cash flow	Discount factors 12%	Discounted cash flow	Discount factors 14%	Discounted cash flow
£100 000	0	—	1·000	−100 000	1·000	−100 000
	1	40 000	0·893	35 720	0·877	35 080
	2	40 000	0·797	31 880	0·769	30 760
	3	30 000	0·712	21 360	0·675	20 250
	4	20 000	0·636	12 720	0·592	11 840
£100 000		130 000		+1 680		−2 070

$$
\text{Therefore, Yield} = \left[\frac{1\,680}{3\,750} \times 2\right] + 12 = 12.90 \text{ per cent}
$$

Expected cash flow	Inflation factor at 3% p.a.	Real cash flow
−£100 000	1·00	−£100 000
40 000	1·03	38 840
40 000	1·03² = 1·061	37 700
30 000	1·03³ = 1·093	27 447
20 000	1·03⁴ = 1·126	17 762

The NPV in real terms:

$$\begin{aligned}
\text{NPV}_{RT} &= 38\,840 \times 0.893 + 37\,700 \times 0.799 + 27\,447 \times 0.712 \\
&\quad + 17\,762 \times 0.636 - 100\,000 \times 1.000 \\
&= 34\,684 + 30\,122 + 19\,542 + 11\,297 - 100\,000 \\
&= \underline{\underline{-£4355}}
\end{aligned}$$

Therefore, in real terms, the investment will not bring the company a return of 12 per cent. The yield can, however, be checked:

−ive cash flow		+ive cash flow	Discount factors 10%	Discounted cash flows	Discount factors 9%	Discounted cash flows
£100 000	0	—	1·000	−100 000	1·000	−100 000
	1	38 840	0·909	35 306	0·917	35 616
	2	37 700	0·826	31 140	0·842	31 743
	3	27 447	0·751	20 613	0·772	21 189
	4	17 762	0·683	12 131	0·708	12 574
				−810		+1 122

Yield is therefore $10 - \left[\dfrac{810}{1932}\right] \times 1 = \underline{\underline{9.58 \text{ per cent}}}$

This result could have been obtained more quickly by applying the factor for inflation to the yield calculated in the first instance on the money terms. The expression $[(1 + y)/(1 + i)] - 1$ gives the real rate of return where

y = yield calculated in money terms
i = rate of inflation

In this case,

Real rate of return $= \left[\dfrac{1 + 0.129}{1 + 0.03}\right] - 1 = 0.0961$ or $\underline{\underline{9.61 \text{ per cent}}}$

(The slight difference between this and the return calculated by the trial and error method arises from approximation in the calculations and linear interpolation of rates.)

4.13 Comparison of appraisal methods

A number of different methods of appraising investments and the application of these methods to a number of different situations have been examined. There is no general agreement about project appraisal methods, as to which method of analysis produces the best practical answer. Three discounting methods, those of annual cost, NPV and Yield have been described, and if correctly used, any one of these three methods will lead to the same accept or reject decision. All three methods place emphasis where it should be, on the cash flows to a time base. However, the requirement in an appraisal of an investment situation is rarely to know whether a project is likely to be profitable or non-profitable. It is more likely that a number of investments are being appraised and the requirement is for these to be ranked in an order representing the relative profitability of each.

Of these three methods, annual cost tends not to be widely used except where a forecast can be made of the possibility of fairly constant cash flows in the future. For example, in large nationalized industries, such as the electricity industry, forecasts of power use can be made over a number of years ahead and the cash flows arising tend to be reasonably constant. In situations where there is an unevenness of cash flow, the annual cost method tends to create a large amount of calculation over and above that of the other two discounting methods.

In comparing NPV and the Yield method, different questions are answered by the different methods. In the case of NPV, the answer obtained is in absolute terms and represents a surplus amount of money after discounting and subtracting the initial investment. This will accrue to a company at the present time as a result of making the investment. In such calculations, the minimum rate of return that is required must be assumed before the calculation can be completed. The Yield method seeks to establish what the rate of return will be if the cash flows, as predicted, finally take place. The result, therefore, is not in absolute terms of profit, but in terms of a percentage which represents the yield or return on the investment.

Comparing Yield and NPV calculations, Yield does present difficulties where a decision is required between mutually exclusive projects. It is necessary to employ the incremental method of yield analysis and this in itself is an added complication and extension of the calculation. The calculated yield of a series of investments does not enable the projects to be ranked directly in an order of acceptability unless certain conditions apply, and hence NPV is the method which is and should normally be employed on engineering studies for the analysis of different methods of carrying out work, such as production systems. In both NPV and Yield calculations, the lives of all the alternatives under consideration or for comparison should be similar, otherwise some compensation must be incorporated in the cash flows in order to convert them to equivalent sums over similar periods of time.

Yield methods, however, do relate to the risk involved in undertaking a project. In looking at the mechanics of a Yield calculation, it will be noted that it is dependent on the amount of money that is invested for a specific length of time. In general, the longer period for which money is invested before the return is obtained the greater the risk in the project. Risk is directly proportional to investment life. In addition, in favour of Yield methods, it can be said that a percentage is often more meaningful to an investor than the lump sum which results from the NPV calculation. On the other hand, the Yield method does suffer from the possibility of multiple yields when there is a large negative cash flow after the early stages of the investment life. If there is any danger of this, some check must be carried out to ensure that a multiple yield situation does not exist. Where negative cash flows do occur after the early stages of the life of the project, then the NPV method is clearly the one to use, since this is not affected by such complications. Both NPV and Yield methods are essentially of value in identifying the margin at which an investment becomes profitable.

In summary, the more sophisticated and mathematical methods of investment appraisal, particularly Yield and NPV, can have extremely useful applications so long as they are used appropriately. Users must have an appreciation of the techniques' limitations and they must be used in the correct method. Such methods take account of the time-value of money and it is fundamental to proper investment appraisal that this should be so. In particular, Yield and NPV are both methods which are suitable and both methods which will give similar accept/reject support. NPV has the slightly wider application as a method and is sometimes more flexible in its application. Both methods encourage the accurate forecasting of the future cash flows to be associated with a project investment and in itself this is a fundamental requirement of understanding an investment situation. Even if the attempted use of such methods has led only to a more thorough and detailed analysis of the situation, a considerable step forward will have been taken.

4.14 Inflation and project appraisal

Whilst in some instances the effects of inflation (inflation being the declining purchasing power of money over time) do not influence the calculations in an investment appraisal, in most instances inflationary tendencies should be considered especially at times of high inflation. A simple instance of where the effects of inflation are not felt too severely can arise in the calculation of net cash flows arising as between the cost of producing an article having wide sales and low unit cost, and the revenue arising from the sales of these units. This depends, however, on the company producing the unit being able to adjust its pricing structure to take account of increasing production costs. In general terms, where incremental cash flows are being considered, if income rises in proportion to costs then the differences will remain approximately constant.

Inflation will have an effect not only on the cash flows of a project, but also on the rate at which the cash flows need to be discounted, since this rate reflects the cost of capital to a company. As far as the cash flows are concerned, the treatment must be consistent for all the cash figures throughout the appraisal. If it is decided to estimate cash flows on the basis of the present-day purchasing power of money, then this must be used consistently. On the other hand, if it is decided to apply to each future cash flow a factor based on future rates of inflation, then detailed consideration must be given individually to each flow.

When considering the cost of capital and how inflation affects it, it will be seen that the NPV method of appraisal offers some flexibility on this score.

Equation (4.5) reads as follows:

$$P = \frac{A_1}{(1+i)} + \frac{A_2}{(1+i)^2} + \frac{A_3}{(1+i)^3} + \ldots + \frac{A_n}{(1+i)^n}$$

In this equation, a different value of i can be used for each year if necessary, when the equation will read:

$$P = \frac{A_1}{(1+i_1)} + \frac{A_2}{(1+i_1)(1+i_2)} + \frac{A_3}{(1+i_1)(1+i_2)(1+i_3)}$$

$$+ \ldots + \frac{A_n}{(1+i_1)\ldots(1+i_n)}$$

If, however, inflation is expected to increase by a uniform percentage each year, d, the present value of cash flows A_n is:

$$P = \frac{A_n}{(1+i)^n(1+d)^n} = \frac{A_n}{[(1+i)(1+d)]^n}$$

therefore, the effective rate of inflation $= (1+i)(1+d) - 1$

Example 4.15

If the current cost of capital is 8 per cent per year, and inflation is expected at the rate of 3 per cent per year, at what rate should cash flows be discounted in order to take account of both factors?

$$\begin{aligned}
\text{Effective rate of inflation} &= (1+i)(1+d) - 1 \\
&= (1.08)(1.03) - 1 = 0.1124 \\
&= \underline{11.24 \text{ per cent}}
\end{aligned}$$

$(1+i)(1+d) - 1 = i + d + id$ where $(i + d + id)$ is known as the *nominal discount rate* that corresponds to the actual discount rate, i, the rate of interest or required rate of return. Where inflation is reasonably constant from year to year, and at a relatively low rate, the use of a nominal discount rate of $(i + d)$ will normally be sufficiently accurate. Where inflation rates are assumed to be an integer, the approximation means that interest tables can be used in the solution of problems.

Example 4.16

If inflation is assumed to average 4 per cent per year over the next 8 years and the discount rate is 6 per cent, calculate the present value of £100 in nominal terms which will be paid at the end of Year 8.

$$\text{Nominal discount rate} = (4 + 6) = 10\%$$
$$\text{Approximate present value} = 0.4665 \times 100$$
$$= £46.65$$

A more accurate calculation would be:

$$\text{Discount rate} = (1 + i)(1 + d) - 1$$
$$= (1.06)(1.04) - 1 = 0.1024$$

$$\text{Present value} = \frac{100}{(1.1024)^8} = £45.84$$

Example 4.17

The future cost of heating fuel is expected to increase by 2 per cent per year in relation to general price level inflation. At what rate should they be discounted if the real discount level required is 10 per cent?
   Combining the relative price change with the real discounting requirement:

$$\left[\frac{(1 + r)}{(1 + z)}\right] - 1 = \left[\frac{(r - z)}{(1 + z)}\right] \quad \text{where } z \text{ is the rate of cost increase relative to general price levels.}$$

Therefore

$$\text{Rate of discount} = \frac{1.10}{1.02} - 1 = 0.0784$$

$$= 7.84 \text{ per cent}$$

The approximate rate of discount could have been calculated from

$$(r - z) = 10 - 2 = 8 \text{ per cent}$$

4.15 The effect of taxes and grants

In general, tax and grants will have an effect on the cash flows for an investment project. However, at the time that an investment appraisal is made, the future of a country's tax structure and the detailed rates of tax are probably less certain than the future cash flows deriving from the investment under examination. This is not a reason for ignoring the implications of tax but rather a case for creating an awareness of the general implications for investment analysis. The subject is complex and certainly expert opinion should be sought on the matter in this country, as well as overseas, where some projects under examination may be sited. It often happens, however, that the basic principles of tax structure are similar from country to country while only the detail may be more or less complicated.

The profits of a business are normally subject to tax and for a corporation in the UK, such tax is known as *corporation tax*. It has been shown in Chapter 2 that profit is what remains of income once the deduction of certain expenses and costs has been made. Tax is yet another deduction and, therefore, it will have an effect on the size of cash flows and also on their timing. Tax is usually paid in arrears, currently just over nine months after the end of a company's financial year.

The amount of tax to be paid will be affected by the *allowances* and *incentives* which are an important feature of all tax systems. The features exist so as to encourage companies, for example, to invest in new production equipment and new factories. A certain proportion of the capital expenditure is allowed as a deductible against tax thus reducing the overall tax bill. Another important allowance is that against the depreciation of fixed assets, a subject which has already been broached in Section 4.4 of this chapter. It is important in this context to have a clear conception of the difference between *depreciation* as it appears in a company's accounts and *writing-down allowances* providing relief of tax on the declining value of capital equipment, buildings, etc. In the case of the former, depreciation is allowed against the income of a company before tax is deducted in recognition of the reduction in value of assets as a result of use in a business. The rate at which depreciation is calculated and then incorporated in the firm's accounts is a combination of practice, custom, need and assessment exercised by the directors of the company, who have a duty to ensure that the value of the assets shown in the books are a fair and reasonable estimate of their true worth. The rates of depreciation may vary widely for different assets. Writing-down allowances are set by the Inland Revenue as fixed percentages per year for different types of fixed assets, for example, 25 per cent for plant and machinery. The amount deductible against tax is then calculated for a particular year by taking this percentage of the residual value for that year. It can be seen, therefore, that before tax is calculated on a company's profit, the value for depreciation which has already been deducted in the profit and loss account is put back into the profit and then the writing-down allowances are calculated and deducted in their lieu.

Example 4.18

A small building contractor shows a taxable profit of £75 000 in 19x4 and £110 000 in 19x5. During 19x4 the company considered purchasing a self-propelled crane on rubber-tyred wheels at a capital cost of £40 000. The crane was estimated to have a residual value of £10 000 at the end of five years. However, the purchase was not made.

Prepare a statement showing what the profit after tax would have been in each of the above two years if the crane had been purchased assuming that all other factors remained the same (see Table 4.16).

Table 4.16

	19x4 (£)	19x5 (£)
(a) Capital value of crane	40 000	
(b) Residual value of crane	10 000	
(c) Total accounting depreciation over five years	30 000	
(d) Accounting depreciation per year	6 000	6 000
(e) Taxable profit	75 000	110 000
(f) Profit plus depreciation	81 000	116 000
(g) Written down value of crane	40 000	30 000
(h) Writing-down allowance	10 000	7 500
(i) Taxable profits	71 000	108 500
(j) Tax	24 850	37 975
(k) Profits after tax	46 150	70 525

Notes on Table 4.16

(a) Purchase price of crane
(b) Residual value of crane after five years
(c) Total depreciation of £40 000–10 000 = £30 000
(d) Company's annual straight line accounting depreciation of £30 000 over five years
(e) Taxable profit in relevant year
(f) Company's depreciation added back into profit before tax
(g) Written down value of crane i.e. capital value less writing-down allowance, both from previous year
(h) Writing-down allowance, i.e. 25 per cent of written down value for that year
(i) Profit plus depreciation (f) less writing-down allowance
(j) Tax at 35 per cent of taxable profits
(k) Taxable profits less tax.

Company depreciation method – straight line
Plant life – five years
Corporation tax – 35 per cent
Taxation writing-down allowances – 25 per cent (reducing balance basis)
Company financial year – 1 January to 31 December

Problems

4.1 Prepare a series of tables showing the position at the end of each year with regard to a loan of £10 000 at 10 per cent interest repaid over five years, by the following means:

  (1) By payment of intcrest only at each year end and the lump sum repayment of the £10 000 at the end of the fifth year.
  (2) By repaying £2000 at each year end plus interest for the preceding year.
  (3) By repaying five equal instalments to include loan and interest.
  (4) By repaying total loan plus accumulated interest in a single payment at the end of five years.

4.2 If £100 is invested at 4 per cent on 1 January 19x5, what is the sum that can be withdrawn at each year end for six years so that the investment fund is empty after the sixth withdrawal?   (£190.8)

4.3 How much will accumulate to the credit of a charitable trust at the end of ten years if it earns 5 per cent interest and £250 is deposited in its favour at the end of each year for ten years?   (£3144)

4.4 If a fixed amount is deposited at $2\frac{1}{2}$ per cent for each of five years, the first payment bcing made on 1 January 19x4 and the last on 1 January 19x8 what must the amount be, in order to accumulate £2000 on the date of the last deposit?   (£381)

4.5 If you wish to make a lump sum investment on 1 January next year so that you will be able to withdraw £500 at the end of each year for six years leaving nothing of the investment, what must the initial investment be? The interest rate obtainable is 3 per cent.   (£2708)

4.6 If £5000 is invested now, £2000 in two years hence and £3500 in five years hence, all at 5 per cent, what will the total amount be in ten years hence?   (£15 566)

4.7 What loan made today at 4 per cent would be paid back completely by payments of £2500 at the end of three, six, nine and twelve years?
                          (£7516)

4.8 Spending £150 now to avoid spending £200 in five years' time is, in effect, securing what interest rate?   (5.99 per cent)

4.9 What annual saving must be anticipated for 15 years in order to justify a present expenditure of £20 000 if interest is at 4 per cent?   (£1799)

4.10 What annual investment must be made at 5 per cent in order to replace a £100 000 structure in 25 years' time?   (£2095)

4.11 What is the equivalent annual cost for 20 years of spending £1000 at the present time, £2000 in six years' time and £10 000 in 15 years' time? Interest is at 3 per cent. (£611)

4.12 Conversion to central heating for a house will necessitate an initial investment of £430. It is estimated that a saving in fuel cost will amount to £50 over one year. How long must this saving continue in order to justify the initial investment? Interest is at 10 per cent.

(20.64 years)

4.13 A small face-shovel is bought new for a price of £23 250. It has an expected life of seven years when it is expected to have no salvage or resale value. Expenditure on maintenance, insurance, taxation, fuel and lubricants is expected to reach £500 by the end of the first year, £700 by the end of the second, £900 by the end of the third and so on increasing by £200 each year. What is the equivalent annual cost of owning this piece of equipment if interest is at 5 per cent? (£5079)

4.14 A contractor has to decide between two types of truck for use in his transport fleet. Type A has an initial cost of £15 000 with a resale value at the end of five years of £3000. Type B has an initial cost of £18 000 with a resale value at the end of six years of £4000. Annual maintenance and fuel costs are similar for both vehicles. Which is the better investment if interest is at 5 per cent? (Type A)

4.15 A development company buys a site near the fringe of an industrial area in a large city for £1 000 000. Annual outgoings on the site for maintenance, fencing, watching, etc. amount to £45 000 per year. It is estimated that the site will not be sold for eight years when the area is due for development. For what price must the site be sold so as to show a profit if the money involved could have been invested at 5 per cent per year during the eight years? (not less than £1 907 210)

4.16 A contractor buys a bulldozer on a time-payment plan making monthly payments of £500 at the end of each month for 60 months. The cash price of the bulldozer is £19 000 less $2\frac{1}{2}$ per cent trade discount. Find the approximate rate of interest per month, and the effective interest rate per year. ($1\frac{3}{4}$ per cent; 23.1 per cent)

4.17 The owner of a quarry agrees to supply roadstone to a contractor over a period of 25 years. The contract arranges for the supply of not less than 30 000 tons at £1.25 per ton each year for the first seven years, not less than 45 000 tons at £1.50 per ton each year for the next 13 years and not less than 50 000 tons at £1.75 per ton each year for the remainder of the contract.

   If payments for the roadstone are to be made at the end of each half year in which they occur, what is the minimum present value of the contract to the quarry-owner on the basis of interest at 3 per cent per half-year?   (£1 817 998)

4.18 Two schemes are under consideration for a water supply installation. The first scheme has an installation cost of £1 250 000 and a life of 25 years. The second scheme has a similar life but requires a greater consumption of electricity. Its installation cost is £950 000. With interest at 6 per cent, how much can be paid for the additional electricity per year before the equivalent annual cost of the second scheme exceeds that of the first?   (£23 470 +)

4.19 In considering the installation of a cooling water discharge system the following alternatives arise:

Pipe diam.	Installation cost	Annual pumping cost
550 mm.	£175 000	£60 000
600 mm.	£198 000	£40 000
750 mm.	£230 000	£27 600

It is estimated that the pipes will remain in service for ten years when the installation will be renewed. The salvage value of each of the alternatives is expected to be 25 per cent of the initial installation cost. Compare the annual costs of each scheme if the minimum attractive rate of return is 7 per cent before tax is paid.

(500 mm: £81 850 p.a.; 600 mm: £71 650 p.a.; 750 mm: £64 370 p.a.)

4.20 A consulting engineer has designed the layout of two road schemes in mountainous country. With the exception of one particular length of the road the estimated costs for the two schemes are similar. In this area the choice of schemes lies between a cutting combined with a tunnel and the construction of an embankment combined with a bridge. The engineer has estimated the costs and the lives of the alternative works as follows:

	Life	Initial cost	Maintenance
Location A			
(a) Embankment	perpetual	£15 000 000	£200 000 p.a. for the first three years and thereafter £50 000 p.a.
(b) Bridge	40 years	£17 500 000	£120 000 p.a.
Location B			
(a) Cutting	perpetual	£ 7 200 000	£ 50 000 p.a.
(b) Tunnel	perpetual	£29 000 000	£175 000 p.a.

Using the method of annual costs, determine which scheme has the financial advantage if the interest rate is assumed to be 5 per cent.

$$(AC_A = £2\,140\,250; \ AC_B = £2\,035\,000. \text{ B is just the cheaper.})$$

4.21 A civil engineering contractor operates a fleet of dumpers. From past experience he has found that a dumper normally has a useful working life of five years. Such a machine has an initial capital cost of £21 500 and at the end of the five-year period has a salvage value of £3500. The cost of maintenance for each dumper amounts to £1500 for the first year and this increases by £750 for each succeeding year. If the current interest rate is 6 per cent, what is the equivalent annual cost of owning and maintaining each dumper? If the contractor can sell the dumpers for £7500 each at the end of the fourth year should he be advised to do so? $(AC = £7396; \text{ Yes.})$

4.22 The sole owner of a small business in a developing locality assesses that his profit for the next three years will amount to £13 000 per year. This profit is estimated to reach £17 500 per year for the next five years after that and then to settle at a figure of £22 000 per year during the foreseeable future. For what price should the business be sold today if a return of 8 per cent is required, based on the profitability of the business over the next 12 years?   (£128 340)

4.23 An initial investment of £375 600 is envisaged for a sewage pumping scheme. The future annual maintenance costs are assessed at £36 000 per year, the pumps are presumed to need replacement every 20 years and some pipework to be replaced every 25 years. The cost of replacing the pipes and the pumps is £30 000 and £45 000 respectively. The service life of the scheme is assumed to be perpetual. Compute the capitalized cost of the perpetual service using an interest rate of 6 per cent.

(£1 002 860)

4.24 What is the present value of the cost of nine years' service from a roller having an initial cost of £25 500, a life of nine years, an estimated salvage value of £3000 at the end of its life and an annual operation and maintenance cost of £3000? Assume an interest rate of 6 per cent.

(£44 129)

4.25 Calculate the yield in each of the following cases:

Initial outlay	Return
(a) £10 000	£3050 for Years 1-5 inclusive
(b) £5 000	£802 for Years 1-7 inclusive
(c) £25 000	£1920 for Years 1-20 inclusive

((a) 16 per cent; (b) 3 per cent; (c) $4\frac{1}{2}$ per cent)

4.26 Calculate the yield for the following project cash flows:

(a)		(b)	
Year 0	(10 000)	Year 0	(10 000)
1	2 000	1	(2 000)
2	2 500	2	4 000
3	3 000	3	4 000
4	3 500	4	6 000
5	4 000	5	6 000

((a) $13\frac{1}{2}$ per cent; (b) 16 per cent)

4.27 If £20 000 is invested in a project, what average annual return is required for each of the first 20 years in order to yield 12 per cent?

(£2677)

4.28 A property developer invests £12 000 000 in a slum clearance site by constructing an office block on it. He estimates that his revenue from office rents will be £2 700 000 per annum and that his total outgoings each year will be £1 300 000. Calculate the prospective yield from the investment based on a life of 25 years. (11 per cent)

4.29 On retirement a man is offered a pension of £11 000 per year for 15 years guaranteed, rather than a lump sum of £120 000. At what rate of interest would he have to invest the lump sum in order to be able to withdraw from the investment an equivalent annual rate?

(4½ per cent)

4.30 Calculate the present value of a series of annual uniform cash flows each of £1000 over a period of ten years, assuming that the interest rate is 5 per cent for each of the first five years and then rises by 1 per cent for each of the succeeding five years. (£7043)

4.31 A man invests £10 000 in the shares of an engineering company. In the first year he receives dividends of the value of £750, in the second year, £1000 and in the third and fourth years he receives £1500. At the end of four years he sells his shares for £9500. What was the yield on his investment? ($10^{1}/_{4}$ per cent)

4.32 A local authority engineer is concerned with making a selection between three types of pumping equipment, each of which will provide the necessary output in the specified conditions.

Manufacturer A presents evidence that his equipment will cost £23 600 per year to operate and will need a major overhaul towards the end of the fourth year of its life costing £6750 in addition to the normal operating costs. His quotation for supplying the equipment is £243 200.

Manufacturer B quotes £305 000 for supplying the necessary equipment and estimates that normal operating costs will be £16 500 per year plus a major overhaul costing £7500 towards the end of Year 6 of its life.

Manufacturer C quotes £269 200 for equipment supply with an estimated annual operating cost of £20 000 ± 5 per cent.

If capital is currently costing the local authority 10 per cent and the life of the equipment in each case is eight years, use the equivalent annual cost and present value methods to make a recommendation for selection.

(Equivalent annual cost – A: £70 050; B: £74 460;
C: £69 460 to £71 450
Present value – A: £373 710; B: £397 260; C: £370 560 to £381 230
Select A in the absence of further definition of cost from C)

4.33 An engineer is designing a heating system for a small hotel. He is concerned about providing suitable insulation against heat loss in the office block and eliminates all but two alternatives.

Scheme A is to spend £240 000 on special insulation and £30 000 on a boiler having total estimated annual running costs of £6750 for each of the first seven years and £9000 for each of the remaining eight years of its useful life.

Scheme B involves spending more on the insulation, £280 000 in total, but less on the boiler installation. The boiler will cost £20 000 and it is estimated that its running costs will amount to £5000 for each of the first five years, £6500 for each of years 6 to 10 inclusive and £7500 for each of the remaining five years of its useful life.

It is expected that in both cases, the salvage value of the boiler at the end of its useful life will be 10 per cent of its initial cost. Ignoring the salvage or resale value of the rest of the investment, that is for the insulation, etc., use the NPV and the Yield methods to assist in making a choice between the two schemes over the period of the useful life of the boilers. Assume a minimum acceptable rate of return of 11 per cent.

How would you approach the economic analysis if the useful lives of the boilers had been different from each other?

(Select Scheme A)

4.34 A company has to decide between five mutually exclusive projects, A, B, C, D and E having investment outlays of £230 000, £270 000, £150 000, £200 000 and £180 000 each, respectively. Each project has a life of 20 years.

Project A produces net annual receipts of £36 000, Project B £39 000, Project C £20 000, Project D £31 000 and Project E £25 000. If the company wishes to minimize its investment and obtain a rate of return not less than 13 per cent, in which project should it invest?

(Project D to maximize return;
Project A is acceptable and maximizes investment.)

Chapter 5
Risk and Uncertainty in Project Investments

5.1 Certainty, risk and uncertainty

In comparing straightforward investment projects thus far, it has been assumed that the chances of obtaining a yield, a specific cash flow at a particular time, a given project life or a minimum acceptable rate of return, has not been in doubt for each of the projects. Alternatively it can be stated that all the assumptions that have been made in each case have been used as though they were *certain*. Most aspects of investment appraisal, however, apart from the actual calculation, involve a degree of forecasting and judgement into which an element of subjective bias is likely to creep. In addition, a great deal of forecasting is based on historic performance from similar or near-similar situations. It is doubtful whether the scale of capital budgeting in many individual companies reaches that number of similar investments where prediction or estimation on historic performance can be used without an element of risk. The investment decision can frequently be one of an isolated 'once in a lifetime' nature which does not have the advantages of being one among a large number, and thus subject to a law of averages.

Investment decisions will normally be made subject to conditions of *certainty*, *risk* or *uncertainty*.

The condition of certainty needs little explanation. As applied to investment appraisal, it is a state of nature which exists when there can be reasonable certainty about a specific outcome. There can never be absolute certainty about investment, but there are situations where nothing short of international economic disaster will influence the outcome of the decision. A decision to invest in government stock bringing a return of 6 per cent is an example of a decision made under certainty. There is little doubt that the return will be received and that, under such conditions as the money might be lent, the invested cash will be returnable when required. Certainty invariably costs something, and within reason, it might be said that certainty and the chance of a very large profit or return are inversely proportional.

Risk has previously been mentioned in Section 3.3. A decision is made under risk when a decision-maker can assess, either rationally or intuitively, the probability of several outcomes of each alternative because of past experiences. A simple example of a decision under risk is that of determining the length of time it will take to complete a specific operation or activity in a construction programme. One knows from previous experience what sort of output is likely to be achieved and the likely chances of this or that going wrong. This is the nature of a large number of decisions in business generally, for example in insurance, and often in construction.

Another simple practical illustration of a problem requiring a decision to be made under risk is that of whether to provide protection against flooding for a factory which is to be built adjacent to a tidal river. Past experience and statistical data concerning the weather, tidal range, the high and low water levels will enable a reasonably accurate forecast of the chance (or probability) of a flood occurring in the future. Further, estimates can be made of its probable frequency and, from its likely magnitude, the probable effect the flood will have on the fabric of the factory and its operations. These probabilities can be converted into monetary terms and a final decision can then be made as to whether to provide complete, partial, or no protection from flooding, using these forecasts as an aid and in the light of a numerical assessment of a disaster recurring. Decisions under risk require past experience, historical data and previous records in order to support them. Usually, therefore, they will be taken in conditions where like circumstances have occurred previously and sufficiently often to have established at least an outline pattern of behaviour.

The third area is that of uncertainty. Again, this type of situation is common in business and is one in which there are not historic data or a previous history relating to the particular situation under consideration which can be used as a guide in arriving at a decision or a judgement. Decisions which are likely to be influenced by government policy can be decisions of this nature. For example, a decision by a construction company to concentrate on the erection of office buildings in a city centre as part of a property development is usually made with regard to demand and known legislation. There can be no certainty with regard to future, long-term central government legislation beyond the immediate future of the project that happens to be at hand at that time. Another example of such a situation is the launch of an entirely new product onto the market. No previous experience of selling the product will be available, no experience of how the customer is likely to react and buy this particular product can be used to predict required production plant capacities, the size of a sales force, advertising methods, etc. Further, in many investment situations, a radical change in Government policy or the tax structure, changes in the trade climate overseas, depression, recession or war may have a far-reaching effect on profitability or viability. Conditions pertaining to such a project will rarely be such that they can be predicted with accuracy or even foreseen at all.

Frequently in decision analysis, an attempt is made to divide problems into those that exhibit the characteristics of risk as described above and those that exhibit the characteristics of uncertainty. However, some statisticians have argued that the division would be more helpful if it were between events which are statistical and those that are not; others suggest the choice should be between events that are repeatable and those that are essentially non-repeatable. The former situation in each case is one wherein the set of outcomes can be described realistically by a probability distribution; the latter situation cannot. There are occasions, however, when it is difficult to decide into which category a particular problem event should fall. One such occasion might be the construction of a bridge pier in a cofferdam constructed in a river bed. There may be ample statistical data available concerning likely weather, rainfall, construction durations, work force absenteeism, ground conditions and so on, but there is little doubt that the event itself is non-repeatable. It would generally be agreed that the problem combines aspects of both risk and uncertainty. The distinction between risk and uncertainty may, therefore, be a useful concept to have in mind in laying out the framework of a decision problem so long as this is not undertaken with a view to categorizing all relevant problems into one or other water-tight boxes.

Situations where appraisals are made under conditions of certainty present little problem except, perhaps, in so far as there may be so many different alternatives to consider that mathematical methods must be employed to narrow the field and eliminate all but the real contenders.

5.2 Probability theory

It is commonplace to use such expressions as 'probably', 'improbable', 'most likely', 'not likely', 'possibly' and so on, in our everyday description of the chances that a particular event will take place. Such descriptions are subjective by nature and have different meanings (though often only slight differences) to each user. They are subject to a range of interpretations that the user wishes to put on them. In order to give such expressions a quantitative classification, statisticians use a scale of *probability* ranging from zero to one.

The likelihood of every or any event occurring can then be assigned a value on this scale and a common language has been established which is universally understood by those who use it.

The probability of 1 ($p = 1$) on the scale represents the condition of absolute certainty. This is a situation which rarely pertains in business, though when money is deposited in the vaults of a major international bank, the probability that it will eventually be returned is often assumed to equal to 1. At the other end of the probability scale we have the figure of zero ($p = 0$). This represents the absolute certainty that an event will not occur.

When a die is rolled, assuming that it is a fair one and not in any way biased, it is possible for it to come up on any one of its six sides with an equal chance. The probability of any one specific number turning up, say a five, is 1/6 or 0.167. When a coin is tossed, it is already known before it lands that it must have one of two sides uppermost (excepting freak circumstances). The probability that it will be either one or the other is equal to 1, and since there is an equal chance of either a head or a tail being uppermost, the probability of a head showing is $\frac{1}{2}$ or 0.5 and similarly for the chance of a tail uppermost. Such probabilities as these can be calculated beforehand out of consideration of the well-defined possibilities and good commonsense. There is no need actually to toss the coin or roll the die in order to establish p for each possible event. Probabilities of this nature, that can be calculated without undertaking the trials, are known as *a priori* probabilities.

There will always be situations where it is possible and perhaps necessary to establish probabilities by observation or practice. In some cases no amount of thinking and commonsense will enable the establishment of an *a priori* probability. For example, if we observe a bricklayer laying bricks and his output on each day for 100 days is recorded, it is then possible to assign an *empirical probability* of the bricklayer's output reaching a certain figure on a future day.

Table 5.1 lists the output as recorded:

Table 5.1

No. of days	Output in bricks per day	Relative frequency of occurrence
2	210	0·02
8	220	0·08
30	230	0·30
50	240	0·50
5	250	0·05
4	260	0·04
1	270	0·01
100		1·00

If the number of days on which a single specified output is achieved is divided by the total number of days on which outputs are recorded, the *relative frequency* of each output is obtained. This is listed in the third column of Table 5.1. The information about the random variable, *output*, and the data in the relative frequency column, together enable a *relative frequency distribution* to be drawn. In the absence of other data concerning this output, it may be assumed that the relative frequencies set out above can reasonably represent the *probability distribution* of output.

It may be expected from looking at the table that the probability of obtaining an output of 240 bricks laid on the next working day will be 50/100 or 0.50. Similarly, the probability of obtaining an output of 260 bricks laid on the next working day is 4/100 or 0.04. The probability of each output occurring can be calculated and it should be noted that the total relative frequencies, and hence the total probabilities for each group, amount to 1. The longer the period over which the recording is carried out, the more likely it will be that recorded relative frequencies will give an accurate forecast of the probability of future similar events.

The empirical probability of an event can be calculated by:

$$\text{Probability} = \frac{\text{Total number of occurrences of event recorded}}{\text{Total number of trials}}$$

It is empirical probability which lends itself to the calculation of the likely effect of risk in investment appraisal.

5.3 Frequency and probability distributions

It is often easier to obtain an overall appreciation of the classification of a set of figures by displaying them graphically. Frequency, and hence probability, distributions are no exception and there are several ways in which this can be

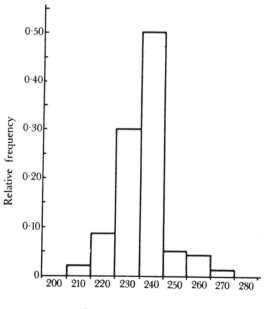

Output – bricks per day

Fig. 5.1 Histogram representing relative frequency/probability.

done. The most useful way is to display the data in the form of a *histogram* wherein the frequency and the value of the random variable are represented by the area of a vertical bar as illustrated in Fig. 5.1. The data set out in Table 5.1 are used for the purpose of constructing this histogram. In Fig. 5.1, it is important to note that each vertical bar is of constant width and therefore the height is proportional to the relative frequency/probability.

It is often of greater practical value to have a *cumulative distribution* of a random variable. Such distributions can give values of the frequency or probability for *all values less than, equal to or less than, greater than*, and *equal to or greater than* a specified value of the random variable. A cumulative frequency or probability distribution can readily be constructed from the information contained in Table 5.1. This is demonstrated in Table 5.2. The distribution is illustrated graphically in Fig. 5.2. It will be seen that the calculation of a cumulative frequency of that value of the random variable, or less, is quite straightforward. For example, the relative frequency of an output of 220 bricks per day is 0.08. There is only one recorded output less than this – 210 bricks per day with a relative frequency of 0.02. The two frequencies are therefore added to give the required cumulative frequency and so on.

A convention is sometimes used in drawing a cumulative frequency diagram to indicate which of the two values of the random variable is

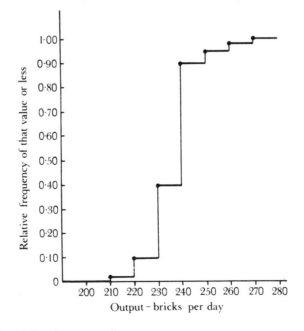

Fig. 5.2 Cumulative frequency diagram.

Table 5.2

Output in bricks per day	Relative frequency	Cumulative frequency of that value or less
210	0·02	0·02
220	0·08	0·10
230	0·30	0·40
240	0·50	0·90
250	0·05	0·95
260	0·04	0·99
270	0·01	1·00
	1·00	

associated with the appropriate relative frequency on the *y*-axis. For example, the relative frequency of 0.40 in Fig. 5.2 can, in the absence of other indications, refer to the value of output of 230 *or* 240 or any value between these two. A dot is used, as in the diagram, to make the required distinction. In the example quoted the dot refers to the value of 230 and this can readily be checked against the table. If the *y*-axis is labelled 'Relative frequency less than that value', the dot would be correctly positioned at the other end of each horizontal line in the cumulative frequency diagram.

5.4 The characteristics of a distribution

Given a set of numbers or data, it is desirable to express the characteristics of the set by using a *representative* figure. One such representative is the *average*, of which there are a number of varieties, the most frequently used being the *arithmetic mean*. An arithmetic mean is calculated by dividing the sum of all the individual values in a set by the total number of values in that set. It is a measure of central tendency but can be misleading and misinterpreted if it is not calculated with discretion and subsequently employed with care. The arithmetic mean is usually denoted by $\bar{x} = (\Sigma x)/n$ where x represents each of the values and n represents the number of individual values.

In the same way that groups of figures or values can be represented by a measure of central tendency, a distribution can be similarly represented. It will be appreciated that a relatively small number of values at one extreme of the range of values for which an arithmetic mean is to be calculated, can seriously affect the true representivity of the arithmetic mean. Alternative or additional measures of the characteristics of a set of data have therefore to be used in order to establish a comprehensive picture of the relationship of each with the others. One such measure is known as the *mode*. This is the most commonly occurring value and as such may be considered to be a *typical* value of a distribution. Referring to the histogram of Fig. 5.1 the modal value of the distribution shown is clearly 240 bricks per day.

A second alternative means of measuring the central tendency of a distribution is the *median*, being the individual value above and below which are equal numbers of values in the distribution. The median value of the histogram of Fig. 5.1 is also 240 bricks per day.

Having calculated the values of a median, a mode and a mean, still they may not provide a comprehensive specification of the attributes of a particular distribution of values. For example, neither singly nor collectively do they give a measure of the *range* of values (from 210 to 270 bricks per day in the example). In addition, they do not give a precise definition of the *skew* or lop-sided shape of a distribution where this is the case. An example of skew is shown in the histogram of Fig. 5.1 where the distribution is not symmetrical about its central value. Figure 5.1 shows an example of *positive skewness*, a situation in which the longer tail of the distribution stretches out to the right, as the histogram is drawn, where the values of the variate (the quantity which varies) are higher than those of the left-hand end of the scale. If the shape of the distribution is reversed about its central value from that of Fig. 5.1 the skewness is referred to as being *negative*. Consideration has only been given, thus far, to distributions which consist of *discontinuous* variates, that is, variates which only assume discrete values. There are many examples of such variates which are commonly collected for statistical purposes, for example statistics concerning the number of cars per family. Certain percentages of the population will have none, one, two, three cars, and so on. The number of cars per family can only assume a discrete number (though undoubtedly there will be statistically awkward families who claim to have a part-share with another family in a particular car). The distribution in such cases must therefore assume the stepped shape which is typified by the histogram in Fig. 5.1. In other cases the variate will be *continuous* and it can assume any value within a given range. For example, the daily output of a labourer digging a trench can be expressed in units or parts of units and would not necessarily be a discrete number of cubic metres.

Figure 5.3 illustrates (a) a symmetrical and (b) a skew continuous distribution with the three measures of central tendency superimposed. Where the skewness of a distribution is not excessive there is an approximate relationship between the mean, median and mode. The difference between the value of the variates at the mean and the median is three times the difference between those values at the mean and the mode.

Having considered the measure of central tendency for a distribution, further consideration needs to be given to the range of a variate. The extent to which a range of a variate is spread is known as the *scatter* or *dispersion*

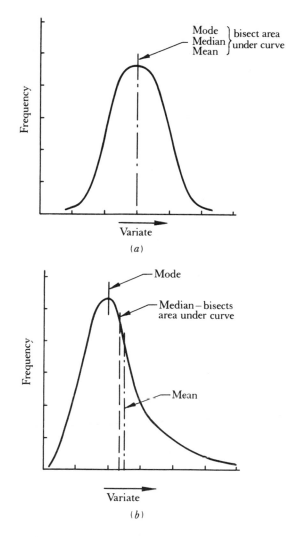

Fig. 5.3   Two typical continuous distributions.
     (a) Symmetrical continuous distribution.
     (b) Positively skew continuous distribution.

and the most important measure of this characteristic is the *standard deviation*. The formula for calculating standard deviation, *s*, is

$$s = \sqrt{\frac{\Sigma(x - \bar{x})^2}{n}}$$

where x, $\bar{x}$ and n are used as before.

Example 5.1

Calculate the standard deviation for the set of values 12, 6, 8, 9, 10, 7, 5.

x	$(x-\bar{x})$	$(x-\bar{x})^2$
12	3·86	14·90
6	−2·14	4·58
8	−0·14	0·02
9	0·86	0·74
10	1·86	3·46
7	−1·14	1·30
5	−3·14	9·86
Total 57	$\Sigma(x-\bar{x})^2 = 34\cdot86$	

$$\bar{x} = \frac{57}{7} = 8.14, \text{ therefore } s = \sqrt{\frac{34.86}{7}} = 2.23$$

The significance of a standard deviation may not be immediately obvious. This is best illustrated when applied to either a symmetrical or near-symmetrical distribution of the type already discussed. Figure. 5.4 gives the relationship between the dispersion and standard deviation as measured about the central value of a normal (that is a particular, symmetrical) or near-normal distribution.

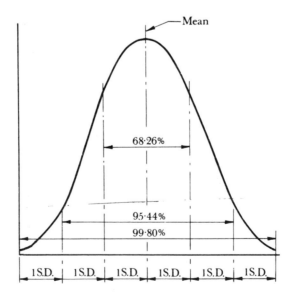

Fig. 5.4 Standard deviation as a measure of dispersion.

It will be noted that these measures are made in terms of the units in which the variate is measured. Just over 68 per cent of the area under the distribution lies within an area of $\pm$ one standard deviation from the mean, 95 per cent lies within $\pm$ two standard deviations from the mean and almost 100 per cent of the distribution lies within $\pm$ three standard deviations from the mean.

Whilst a standard deviation gives a measure of the dispersion of a distribution about its central value, it still does not allow a ready comparison of the attributes of one distribution with another where the variate is expressed in different units. To enable relative variability to be measured, the *coefficient of variation* (Karl Pearson) is most commonly used. This coefficient of variation is obtained by expressing a standard deviation as a percentage of the mean value of the distribution. If the coefficient of variation is denoted by v, then

$$v = \frac{100s}{\bar{x}}$$

The larger the value of v obtained from this expression, the larger is the relative reliability of the data to which it refers.

5.5 Empirical methods of dealing with risk and uncertainty

Before proceeding with the application of probability theory to risk analysis in an investment situation, it is helpful to look at the empirical methods by which risk can be taken into account.

One of the most common approaches remains entirely within the subjective judgement of the decision-maker. The individual making the decision uses his own experience and knowledge to assess whether one project is *riskier* than another. Having evaluated the project using one of the techniques already described, the ratings of the projects are compared. If NPV is the method of comparison chosen, then where similar NPVs are obtained as between projects, the riskier one will be rejected. It is not possible to identify or specify a rule for trade-off between risk and NPV, if this is the chosen method, and much will depend on the nature of the projects chosen in the eyes of the decision-maker.

A move towards a quantitative approach to risk analysis is that of adopting the need for varying discount rates or varying minimum acceptable rates of return for projects which are believed to involve different degrees of risk. On the basis that a firm has computed its cost of capital as being, for example, 11 per cent, it might adopt this rate as the minimum acceptable rate of return for projects which, from experience, are known to bear no more than average risk. Such projects may be extensions to the well-established business of the firm. Perhaps market trends indicate an increasing demand for the products from existing plants and the costs of plant extension and increase in product revenue can be estimated with considerable accuracy. However, if the firm decides to diversify into a business of which it has little or no experience, there may be some justification for using a discount rate of, say, 16 per cent or more to compensate for the obvious risks and unknown factors which may be involved. At the other end of the risk scale are probably the inevitable series of cost reduction exercises which are in themselves almost risk-free, for example, the substitution of mechanical-handling equipment for manual labour in what is currently a labour-intensive operation. The return in such cases is usually relatively risk-free allowing the use of a discount rate slightly lower than the computed cost of capital, say 9 or 10 per cent in this example range of rates. The discount rate can be looked upon as being of two parts; one the rate for a risk-free investment, the other to compensate for the risk element. The difficulty with this concept is probably the determination of a suitable rate for a risk-free investment.

In another way, the payback method, referred to in Chapter 4, can be considered to be a means of evaluating risk for a manager. Cash flows beyond the payback period as calculated are considered to be so uncertain as not to warrant consideration. However, in situations where the data used in making the evaluation prove to be incorrect in practice, particularly as far as its timing is concerned, the results can be, in themselves, very risky. This is never more so than in the situations where, on the one hand, the larger annual returns are expected in the latter years of the payback period, and on the other, they are expected to be received early on. For two projects with similar paybacks and these characteristics, the risks can be quite different.

It is not good policy on the part of a firm to attempt to assess single projects in isolation from all the other activities of the firm. The decision to undertake a project which is considered to be risky must be taken out of consideration of the risk which the firm has already undertaken. A balance of risky to risk-free projects is required at any one time if the future of the firm is not to be put in jeopardy. Many firms have become troubled because they have chosen to accept only risk-free (or nearly so) projects and as a result have ignored those projects which are necessary to the future well-being of the firm. The development of new processes, research and development as a whole, and the marketing of new products are all ventures which cannot be undertaken in a commercial environment which is entirely free of risk. A proper balance of projects of this nature with others of a more mundane nature must always be part of a broad investment programme.

5.6 Risk and probability distributions

The difference between a *risky* project and a *risk-free* project can be said to be dependent on the variability in the return which is likely to arise from each. A relatively risk-free project is one in which the rate of return (to use a quantitative measure) can be forecast with some certainty and thus some accuracy. For example, an investment in the loan stock of a large international corporation at fixed interest rates is reasonably risk-free, since considerable dependence can be placed on the receipt of interest payments year by year. On the other hand, a similar investment in a company set up for the sole purpose of mineral prospecting in the forests of an undeveloped country may produce a complete loss of the sum invested or, if the prospecting is highly successful, a return many thousands of times the original investment. Such an investment is of very high risk. The possible variability of the outcome is therefore likely to be a measure of the riskiness of the investment.

When evaluating a project investment, it is necessary to estimate the net cash flows for a number of periods over the project life. The establishment with certainty of a precise value for a variable in a decision problem of this nature is a difficult process. Often, at best, the answer can only be an enlightened guess and, frequently, conditions are such that it has even less foundation. There is always a much greater probability that a single estimated value for a cash flow, or of another estimated variable, will not be achieved, than that it will be achieved exactly. It is simpler to make an estimate of a range of values rather than a single figure. To do so will breed confidence and tend to encourage a greater depth of thought. It is possible, by careful and intelligent questioning, to elicit the required information from an expert or a knowledgeable source, over a range of figures together, with some expression of their confidence that the results will be achieved. It is convenient to use a statistician's scale of probability in order to express the likelihood of a particular value of the variable being achieved or of the achievement of a particular event.

The result of estimating a range of values, and the probability of the achievement of each one, can be represented graphically by a probability distribution. If, for the sake of example, it is estimated that the net cash flow for a particular year is £5000, this in itself does not express any measure of risk. By examining the probability distribution of the possible range of values, comparative risks may be assessed.

Figure 5.5 illustrates two possible probability distributions, both normal or symmetrical in this case, and both based upon an *expected cash flow* of £5000. Since the area under each distribution must equal 1.00, the probability of achieving the expected cash flow in the case of the estimate with the shorter range of values will necessarily be larger than in the other instances. Since, in normal and near-normal distributions, the standard deviation is a measure of the range of values involved, it can also be a measure of the risk or uncertainty involved. The larger the standard deviation, the greater the likely variability of the net cash flows and hence, the greater the risk of achieving that particular value. In

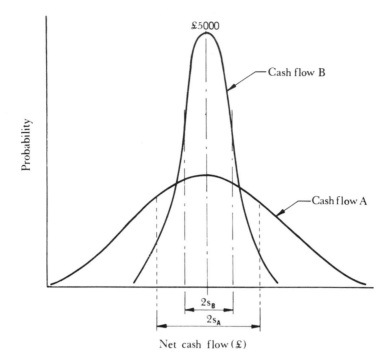

Fig. 5.5   Cash flows at different risks.

the example illustrated in Fig. 5.5, 'Cash flow A' is typical of a higher risk cash flow than 'Cash flow B'.

A number of general conclusions as to the certainty or otherwise of an estimator's forecasting can be drawn from examining the shape of the probability distribution which arises from the range of values so estimated. Three examples are illustrated in Fig. 5.6. Curve A shows a confident, certain and relatively risk-free estimate. Curve B illustrates a skewed distribution which indicates that the expected value is more likely to be exceeded than not to be attained.

Curve C of Fig. 5.6 illustrates a cash flow pattern in which there is a greater probability that the estimate will be underrun than the reverse.

In some instances, a comparison of the risk involved between different net cash flows, as in the above examples, using the standard deviation as a basis, does not give a complete answer. Cases can arise where two cash flows can have distributions of similar shapes and dimensions but different expected values. Referring to Fig. 5.5, assume that 'Cash flow B', with its expected value for a net cash flow of £5000, is being compared with a cash flow having a similar distribution shape but with an expected cash flow of £2000. In terms of the illustration it simply means that the distribution having the smaller

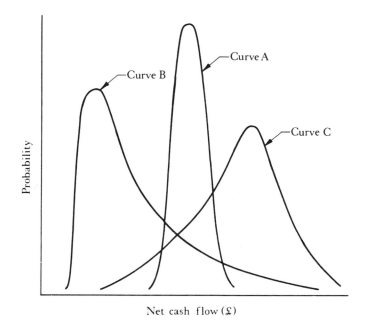

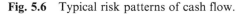

Net cash flow (£)

Fig. 5.6 Typical risk patterns of cash flow.

expected net cash flow appears nearer to the *y*-axis of the graph; it is therefore wholly transposed to the left. Both distributions may have similar standard deviations, say £350. However, as a measure of variability the standard deviation does not give a true comparison because the percentage variation of one standard deviation to the mean value is different in each case. It is in such cases as these that the coefficient of variation should be used. In the case of the cash flow of £5000,

$$v = \frac{100 \times 350}{5000} = 7 \text{ per cent}$$

and in the other case,

$$v = \frac{100 \times 350}{2000} = 17.5 \text{ per cent}$$

The risk is therefore greater in the lower expected cash flow.

Only in the case of Curve A of Fig. 5.6 will the expected value (the mode) of the frequency distribution of cash flows give a reasonable answer in the appraisal of an investment, when taken along with all the other factors affecting the likely outcome. To overcome this deficiency, a simulation technique for risk analysis can be used. The general basis of the simulation is that all possible combinations of probability values of the variables are examined, though in practice the range of combinations must be limited to avoid excessive processing costs. An estimated subjective probability distribution (either continuous or discrete) for the levels of achievement for each of the key variable factors relevant to the appraisal is prepared.

It is then necessary to select, by some random procedure, one sample value from each of the distributions of the relevant variable factors. (Monte Carlo sampling is the usual procedure to be used for the random selection from frequency distributions of variables.) The values of the variable so obtained are then combined to calculate a rate of return for the project and the sampling and calculation process is then repeated a large number of times. Eventually, a probability distribution for the rate of return can be drawn up and the probability of achieving any one particular return in the range of the final distribution can be established. A simple example will illustrate the principles.

Example 5.2

Three estimates of the annual operating cost of a proposed new plant are given together with the associated probability that any one of them will be achieved. (All costs are in £000.) The costs are £1000, £1100 and £1200 having probabilities of 0.30, 0.50 and 0.20, respectively. Annual revenue is estimated to be £4400, £4500, £4600 or £4700 having probabilities of achievement of 0.10, 0.30, 0.40, and 0.20, respectively. In estimating the capital cost of the plant, the engineer in charge of the work believes that there is a 0.30 probability of it being £14 000, a 0.60 probability of it being £15 000 and a one in ten chance of it being £16 000. When consideration of the life of the plant is made, the three lives and their associated probabilities are five years and 0.40, six years and 0.30, and seven years with a probability of 0.30. How can this information be used in order to assess the risk in making the investment?

A histogram or probability distribution can be drawn for each of the variable factors to be considered. In the case of the estimated figures for annual revenue, the distribution is shown in Fig. 5.7(a). In Fig. 5.7(b) is illustrated the cumulative distribution derived from it. Single values of the variable factor (annual revenue in this case) can now be selected from the cumulative distribution in a random fashion by the use of Monte Carlo methods.

The first step in applying Monte Carlo methods is to assign a series of numbers (in this case they are two-digit numbers) to each of the discrete revenue probability intervals. In the case of an annual revenue of £4400 the probability of achieving it is 0.10. Since there are 100 two-digit numbers in the scale 00–99, 00–09 are allocated to this revenue. 10–39 are allocated to an annual revenue of £4500 and so on, so that the number of different two-digit numbers in the range allocated is proportional to the probability of achieving its associated revenue. Using a table of uniformly distributed two-digit

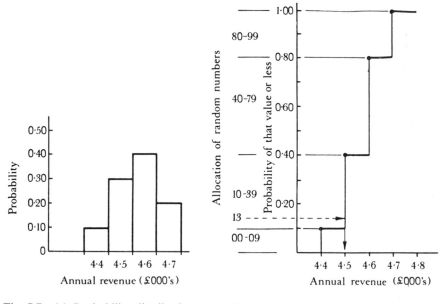

Fig. 5.7 (a) Probability distribution for annual revenue.

Fig. 5.7 (b) Cumulative probability for annual revenue.

random numbers (such as those in the *Cambridge Elementary Statistical Tables:* Lindley & Miller) it is possible to select random values of the annual revenue, together with their associated probability.

The first four random numbers from the table are 13, 25, 62 and 09. 13 falls between 10 and 39 in the allocation; if the left-hand side of the cumulative distribution at the appropriate point is entered, the horizontal line cuts the distribution at a value of revenue of £4500. Using the remaining three random numbers in similar fashion, revenues of £4500, £4600 and £4400 respectively are selected. If the simulation is carried out over a limited range of four trials for each of the variable factors in the appraisal, using a similar sampling procedure to that above, the values in Table 5.3 are obtained.

Table 5.3

Sample number	Annual operating cost (£)	Annual revenue (£)	Capital cost (£)	Project life (years)	Net annual revenue (£)	Rate of return (%)
1	1100	4500	15 000	5	3400	4½
2	1100	4500	16 000	7	3400	11
3	1000	4600	14 000	7	3600	17
4	1000	4400	15 000	5	3400	4½

From the random samples, the net annual revenue can be calculated and for sample numbers 1 to 4, a rate of return can be established.

In the case of sample number 1 the calculation is as follows:

$$\text{Uniform series capital recovery factor} = \frac{3400}{15\,000} = 0.227$$

Therefore, with a life of five years the rate of return or yield will be $4\frac{1}{2}$ per cent.

If a large number of samples are combined in this way (certainly not less than 100 but the more the better) a frequency distribution for rate of return can be drawn. This will give a range of rates of return and the probability (or relative frequency) of each being achieved. The final decision about the project can then be made, not only by considering a single yield, but by

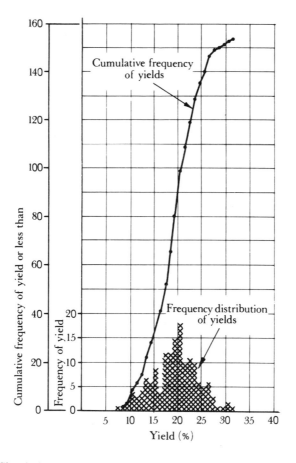

Fig. 5.8   Simulation results.

having a measure of the chance that each or any yield rate in the range will be achieved.

In practice, such a solution is not reasonably possible by using manual simulation and it is necessary to make use of a computer for the purpose. The simulation in practice is even more complex than the above example, because many more variable factors than those chosen usually need to be considered. Some further examples are tax rates, product price, raw material prices, labour costs, overheads, resale or residual values, cost of capital and rates of inflation. Figure 5.8 illustrates the form of a typical output from such a method giving a measure of the probability of achievement for each rate of return in two forms, the number of results and the cumulative frequency.

Having obtained the output as illustrated in Fig. 5.8, the answer to many additional questions can be obtained. For example, if a return of at least 15 per cent is required, what is the probability of it being less than this? From the cumulative frequency curve this will be seen to be 0.29. The expected rate of return is approximately 20 per cent.

Example 5.3

A company reviews its past investment programme and finds that the type of investment projects which it has carried out tend to produce results which fall into one of three categories. In Category A the yield from each investment has an equal probability of being either – 5 per cent or + 10 per cent per year in any one year. (The yield in this example is an expression of annual net profit over the initial capital investment.) In Category B, the yield has a probability of 0.10 that it will be 2 per cent per year, 0.40 that it will be 5 per cent per year, 0.30 that it will be 8 per cent per year and 0.20 that it will be 12 per cent per year. For Category C, the probabilities and returns are 0.20 for 5 per cent, 0.40 for 9 per cent, 0.30 for 12 per cent and 0.10 for 18 per cent yield. All projects are considered to have a life of ten years.

(a) Assume that the firm has £100 000 to invest in five projects of value £10 000, £10 000, £20 000, £30 000 and £30 000. The two projects with a value of £10 000 are expected to fall into Category A as far as yield is concerned, the £20 000 and one £30 000 project are expected to fall into Category B and the remaining project to fall into Category C. Simulate the results from these five projects over the first five years of their life.

(b) Simulate the total results from the five projects during the first five years of their lives if all turn out to be Category A projects.

(c) If the projects are all Category A, what is the probability that the company will not make a profit in any one of the ten years of the project's lives?

Frequency distributions for Category A projects:

Yield % p.a.	Probability	Cumulative probability of yield or less
−5	0·50	0·50
+10	0·50	1·00

Frequency distributions for Category B projects:

Yield % p.a.	Probability	Cumulative probability of yield or less
2	0·10	0·10
5	0·40	0·50
8	0·30	0·80
12	0·20	1·00

Frequency distributions for Category C projects:

Yield % p.a.	Probability	Cumulative probability of yield or less
5	0·20	0·20
9	0·40	0·60
12	0·30	0·90
18	0·10	1·00

The above cumulative distributions are illustrated in Figs. 5.9, 5.10 and 5.11. Random numbers from 00 to 99 can now be assigned to each cumulative distribution in proportion to the probability of any or each yield being achieved. These are shown at the left-hand side of each of Figs. 5.9, 5.10 and 5.11. Given a means of generating two-digit random numbers, we can now sample from the distributions as required.

The simulation called for in (a) can be set out as shown on page 137.

The probability is therefore 5/10 or 0.5, which is what one would expect if the sample were large enough, since there is an equal probability of making a profit or a loss.

Simulation as a technique in capital investment, as elsewhere, has the danger that a problem may be over-generalized in the way that the interaction of so many variables, one with another, is concealed beneath the

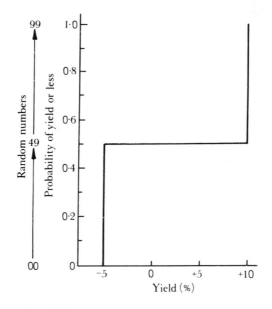

Fig. 5.9 Category A projects.

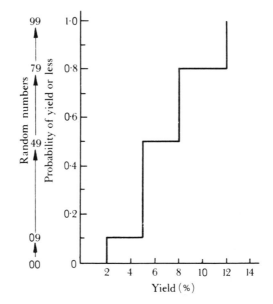

Fig. 5.10 Category B projects.

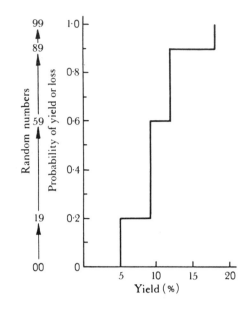

Fig. 5.11   Category C projects.

processing operation. It may, therefore, need supplementing by other methods such as sensitivity analysis for a single variable, which provides a wider base on which to take a decision. It has the distinct advantage, however, that probability distributions can be built up as either continuous or non-continuous and that they do not have to conform to standard shapes or mathematical equations in order to be of use. The simulation model, however, is no different from other models in that an accurate answer is dependent on the accuracy of information put into the system.

The accuracy of input must, in its turn, be the subject of economic appraisal, the trade-off being between the cost of estimation and collection against the fineness of detail of the outcome.

5.7 The use of expected values

The term *expected* with reference to the most probable value of a frequency or probability distribution has already been used. It is the value of the variate that has the greatest probability of occurring and which corresponds to the mode of the distribution. The slightly difference concept of *expected value* can be used in arriving at a decision in investment and other spheres.

Most decision situations can clearly have more than one outcome. If it were possible to assign a probability of each outcome being achieved, then a weighted average of the whole range of outcomes can be calculated. It is this

(a)

Year	Category A			Category B			Category C			Total Profit
	Random No.	Yield (%)	Profit (£)	Random No.	Yield (%)	Profit (£)	Random No.	Yield (%)	Profit (£)	(£)
1	48	−5	(1000)	73	8	4000	85	12	3600	6600
2	36	−5	(1000)	78	8	4000	41	9	2700	5700
3	75	10	2000	04	2	1000	76	12	3600	6600
4	66	10	2000	50	8	4000	62	12	3600	9600
5	26	−5	(1000)	90	12	6000	53	9	2700	7700
									Total	£36 200

(b)

Year	Category A		
	Random No.	Yield (%)	Profit (£)
1	25	−5	(5 000)
2	83	10	10 000
3	35	−5	(5 000)
4	66	10	10 000
5	61	10	10 000
		Total	£20 000

(c) The remainder of the table in (b) is:

	Random No.	Yield (%)	Profit (£)
6	78	10	10 000
7	14	−5	(5 000)
8	93	10	10 000
9	21	−5	(5 000)
10	25	−5	(5 000)

weighted average that is known as the *expected value*. For example, if three possible yields for an investment are 7 per cent, 11 per cent and 14 per cent and the probability that any one of these will be achieved is 0.30, 0.40 and 0.30 respectively, the expected value of this yield,

$$EV = (0.30 \times 7) + (0.40 \times 11) + (0.30 \times 14)$$
$$= \underline{10.7} \text{ per cent}$$

In general terms, if the probabilities of outcomes $O_1, O_2, O_3, \ldots O_n$ are $p_1, p_2, p_3, \ldots p_n$, respectively

$$EV = p_1 O_1 + p_2 O_2 + p_3 O_3 + \ldots + p_n O_n \qquad (5.1)$$

From the previous discussion of probability, it is known that the sum of the individual probabilities must add up to 1.00, since one or other of the outcomes must necessarily occur.

Not in every case will the outcome be positive and due allowance must be made in the arithmetic for positive or negative results. An example of this situation arises in the contracting business where a contractor knows full well from his experience that he will only obtain a certain proportion of the work for which he submits a tender. Supposing that he believes that this proportion is 1 in 10 for a particular section of the industry. The probability that he will be successful with any one tender is 0.10 and the probability of failure is 0.90. If the tender costs £10 000 to prepare and he includes £100 000 profit in each similar tender which he submits, then his expected return from *each* submission,

$$EV = (0.10 \times £100\,000) - (0.90 \times 10\,000)$$
$$= 10\,000 - 9000 = +\underline{£1000}$$

It is as well to examine the concepts behind this calculation. First of all, it is not *expected* that any one situation will arise in which the contractor will actually receive £1000 return. At each point where a decision is made to tender for work, it is expected that the outcome will either be a loss of £10 000 (as the abortive cost of preparing the tender) or a profit of £100 000 as a result of being awarded the contract. Clearly, the evaluation of expected value can only be used with effect where the decision situation to which it applies is one of many similar situations of like character. The theory of probability is derived on the basis that a large number of trials will take place and expected value is a weighted average of the outcomes. Expected value, therefore, depends on the repetition of the situation under examination. In fact, a businessman intuitively makes this evaluation in this contracting situation when he spreads the cost of abortive tenders over all his tenders in proportion to his likelihood of being successful. On a broader basis, he expects the successful tender to support the failures.

The second point to be made in this situation is that the tendering opportunities and their outcomes must be independent of each other. The tendering situation, for example, would not lend itself to the expected value concept if the award of a series of contracts were the result of, and dependent on, the award of the first. Neither would it be realistic if a large proportion of them were for one client who might be influenced by the outcome of one or other tenders in his treatment of future awards. It will be seen that the accurate calculation of the probability is very important.

5.8 Expected values and utility in investment decisions

If one is faced with a choice between investing in one of two projects, both of which have similar capital investment costs, and for each of which the probability of specific returns can be estimated, expected value can be used as an aid to the investment decision. Project A involves an initial investment of £100 000 and the present value of the total returns which are likely to arise from it are estimated to be over a range which includes £90 000, £100 000, £120 000 and £200 000 with probabilities of their occurring of 0.10, 0.25, 0.55 and 0.10, respectively. Project B involves a similar capital investment but with present values of the total returns falling in the range of £100 000, £110 000, £120 000 and £125 000 with probabilities of achievement of 0.60, 0.25, 0.10 and 0.05, respectively.

The expected value of Project A is therefore:

$$(90\,000 \times 0.10) + (100\,000 \times 0.25) + (120\,000 \times 0.55)$$
$$+(200\,000 \times 0.10) = \underline{£120\,000}$$

In the case of Project B the expected value is equal to £105 750.

If expected value were the sole consideration in this case, there would be no reason for choosing other than Project A because it has the greatest expected value. However, decisions in such matters are always accompanied by other considerations many of which are reflections of the risks that are inherent in any situation. In Project A, for example, it is not every businessman who can afford to ignore a 0.10 probability that a project will show a £10 000 loss on present value. To a cautious decision-maker this alone may rule out Project A. It will also influence those who, at the time, cannot afford to undertake the slightest risk because of a current doubtful or unsteady financial situation. A large, prosperous corporation which could readily absorb a loss of £10 000 (with a probability of only 0.10) would clearly go to Project A, because over a series of such projects the result of choosing one with these characteristics will result in a greater estimated (and expected) profit than those of Project B.

There are two further considerations that may be taken into account in such decision-making situations:

(a) the use-value of the outcome, called the *utility*, resulting from the attitude of the decision-maker at the time that the decision is made;
(b) the amount of the outcome, relative to the current overall financial situation of the decision-maker.

All individuals have an attitude to the use-value for money; seldom will that attitude be exactly similar from person to person. Many attempts have been made to measure this use-value or utility in the case of individuals, but it is sufficient for the time being to describe it in the general terms of a graph of utility plotted against costs or money values. *Cardinal utility theory* supports the drawing of such a graph and the expression of utility (or satisfaction) in units of measurement. (*Ordinal utility theory* does not use units of measurement.)

Such a graph is illustrated in Fig. 5.12. The relationship shown describes the attitude of a person who would derive a great deal of satisfaction from having the use of an initial specific sum of money, but that satisfaction gradually diminishes per unit as more and more money becomes available to him. The total utility will always tend to increase directly with the quantity of money available. In a practical sense, this is explained by the fact that once having bought many of the things that have persistently been sought after as being those that will give the greater pleasure and satisfaction, there will be a tendency for each successive unit to give less satisfaction the more that become available.

A number of theories concerned with utility have been developed within decision theory. One approach to the problem of the differing views on the utility of expected values, such as above, is concerned with ranking outcomes using what is known as the *standard-gamble technique* (developed by J. Von

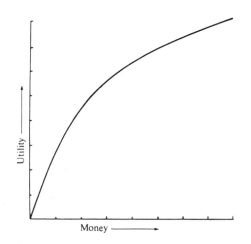

Fig. 5.12   Utility of money.

Neumann and O. Morgenstern). This technique is concerned with providing a numerical method of dealing with decision-making under conditions of uncertainty by establishing a *utility index*. As such it is a means by which the relative value of monetary returns to a decision-taker can be measured. The mechanism for doing so is by establishing the point at which two alternatives become of equal attractiveness to the decision-maker concerned. One alternative is a sum of money for certain; the other is a gamble, at stated probabilities, of a certain gain or loss.

The utility value determined by the method rises and falls as pay-offs rise and fall in acceptability. In other words, if the decision-maker prefers the outcome or pay-off X to that of Y, his particular utility index shows a greater utility of X to that shown for Y. In addition, if a decision-maker is faced with a proposition in which the pay-off of X has a probability of occurrence of p and the pay-off of Y has a probability of occurrence of $(1-p)$, then the utility of the proposition is the expected value of the utilities of the pay-offs. Therefore, if the utility of X is u(X) and the utility of Y is u(Y), the utility of the proposition, u(Z) is:

$$u(Z) = u(X)p + u(Y)(1-p)$$

Using Projects A and B above, the composite list of probable outcomes of both projects and the probabilities of each being achieved are first ranked in decreasing order of magnitude as in Table 5.4.

The attitude of a decision-maker towards each of these particular outcomes in relation to another is then tested. The tests carried out in this example take the form of a rating on a scale of preference or choice similar to that used for probability, as between a certain ($p = 1$) outcome of one value and that of a higher-valued outcome at some degree of risk. The rating or utility is assigned a value between 0 and 1 and represents the point on the scale of utility at which there is complete indifference between the two choices available. It is purely a matter of convenience that the scale of utility is used from 0 to 1; any other numerical scale, such as 200 to 500, will be equally satisfactory since the values of utility are only used in a relative fashion.

Table 5.4

Estimate present value of return	Probabilities	
	A	B
£200 000 (A)	0.10	–
125 000 (B)	–	0.05
120 000 (A & B)	0.55	0.10
110 000 (B)	–	0.25
100 000 (A & B)	0.25	0.60
90 000 (A)	0.10	–
	1.00	1.00

In Table 5.4, starting at the top of the table, an individual is given the choice of whether he prefers a certain outcome of £125 000 (i.e. a profit of £25 000 with a probability = 1), or £200 000 (a profit of £100 000) with a probability of income, p, of less than one, implying an alternative outcome of £0 with a probability of $1 - p$). The probability of the larger outcome, p, is then adjusted up or down until the individual's point of indifference between the two choices is reached. This may, for example, be at 0.60. That is to say, the decision-maker values equally, with complete indifference, a certain return of £125 000 to the larger possible return of £200 000 with a probability of occurrence of 0.60 (giving a zero outcome at a probability of 0.40 as the other risk in the choice). This is based on the premise that the decision-maker is really being presented with a choice between a certain £125 000 and a gamble between the £200 000 at a probability of p or £0 at $(1 - p)$. If p is given a value of 1.00, that is certainty, clearly the decision-maker will opt for £200 000. If it is given a value of 0, that is no chance of receiving anything, the decision-maker will clearly opt for the £125 000. At some point the preference changes and by adjusting the value of p, this point can be located.

The next step is to move one place town Table 5.4, and in similar fashion, to choose between a certain income of £120 000 and an outcome of £125 000 having a probability of less than 1.00 of being achieved. Again, the probability of achievement of the higher figure is adjusted until the decision-maker is completely indifferent to either outcome. When the probability of achievement of the larger return at which the decision-maker is indifferent between that and the smaller but certain sum is settled, the utility of the smaller sum can be calculated. The scale is now one of 0.60 at the upper (for a return of £125 000) and 0 at the lower (for a return of £120 000). The decision-maker's utility for the return of £120 000 will, therefore, be $(0.60 \times p)$ where p is the probability assigned to the larger return. The procedure is followed until each outcome has been assigned a utility value.

The utility scale (not to be confused with probability) measured from 1 to 0 is used for assigning values to all the outcomes in terms of preference of that outcome to the upper one. In other words, the most advantageous outcome assumes a value at the top of the chose scale of utility (in this example 1.00) and the least advantageous outcome assumes the value at the bottom of the utility scale (0 in this case). The utility values between the top and bottom will gradually fall, since, when two returns are compared, the smaller one is assigned a certain achievement ($p = 1.00$) and the other is assessed on an adjusted probability of less than 1.00. The utility value assigned is therefore calculated as a fraction of the utility value previously allotted to the upper of the two values in the comparison. For example, in deciding the utility value for a present value of £120 000, the comparison is made between that value with certainty and £125 000, at some lower probability. If the point of indifference is found to be where p for £125 000 is 0.93, the utility value of £120 000 is $0.093 \times 0.60 = 0.56$. For one individual decision-maker the

Table 5.5

Estimated present value of returns	Utility
£200 000	1.00
125 000	0.60
120 000	0.56
110 000	0.45
100 000	0.32
90 000	0

utilities might be as set out in Table 5.5. The individual's utility curve can be plotted from these figures.

The view of an individual making the assessment of utility can now be evaluated by multiplying each of the original probabilities of achieving the present values by the established utility values to give the *expected utility value* (EUV) as follows:

Project A $(0 \times 0.10) + (0.32 \times 0.25) + (0.56 \times 0.55)$
$+(1.00 \times 0.10) = \underline{0.488}$

Project B $(0.32 \times 0.60) + (0.45 \times 0.25) + (0.56 \times 0.10)$
$+(0.60 \times 0.05) = \underline{0.391}$

In this case the decision-maker should opt for Project A because it is calculated to have the higher utility value *to the decision-maker*.

Curve A in Fig. 5.13 shows the general shape of the utility curve for the decision-maker who has expressed his preferences in the pattern of Table 5.5. It is a typical curve of a decision-maker who wishes to avoid the risk of a loss and adopts a cautious attitude to risky projects – probably because his business is such, at the time of the decision that a loss, even at a low level, would make acute problems for his organization. The rate at which the decision-maker's utility value falls gets faster as the present value of the returns falls towards the initial investment and below. This can be inspected in Fig. 5.13 by checking the slope of Curve A to the left of zero NPV. It falls 0.32 utility values between zero NPV and a negative NPV of £10 000; to the right of that line, Curve A rises 0.13 for an increase in NPV of £10 000.

Curve D represents the attitude of a decision-maker which is the reverse of that portrayed by Curve A. It is the curve of a gambler who favours long odds in an attempt to scoop a big return. It portrays an attitude that suggests only a large return will be acceptable and that a loss can hardly make matters worse.

Curve B represents a standard against which other decision-makers can be measured. The decision-maker consistently believes in making some insurance against risk. He is indifferent towards a loss or a gain and therefore probably has a well-financed and sound organization.

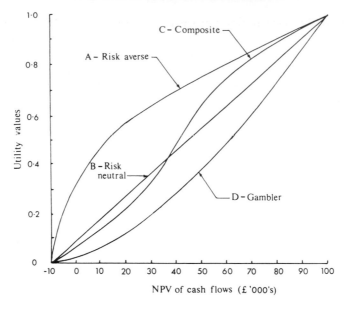

Fig. 5.13 Attitudes to risk.

Curve C shows a combination of attitudes. The decision-maker is risk aversive when there is little likelihood of a loss and even more so as he approaches the point where there is a danger of making a loss. When nearer to the loss situation and continuing into it, he becomes a gambler and is prepared to take greater risks in order to attempt to make a profit.

5.9 The evaluation of an investment to avoid risk

When an investment is made, whether it be in some commercially productive unit, such as a new factory to produce consumer goods, or in a highly desirable asset for the benefit of the community, such as a reservoir or a sewage works, a decision must be made, amongst others, as to the extent to which the risk of failure of the asset should be reduced. The sort of natural hazards which are likely to affect such investment are fire, flood, earthquake, explosion, frost and gales. At one end of the scale the design of the unit can be made so substantial that all the foreseeable hazards can be resisted. At the other, considerable risk can be involved by taking chances with the degree to which provision against hazards is provided.

Some investment to reduce risk is compulsory under construction legislation and local ordinances. The insistence on a minimum standard of fire-resistant construction is one example. Other investment might be conditional on the need to insure the asset against hazard; the compulsory installation of a burglar alarm against theft, for example. Where an investor considers providing protection or insurance against such hazards in excess of the minimum required for structural stability, for example, the test will be that of capital recovery plus adequate return out of the savings from avoiding losses due to the hazard. The following examples illustrate the application of expected cost to investments which are intended to reduce risk.

Example 5.4

In constructing the foundation for a river bridge pier, a decision has to be made as to the height of the cofferdam in which the work will be carried out. Because of other construction considerations, two alternative heights are taken into account. By reviewing data concerning the level of water in the river over the past 100 years, it is determined that the probability of it rising above the lower of the two levels of sheet-piling is 0.10 or once in every ten years. To raise the height of the cofferdam to the upper of the two levels will require longer sheet-piles and these will cost an additional £90 000. Flooding of the cofferdam is likely to delay work by two weeks at a cost of £200 000 per week. What decision should be made if the work is to be carried out through one flood season?

Total cost of delay because of flooding – £400 000
$$\text{Expected cost of flooding} = (0.1 \times 400\,000) + (0.9 \times 0)$$
$$= \underline{£40\,000}$$

Since this is less than the cost of the additional piling, on the basis of expected cost the lower level of cofferdam should be used.

Example 5.5

A building is being designed for an area on the fringe of an earthquake zone. A design to make the building fully resistant to earthquake shock will cost £2 000 000 more than one built to normal standards. Based on historic data, seismologists estimate the chance of an earthquake severe enough to damage the building structure to be 1 in 12 in any one year. The cost of repair to damage caused if an earthquake does occur is estimated to be £1 200 000. If the owner's cost of capital is 10 per cent and the economic life of the building is considered to be 30 years, which design should be selected?

Expected annual cost of normal design
$$= \frac{1}{12}\,(1\,200\,000) + \frac{11}{12}\,(0) = \underline{£100\,000}$$

Annual cost of additional work in structure
$$= (2\,000\,000)\,(A/P, 10\%, 30) = 2\,000\,000 \times 0.106$$
$$= \underline{£212\,000}$$

Therefore, on the basis of expected cost, no additional work will be carried out.

5.10 Introduction to decision trees

Many investment decisions are not isolated as single instantaneous decisions but are the first of a series and will lead to the necessity for a further decision(s) at a later date depending on the outcome of events resulting from the first. Many of these events will be quite beyond the control of the decision-maker and will be the result of chance to a greater or lesser extent. Examples of such chance events are rife in the world of fashion where it is not at all easy to predict the outcome of a change in style. Prospecting and well-boring are two related areas where chance events will influence the final outcome of a decision to spend money. The financing of a new development or invention might lead to diversification in entirely new industries with their associated decisions. Where a sequence of investment decisions has to be made over a period of time, a convenient and useful method of illustrating the possible outcomes of these decisions and of facilitating the analysis of the factors involved on the basis of expected value is the *decision tree method*.

A decision tree consists of a number of branches springing from each fork or node as a result of several alternative conditions which may arise between nodes. The notation of a square in the diagrammatic decision tree (Fig. 5.14) represents a point at which a decision is required and a circle is used where the outcomes represented by the branches spring from it are the result of a chance or uncontrollable event. Figure 5.14 represents a simple decision tree for the situation evaluated in Example 5.4.

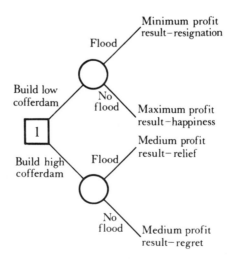

Fig. 5.14   Simple decision tree.

At Decision Point 1, a decision must be made as to whether to build a high or low cofferdam with the attendant risks of each. The branches leading out of this point represent the two possible alternative decisions. (It is by no means essential that only two outcomes should be considered at each decision – as many alternatives as are reasonably available and which the decision-maker wishes to consider can be included.) The occurrence or not of a flood is then a chance event over which the decision-maker has no control. A chance event is represented by a circle and the outcomes which might result from the chance events occurring are shown as further branches. In a situation of uncertainty, a decision area will always have two constituents – decisions which must be made in order to initiate action, and events or outcomes which are entirely or partially matters of chance or are circumstances outside the control of the decision-maker.

If a flood does occur and the low cofferdam has been constructed, then considerable cost will be incurred. If it does not, having made a similar decision, then the maximum profit will be made. The branches arising from a chance event can generally have a probability of the outcome assigned to them. In the example illustrated in Fig. 5.14 the probability of a flood occurring is 0.10 and therefore the probability of no flood is $(1 - 0.10) = 0.90$. The total of all probabilities leading from a chance event must always equal 1.00. This is a simple example for the purpose of illustrating underlying principles. Figure 5.15 presents a more complex decision concerned with investment and is described in Example 5.6.

Example 5.6

A property developer is offered a vacant site at no cost so long as he is prepared to develop it for office accommodation. The site is on the outskirts of a city area and is part of a major redevelopment programme scheduled for office accommodation and high technology industries. There is no restriction on the size of building which can be built on the site subject to the usual consent procedures being followed. Because no public transport systems presently exist to carry the future office workers to and from both the buildings on this site and neighbouring sites, the local authority is investigating, without commitment, the possibility of constructing a monorail system to serve this purpose. The final decision in this respect will not be made for another two years by which time the office block has to be built. However, a decision has been taken that requires the developers of the sites to subscribe to the capital cost of such a transport system and in the case of this developer, the contribution will be £4m. The current probability that the monorail will be built is assessed at 0.60. Should the developer decide at any future time before construction starts that he will not develop the site, he is required to hand it back to the original owner at no cost.

The developer's plans for the site include several alternative schemes associated with an estimate of the return for each. These include alternatives which hinge upon the decision as to whether the monorail is to be built or not. The two main alternatives to be considered from the outset are whether to build a small or a large office block at a cost of £7m and £14m, respectively. Assessments are made of the likely average net annual costs that should arise from the various combinations of assets to be created. At this stage the additional alternative of returning the site to the original owner is still available to him. If the developer initially builds a large office block his future course of action depends on whether the monorail is built or not. If it is built, there will be a capital outlay of £4m whichever subsequent course of action is followed. The alternative, if the monorail is built, are seen as either leaving the large office block at the size that has already been built or extending its size for £5m. Extending the large block will necessitate the purchase of some adjacent land at a cost of £5m. If the land is purchased and the initial large building is extended, it is estimated that there is a probability of 0.55 that the percentage occupancy will be such that a net income of £5m per year will accrue and a probability of 0.45 that the occupancy will be such that a net annual income of £4.30m will accrue. If the large building is not extended at this stage there is a probability of 0.35 that the annual net income will be £3.75m per year and a probability of 0.65 that it will be £3.45m. The probabilities are established by the developer and his staff from their previous experience in such situations.

If the developer decides to build the large office block initially and the monorail is not then built, the developer feels that he has two alternatives from which to choose. The first is to provide a transport system for the occupants of the offices at a capital cost of £2m; the second is not to provide such transport arrangements and to leave the office occupants to make their own. In the former case the developer estimates an occupancy of approximately 90% having a net annual revenue of £3.40m with a probability of 0.45 and a net annual revenue of £3.20m with a probability of 0.55; in the latter alternative the occupancy may fall considerably and a net annual revenue of £3m is likely having a probability of occurrence of 0.50 along with a probability of 0.50 that the net annual revenue will fall to £2.80m.

The developer then turns to the alternative of building a small office block in the first instance with a capital cost of £7m against the £14m for the larger block. Again the question of with or without a monorail needs to be considered as a chance event over which he has little or no control. If the monorail is built he sees his options as being to extend the building within the existing site at a cost of £9m or not to extend it. If he extends it he estimates that there is a probability of 0.70 that he will achieve a net annual revenue of £3.70m and a probability of 0.30 that a net annual revenue of £3.50m will be achieved. If the small building is not extended, net annual revenues of £2.70m and £2.50m are estimated with probabilities of 0.60 and 0.40, respectively.

If the small office block is constructed and the monorail system is not installed, the £4m levy on users will not be needed but the developer will need to find £2m to provide his own transport system if he wishes to do so. The developer decides to examine two options in the case of no monorail and a small initial building. The first is to provide his own transport system and to extend the small building at a cost of £9m; the second is neither to provide a transport system nor to provide an extension to the building. In the first situation it is estimated that the net annual revenues will be £3.65m and £3.50m with probabilities of 0.60 and 0.40, respectively; in the second situation the equivalent figures will be £2.65m and £2.45m with probabilities of 0.70 and 0.30, respectively.

The developer's cost of capital is 12 per cent and he decides to evaluate his decision on the basis of ten years' income. Furthermore, he assumes that the income will start with the completion of the building which coincidentally coincides with the decision concerning the monorail. He further assumes that all capital costs of the various projects may be assumed to occur at the date at which the income commences.

At chance event nodes, the expected outcome of that event is calculated by taking into account all the branches which spring from it. For example, in Fig. 5.15 the expected outcome from Chance Event Node C is $(0.55 \times 10 \times 5) + (0.45 \times 10 \times 4.3) = £46.85m$ (if, for the time being, the time-value of money is ignored). It should be noted that the probabilities assigned to all the outcomes emanating from a single node must add up to 1.00.

Figure 5.15 shows the developer's problems set out in the form a decision tree. One of the principal advantages of such an illustration is that it enables a clear and precise overall view to be taken of the problem and the probable consequences of following each path through it. The decision tree is simply a statement of what is known about the problem with estimates of the likely outcome of each happening. Nothing more is included in the diagram than that which is known immediately prior to drawing the tree. In practice, one of the major disadvantages of using a decision tree is that it can quickly become complex and the subsequent calculation is then confused. It is as well to bear in mind that the tree is not intended to portray absolutely every possible outcome or solution to a problem, but only to enable the required outcomes to be examined. For example, in enumerating the branches from a chance event node, not every possible chance outcome is portrayed since, in some instances, there could be a very large number. Only those which are believed to be probable and hence can have a probability of occurrence assigned to them are included. In similar fashion, only the alternative decisions which are considered feasible are drawn emanating from a decision box. As often as not these alternatives are likely to be restricted to two in number hence contributing to the simplest form of tree.

The analysis of the decision tree is carried out by working backwards in time from right to left. This is something referred to as *rolling back* or the *roll-back* technique. By so doing it is possible to eliminate each decision point

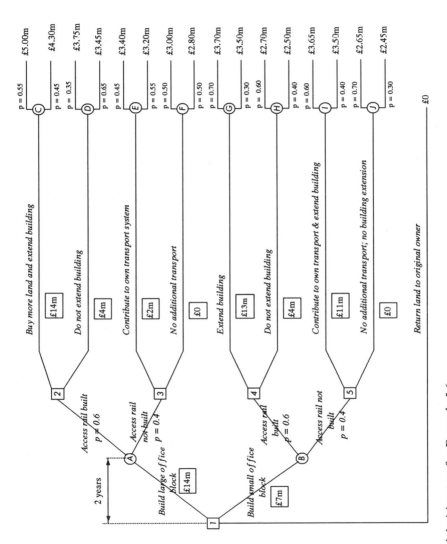

Fig. 5.15　Initial decision tree for Example 5.6.

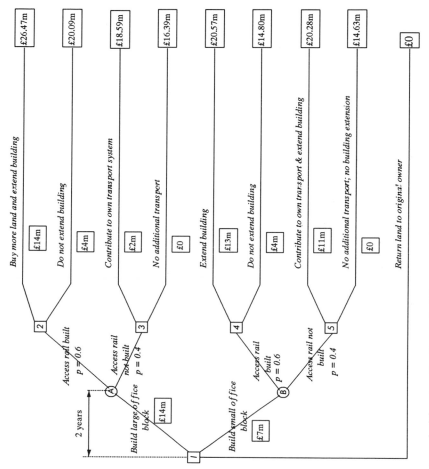

Fig. 5.16 Second stage decision tree for Example 5.6.

starting with that which will be the latest on the timescale. Of course, not every decision point will necessarily be reached in practice, depending on the most desirable path up to and preceding that point. In the analysis, however, each decision must be evaluated in case it happens to be on the appropriate final path and so that the complete picture of the problem and its possible outcomes can be obtained.

Referring now to Fig. 5.16, rolling back from right to left, the chance event nodes can be evaluated over the ten year period of the study. Event C has the following expected outcome:

$$0.55[5(P/A, 12\%, 10)] + 0.45[4.3(P/A, 12\%, 10)]$$
$$= (0.55 \times 5 \times 5.6502) + (0.45 \times 4.3 \times 5.6502) = £26.47m$$

Expected outcomes for each of the remaining nodes, D to J, are shown in Table 5.6.

Decision Point 2 may now be evaluated as a choice between a surplus of $(26.47 - 14.00) = £12.47m$ and one of $(20.09 - 4.00) = £16.09m$. The decision at Point 2 should therefore be *not to extend the building*. At Decision Point 3 the choice is between $(18.59 - 2.00) = £16.59m$ and £16.39m when the decision should be to contribute to the developer's transport system.

The Decision Point 1 choice of *build large office block* can then be evaluated as a net present value of

$$(0.6 \times 16.09) + (0.4 \times 16.59) - 14.00 = £2.29m$$

Decision Point 4 can be evaluated as a choice between $(20.57 - 13.00) = £7.57m$ and $(14.80 - 4.00) = £10.80m$. The latter is the worthwhile choice.

Evaluating Decision Point 5 in a similar manner gives a choice between $(20.28 - 11.00) = £9.28m$ and £14.63m, making £14.63m the choice.

The option of *build small office block* at Decision Point 1 can now be evaluated as $(0.6 \times 10.80) + (0.4 \times 14.63) - 7 = £5.33m$

Decision Point 1, therefore, has three paths emanating from it. One has an NPV of £2.29m, one of £5.33m and the other, *return the site or do nothing*, has

Table 5.6

Chance event node	Expected outcome £(m)
C	26.47
D	20.09
E	18.59
F	16.39
G	20.57
H	14.80
I	20.28
J	14.63

zero NPV. The decision should therefore be to build a small office block for £7m in the first instance, and if the access rail is built, not to extend the building. If the access rail is built, the decision should be not to provide additional transport or to build an extension to the initial small block.

5.11 Sensitivity analysis

One essential part of risk analysis is the estimation, not of a single value, but of a range of values or alternatives for the variables which form part of the whole project investment picture. A further aspect of such analysis is the evaluation of its sensitivity to possible errors in estimation of each or some of the variables in the first place. There is a need to provide a decision-maker with information concerning the effect on the overall outcome of changes in the value of those variables which form an essential part of the analysis.

It has already been seen that if the values of the variables in the investment are estimated correctly, then there need be little doubt about the correctness of the application of discounting methods. However, such a situation rarely occurs in practice and it is thus advantageous to have a measure of how *sensitive* the calculated outcome is in relation to possible errors in the initial estimation. A sensitive project is one in which the ultimate outcome as calculated is likely to change significantly following a relatively small change in the value of one or more of the variables involved. If the values of the variables can be changed over a wide range without altering the ultimate outcome, then the project is said not to be sensitive.

Clearly, it is desirable that a decision-maker is aware of the most sensitive areas of his project. If the outcome of a particular situation is highly sensitive to a particular variable then more information concerning the probability of a change in practice of that variable must be sought. If a variable, say material costs, is not sensitive for the project, then there is little point in expending effort to produce highly detailed forecasts of the movement of prices in that area.

An example will illustrate how use can be made of a sensitivity analysis.

Example 5.7

An item of mechanical plant is purchased for a capital cost of £10 000. It is estimated to have a useful life of ten years. If the cost of capital to the company is 10 per cent, illustrate the sensitivity of the annual cost of the equipment to changes in its useful life and the cost of capital to the company. Assume that the residual value of the plant is £1000 at the end of its useful life and ignore the effects of tax.

Figure 5.17 shows a series of curves plotted on the basis of annual cost against life for various interest rates. Taking a ten-year life and an interest

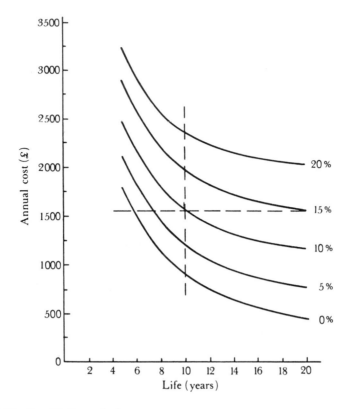

Fig. 5.17 Sensitivity analysis.

rate of 10 per cent as the base case, the horizontal and vertical broken lines through this point give an indication of variation of interest rate with variation of life that is necessary to maintain a constant annual cost. In addition they demonstrate the variation of annual cost with variation of interest rate.

For a given interest rate, as might be expected, a reduction of asset life causes a steep increase in annual cost. An extension of asset life causes a change in annual cost at a reduced rate.

It is estimated that the life of the asset may range from 8 to 12 years and the interest from 10 to 15 per cent, from Fig. 5.17 it may be established that the annual cost will vary between £1425 and £2150 per year. If at an interest rate of 10 per cent it is felt that the annual cost must not fall outside the range of £1250 to £1500, then the life must fall between 10 and 16 years to achieve these costs.

For the same item of equipment, Fig. 5.18 shows the change in annual cost of capital recovery against changes in interest rate for a given life between five and ten years. The broken line takes into account a notional tax rate of

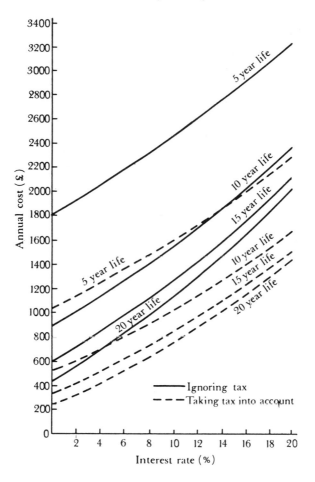

Fig. 5.18 Variation of annual cost with interest rate.

$42^1/_2$ per cent, initial allowances of 60 per cent on capital investment and an annual allowance against tax calculated by a 25 per cent reducing balance method.

Emphasis is frequently on investments which can be recouped in as short a time as possible, particularly where technological change is likely to be rapid. Hence it is often very important that the sensitivity of a project to obsolescence is gauged before the investment is made. Very often it is extremely difficult to estimate the life of an asset which is producing in such an environment, where predictions 10 years or more ahead are virtually impossible to make with any degree of certainty at all.

Further it is extremely difficult to compare the sensitivity of one project to its component variables with that for another when it comes to making a choice as between projects.

An investment project is not only sensitive to the risks which surround the analysis but also to the possibility that there will be errors in the forecast of the magnitude of one or more of the components making up the cash flows which are used in the calculations.

A trivial example will illustrate the drastic effect of changes in the actual initial capital investment, perhaps arising from uneconomic construction, and the resultant cash flows, from those that were anticipated at the appraisal stage. It will be assumed that an initial capital investment of £100 provides a return of £15 per year for each of the succeeding ten years. The anticipated yield is therefore 8.14 per cent. Figure 5.19 shows the drastic variations in yield for variations from the original estimated values in this particular example.

5.12 The application of risk analysis

In the early 1950s almost no sophisticated methods for the appraisal of investments were in current and regular use within companies. Discounted cash flow was, at that time, an advanced concept which was rarely

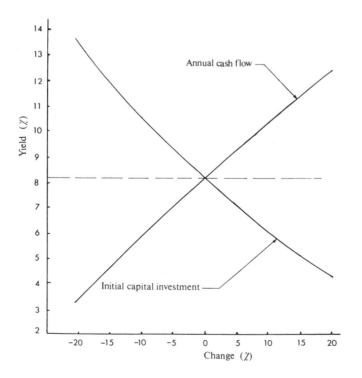

Fig. 5.19   Sensitivity of yield to cash flow and investment.

implemented by companies, except the most forward thinking, for this purpose. The rule-of-thumb methods were in vogue as the decision-aiding tools which supported managers' decisions in a choice of investments. It was not until the early 1970s that the use of discounted cash flow and NPV were regularly used and case studies or project appraisals using these methods began to appear in the literature. Techniques to aid the analysis of risk in the appraisal process, such as the repeated simulation of the calculation of rates of return by random sampling from frequency distributions of the constituent variables, are still not used to any great extent. The tendency is for such matters to be confined to the analysis of the risk in very large projects and for those which are outside the normal type of project for a particular firm.

Various researchers have ascribed the reason for this reluctance to use such methods to a lack of clear understanding of the statistical basis for them. An unfamiliarity with the theory of probability and stochastic variables, together with a lack of facility in interpreting data in the form of frequency distributions, has led to considerable difficulties. This general lack of understanding has also meant that there have been few data collected in an appropriate form. This, therefore, does not allow such techniques to be used without delays while fundamental supporting data are brought together. Managers still tend to put considerable reliance on their experience and managerial feel for the problem of making an appropriate selection of investments from the range that are available. Further, some managers are put off such methods, not just because of the quantity of data that need to be collected in order to support such techniques, but, more particularly, because of the costs of doing so.

Given that risk analysis methods are important tools in the armoury of a manager concerned with investment appraisal, it would appear that the problem of their non-use, to any large extent, is due mainly to the fact that managers of today are generally unfamiliar with the details of the techniques. Unfamiliarity of this kind can often lead to a tendency for individuals to stick with the methods that are well established and with which they are familiar rather than to expose themselves by venturing into unfamiliar territory. One of the ways to aid this lack of confidence is to encourage the use of such techniques in parallel with the well-established methods. By so doing, a basis is established whereby uncertain and risky data are established either in the form of probability distributions or in some other suitable form, even though it may not be used in an analysis which influences a decision. Retrospective comparisons can then be carried out when the subsequent actual data become available, to see how accurate was the prediction, or in what features the original data lacked credibility. This experimental introduction of the techniques will help to improve the use of new methods, not only in their application and the frequency of their use, but also to give considerably more confidence in the reliance that can be placed upon them. In addition, managers using such techniques, even though it be by trial, will begin to develop a feel and experience in the methods of interpretation for given solutions to their problem.

Problems

5.1 Data are collected and sorted into a frequency distribution following a study of the time a batching plant takes to batch the materials for 1 m³ of concrete. The frequency distribution appears as in Table 5.7.

Draw a frequency distribution from these data and from it derive a cumulative frequency distribution. What is the probability that the next time the batcher operates it will take either 1.10 minutes or 1.15 minutes to batch materials?   ($p = 0.40$)

Table 5.7

Time to batch materials (min)	Number of observations
0.95	17
1.00	23
1.05	29
1.10	36
1.15	27
1.20	18
1.25	6

5.2 Calculate the standard deviation for the following set of numbers: 5, 9, 8, 7, 10, 4, 6, 6, 5, 10. Calculate the coefficient of variation for this set of data and determine whether it is larger or smaller than that for the data in Example 5.1.   ($s = 2.05$; $v = 29.3$ per cent; greater)

5.3 A manager is asked to estimate the cash flow concerned with a particular project. His answer is to the effect that it will have a mean of £5000 but that there is a 1 in 3 chance that it will fall outside the range £4500 to £5500. He adds that he expects the distribution of the cash flows to be normal. What is the standard deviation of the cash flow distribution? What would the standard deviation become if the chance of falling outside the range was 50/50?   (Standard deviation = £220; £750)

5.4 From the cumulative distribution derived in Problem 5.1 use the Monte Carlo technique to establish the first ten durations for batching the materials on a random basis.

5.5 A company has the opportunity to invest £100 000. It can invest in its established business with what may be assumed to be a certain return of 8 per cent. Alternatively there is an opportunity to invest the money in a diversification which, if successful, is estimated to bring a return of 17 per cent. If the diversification is not a marketing success it is estimated that it will only bring a return of 2 per cent. What must be the probabilities of the two outcomes of diversification to make an investment in the new venture worthwhile? (*p* success $\leqslant$ 0.40)

5.6 In the design of a refrigeration plant for a small foodstore holding perishable foods, a decision has to be made as to whether a stand-by generator is to be installed so as to ensure against failure of the mains supply. The stand-by generator will cost £2000 to purchase and install. The company's insurers will hold the contents of the store insured to an upper value of £50 000 for an annual premium of £500 if the stand-by generator is not installed, but this premium will be reduced to £300 if the additional plant is put in. If the goods which are stored are ruined because of failure of the refrigeration system it is estimated that consequential losses will amount to £10 000. If the life of the generator is estimated to be ten years with a scrap value of £100 at the end of its life and capital costs 10 per cent, will the installation of the stand-by generator pay? (Hint: base probabilities on insurer's assessment.)

(No)

5.7 Place yourself in the position of the managing director of a company which is thinking about doubling the size of its only producing factory. Draw a decision tree which represents, in your own opinion and experience, the possible outcomes of making such a decision. Put some values on each outcome and use the roll-back method to evaluate the decision.

Chapter 6
Management Information and Control Systems

6.1 Communication

When dealing with management information and control systems, it is important to understand a number of terms that are in frequent use. It is important, for example, to understand the difference between *data* and *information*. Data are facts that have been obtained through research, by measurement or observation and subsequently have been recorded. Data are used as basic material input to a form of processing which may be that of reasoning, or calculation, or both. Data can originate internally, that is from within the organization which uses them or externally, that is from the environment encompassing the organization but not under its control. Data or facts become information when they have been through this conversion process. For example, the details of the man-hours that have been used in order to produce a specific quantity of concrete, are considered to be data; if the man-hours are then processed with other data to become *man-hours per cubic metre of concrete produced*, they are converted to *information*. However, whether the processing has created information depends entirely upon whether its new form is meaningful and of use to a decision-maker or user. Information to be useful to a user must inform the receiver and contribute to the body of knowledge which already exists with that person. Figure 6.1 shows the diagrammatical layout of the process.

Communication is, therefore, an essential and a most important part of the management information process. The process of communication involves an interchange of thoughts, information, knowledge or opinions. To transmit information to an individual can only be part of the total process. It cannot be said that communication has taken place unless the person receiving the information is in a position to understand what the information means and then to respond to it. The response is known as *feedback* and it is an essential part of the total communication process.

When communication is in process, therefore, it is not complete unless it engenders a response from the person to whom it is directed. Information has to be identified by its receiver as being meaningful and to which, if the communication process is to be completed, a response has to be

160

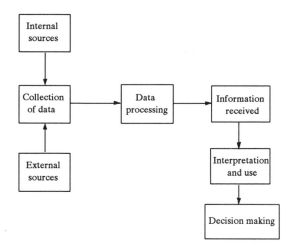

Fig. 6.1   General information system.

made. The design, construction and commissioning of a project can only be successful if good communication exists between all the parties that are involved, and adequate, relevant and high quality information is passed throughout the whole of the project management system. Nevertheless few, if any, project management communication systems are free from serious defects.

Management information is produced as a result of data processing operations; it is relevant knowledge that is acquired for specific purposes. A management information system is a system through which specified data are collected, processed and then communicated in order to support decision-making by those who are responsible for the management of resources, by providing accurate and relevant information concerning those resources at the appropriate time. Because of the vastly improved data processing functions as a result of the use of computers, management information systems are frequently assumed to be synonymous with computer-based information systems. This is not necessarily the case, though clearly, with the undeniable advantages of the rapid and extensive processing facilities that are available by using computers, it is desirable that a management information system of significant size and complexity should, more conveniently, operate in this way. The use of a computer for this purpose, in itself, demands a readjustment of many management philosophies in order to adapt to the facility to process ever-increasing volumes of data and information for the purposes of construction project control.

In order for a manager to be able to request information, have the data collected and processed, and then to have the information transmitted back to him, there has to exist a communication system. Within this communication system will be the individuals who will undertake all the processes that are involved, both in relation to the internal organization of that system and its environment. For example, a forecast of a rapidly increasing labour requirement to undertake the construction work for the next month will not be of much practical use unless it is accompanied by an indication of its availability through contact with the environment of the system. The duties of each of the members of the communication system, with respect to their contribution to the management information system, must be clearly defined.

Jackson identifies four problems that need to be solved within an organization so as to promote better communication:

(a) complete trust between the sender and the recipient;
(b) the establishment of mutual understanding about needs and motives;
(c) fair awards of credit to be given for each contribution to the system;
(d) common agreement about the social structure of the organization.

The system that enables a construction company to operate as a whole and to co-ordinate the various skills and the efforts of all of those individuals involved towards achieving the company's objectives, is its communication system. These communication systems will be at various levels of operation, as between individuals, as between divisions or departments of the company, as between different projects and as between different companies within a group of companies. Without communication systems a company could neither be led by its senior management nor can senior and subordinate managers and supervisors be expected to participate in the decision making appropriate to their particular level within the organization.

Guevara and Boyer undertook a study within nine unionized construction companies in Illinois which was concerned with establishing the possible causes of communication problems within construction companies. The study was carried out by questionnaire and from 122 responses – the response rate for all employee questionnaires was 61.6 per cent – 72 per cent of the respondents indicated that having too much information was a moderate to severe problem. In addition, 53 per cent of the respondents indicated that the withholding of information was a moderate to severe problem; 59 per cent of the respondents indicated that distortion of information was a moderate to severe problem. Only 5 per cent of the respondents indicated that having insufficient information was a moderate to severe problem. The authors of the paper go on to further analyse these responses. In each case they result in insufficient feedback being available and the general communications within the organizations and the interaction between employees need to be examined. When information was withheld, it occurred because there was insufficient feedback or employees were not satisfied with the communication system or both.

6.2 Systems concepts

The system concept has become of fundamental importance to modern management science thinking. It has made a far-reaching impact on the ways in which the functions of management are being both planned and implemented. Its application over a wide and diverse range of activities has provided a modern manager with an analytical framework that can be used to cope with problems and situations which are increasingly complex and difficult to understand.

A system is defined as a set of elements or components which are interconnected, interact, or are interdependent so as to form a complex or unitary whole. By applying the system concept to a specific situation within a defined boundary, it is possible to identify, describe, examine and experiment with the set of components which make up the whole and their interrelationships, in such a way that the working of the system can be understood. A manager can thus examine the components of the system whilst maintaining a perspective of the whole. It will be appreciated that a system in itself may be acting as a subsystem within a set of subsystems and may have interactions with these other subsystems. Systems in this context are usually viewed as block diagrams or flow charts.

The principal parts of a system are the *input*, a *conversion process* of some kind and an *output*. An output results after a conversion process has been applied to an input. If the construction industry is considered as a system, then the input is the resources that are used such as money, men, materials and machines, the conversion process is the construction process itself and the output is the constructed facility. Every system must have objectives for which it strives – where they are quantifiable and set out in specific terms it is possible to measure the effectiveness of the system by comparison of the system output with the objectives. If the objectives are ill-defined and expressed in qualitative terms it is often very difficult, if not impossible, to measure the effectiveness of the conversion process.

A management system will consist of all those parts of the system which come under the control, and can be influenced by, a particular manager or management. Everything outside the boundary will become the environment of the system. If the environment is changed the system will be affected; the system may change some or all of the activities of the environment. The environment of a management system consists of those aspects which influence the system contained within the boundary but which are not directly under the control of the system manager. For example, if the management of a construction site is considered, then part of the environment will usually consist of the operations and requirements of the relevant local government, the activities of the national trade unions, national legislation concerning safety, and so on.

All of these factors, and others, influence the system under the manager's control but are not directly controlled by the manager. It is often difficult to draw a precise and exact boundary round a particular management system since there will be grey areas of interaction between the two. In spite of this, an attempt to define the boundary is essential for an adequate understanding of not only how the system will interact within itself but also with its environment. It will enable the manager to define more clearly the objectives of the system and/or, if necessary, of each part of the system.

In order to maintain the output of a system at the appropriate levels of quantity and quality a form of control is required. A simple management control system is illustrated diagrammatically in Fig. 6.2. In the diagram shown, as with all non-automatic systems, management decision is required at one of the many management levels before corrective action is taken in order to adjust the system in the case of objectives not being met.

A *closed* system is one that is entirely self-contained and controllable. The system's environment cannot influence it. Whilst closed systems are highly convenient most management systems are *open* because they are not entirely controllable and they can usually be influenced by their environment. Elements involving human beings introduce uncontrollable elements into systems and social, economic and political environments all exert an influence on such systems.

If a management system is to be controlled, data must be provided as a feedback in order to form the basis for decision-making. This in turn implies that a system must incorporate the means for collecting, receiving, storing, classifying, analysing and transmitting data as part of the management process as depicted in Fig. 1.1.

6.3 Need for management information

In industry generally, but perhaps more particularly in the construction industry, the emphasis on information systems has tended to be at the operational level. With most of these operational applications now becoming routine and accepted, the demands of higher-level decision-making have been and are being studied. This has emphasized the desirability to place

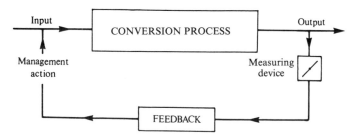

Fig. 6.2 Management control system.

management information systems in the context of the needs of a specific organization. Not only is there a need to consider the particular organization for which the higher levels of a system are required, but also the demands of individual managers on a personal basis and the structure of the decision-making process with its different levels and applications. The management information system then becomes a subsystem within a larger system which also includes organizational development, managerial functions, control systems and decision systems. The designer of a specific management information subsystem must, however, ignore though be aware of the related demands of the other associated subsystems.

The success and profitability of a construction company is usually determined by how well a company's managers achieve the company's objectives. The activities of managers in achieving objectives are broadly similar from company to company though the objectives almost certainly will not be. One important factor that contributes to the efficiency with which a manager carries out his functions is whether or not his information needs are being met. Managers are expected to make decisions and in order to do so they must be supplied with accurate, appropriate, relevant and exhaustive information at the right time.

In order for a manager to identify his own needs for management information he should examine and analyse the following:

(a) his required contribution to the success of his organization;
(b) the particular elements of endeavour that are critical to his contribution being relevant and successfully carried out;
(c) the yardsticks by which these particular elements will be measured and judged;
(d) the means by which the elements can be quantified;
(e) the information that is required to ensure successful achievement of the elements of endeavours.

Without this critical appraisal by a manager of his specific needs for information in order for him to carry out his duties in the most expeditious and effective manner, there is a tendency for those needs to be determined by a systems analyst whose expertise does not lie in the construction management area. This necessity for a user-choice for management information may not be so critical at the lower operational levels of an organization. At these levels users are constrained by more tightly drawn rules and less freedom is available for managerial type decision-making. For example, in a system designed to process payroll information a systems analyst, whether experienced in construction or not, can study the rules and requirements which are clearly specified and one has less room for manoeuvre in the design of the required system.

For the above reasons, when applied to the widely dissimilar methods of operating, organizing and administering construction firms, whether in design or contracting, it is highly improbable that one management information system can be designed so as to satisfy the needs of all firms – or even a majority. Many procedures, such as those found in accounting systems, are well-established in companies and have been tested and tried over long periods of time. Often they have formed an administrative/control base which is identified with successful operations and managers who have grown with the organization are reluctant to see fundamental and far-reaching changes made to them.

In a research thesis (Edwards) a study has been reported of the usage of computer-produced information. During the study, which covered 109 copies of reports which were available to 15 of the company's managers, it was established that 45 per cent of those reports were not used at all. Further analysis showed that the reports were not used during two specific time periods. The first was immediately after the system providing the information reports was installed for two principal reasons. Firstly, because of the general unsuitability of the information that was included in the reports – the reason generally assigned to non-use by the managers who received the reports – and, secondly, a built-in resistance on the part of the potential users of the information. This latter reason was generally thought to be the most important by the designers and developers of the system for producing the information.

The second time period was identified as that period of time following the installation of the system during which many of the actual needs of the relevant managers receiving the information changed; some individual managers had been replaced during this time and therefore the information produced by the systems apparently was not suitably structured and in some cases was not put to any use. In this general context, changes within the organization and the way it was operating, as well as its general environment, also contributed to the reduction in information usage.

In order to overcome these two difficulties it would appear that senior management needs to make sure, when a computerized information system is installed, that every step is taken in order to encourage the use of the information so produced from the moment that it is produced. Senior management also needs to ensure that systems are flexible enough to be modified in order to supply the actual needs of individual managers and/or that the form and nature of the information provided has, in itself, a sufficiently long life within the organization in order to be useful and viable over a reasonable period of time.

6.4 Design of management information systems

With the advent of more advanced technology, the ability to process, edit, accumulate, store, review, manipulate, interrogate and transmit information at high speeds, has had its effect on the management, organization and construction of projects. Activities by management which were seen as being routine and sometimes not very important 25 years ago, are now having to be changed in order to fit with the new facility to create information flows. The construction process is one of a highly variable nature and without control can give rise to the generation of a vast amount of information. This information needs to be used selectively and carefully with the changing management methods if computers processing this information are to be used efficiently and with better and more efficient management as the objective. Different construction managers will always have different criteria by which they judge and then control the activities for which they are responsible. These needs must be taken into account in devising information systems. However, whatever the method of presenting information for control purposes, it is necessary to make sure the facts and basic data are collected in a thorough and accurate fashion before being processed to form the information input to management reports.

The history of management information systems in the construction industry is littered with as many failures as successes. On the whole, however, management information systems have become accepted as a necessary part of the construction management process. In overall concept and design the systems tend to be extremely varied and many lack the necessary flexibility to cope with changes in the structure of an organization and/or with the demands of individual managers.

There are a number of reasons for the failure of such systems. Some of these (either in combination or alone) that have contributed to such failures are:

(a) the failure of experts in computer technology to understand the real needs of a system for management information in the construction industry;
(b) a failure to appreciate the extent of the investment in technology and personnel required to pursue and establish a satisfactory system;
(c) a failure to make allowance for changes in the organization structure and the demands of project managers.
(d) a failure to ascertain the precise needs of the individual managers in respect of information presented in reports and in respect of the timing of this information;
(e) a general incompetence and mismanagement of the system as a whole once it is implemented.

It is important to differentiate between systems which produce routine, record-keeping, transaction-recording processes from those which are intended to support and facilitate decision making and the implementation of those decisions once they are made. Both types of system will very often have some data which are of common use and may be drawn from a common database to be used by more than one system. However, it is likely that the style of data manipulation and report presentation will be different for each. Very often the design of information systems has sprung from the ease with which certain data have been made available and can be recorded. The data have been accepted when their precise future use and value has not been known and the style of the presentation of the records has not always been such that the information concerning totals and ratios when presented have been of use to an intended decision-maker.

Before a system can be designed it is necessary to undertake a searching study of the organizational structure that is to be served by the information system. It is necessary to identify in the organization structure, the functions, the authorities and responsibilities, the decision-making structure, the needs for information and the mode of reporting for relevant levels in the organization and for the personnel involved. In addition, a facility needs to be provided that enables the system to be updated as a routine matter to reflect any changes in the organization, and the individual personnel within it, as time proceeds.

The following seven stages are important in the design of a suitable information system:

(a) identify the organization structure and the functions, authorities and responsibilities of the various levels of the hierarchical structure;
(b) the determination of the objectives of the organization or the department or the project for which the information system is to be implemented;
(c) identify and define each type of managerial decision required in the organization and the significant areas where control of performance will be critical in attaining the objectives;
(d) identify the statistics, ratios, or totals that will be needed to measure progress and performance in the above significant areas;
(e) define clearly the basic data which will need to be collected in order to enable the calculation of these statistics, ratios, totals, etc.
(f) pay particular attention to the way in which the information can be presented to the decision-maker so that it can be assimilated in an efficient manner and be presented in as concise a form as possible;
(g) establish limits of control which will define the variances within which management or decision action need not be taken.

One of the first activities in setting up the structure for a management information system for project management purposes is to break the project down into a hierarchical structure which reflects, at the lowest level, operations which have a short duration and a more or less constant expenditure rate over that duration. Each of these activities, at the lowest level, will then be combined in groups to provide the next higher level in the hierarchical structure – perhaps forming recognizable and convenient elements of a project. The elements will then be grouped together forming components of the project and so on until the grouping reaches the peak of the hierarchy and represents the total project structure. The number of levels in the hierarchical structure that will be used for each and every project is very much a function of size and nature of the project itself.

One outstanding fact is frequently overlooked when developing management information systems for the control of construction. This is that an organization is a dynamic organic structure which changes and evolves as different pressures are put on it. The speed of the reaction to changes is often not immediate but, eventually, the changes will have to be made to meet the new conditions. This is never more true than in the field of construction where different methods of construction, contractual practices, professional organizations, designs and materials are frequently changing, if not necessarily in an extreme or radical fashion. Organizations that change are sometimes inclined to neglect the fact that systems that they have designed some years previously are rarely brought up-to-date to deal with the current situation. In the field of management information systems, however, there are devices which allow this flexibility in construction management.

There is some confusion between the various levels of sophistication of a management information system, what its various functions are, who will use it and how. In construction a limited view of management information systems is frequently taken and a system at the operational level is all too often regarded as all that is required. Operational systems are based on data that are generated internally to the organization, for example, the collection of costs relating to a particular activity or operation with a view to establishing the variance between the estimated cost of carrying out the work and the actual cost. As a result of the comparison, if it is obtained in time, the user of the information will make a decision as to whether or not to change the extent or mode of use of the resources available to do the work. It is necessary in undertaking this calculation to feed into the system some or all of the data collected concerning man-hours employed, rates of pay, incentive payments, materials used, invoiced material costs, measured work carried out, and so on. All such data are generated within the organization and where computer processing systems are employed, much of it will almost certainly be fed into the system from other systems that, for example, prepare purchase orders, pay invoices, prepare valuations with measured work, etc.

Operational systems of this kind are restricted in their concepts. If well-designed they provide a running commentary on the day-to-day operations of a production unit, such as a project site, but do not make a significant contribution to the strategic management of the business as a whole of which the particular project is a part. The higher the level of a management decision-maker in an organization, in general terms, the larger will be the proportion of data and information needed from sources outside that organization. In construction, data will be required on potential markets at home and overseas, trends in such markets, the state of health of the competition, the political stability of overseas countries, future government legislation, the likely trends in inflation, etc. Management information systems at this strategic level are not easy to design – the more so because at these higher-decision-making levels, the greater will need to be the influence of the personality and entrepreneurial skills of a particular manager. Hence the designer of such systems will need to come to grips with these influences to a greater extent than would be the case with a system designed for the operating level.

In establishing the design of a management information system, it is necessary to keep in mind the general relationships between the cost of obtaining data and their value to the user. Those relationships can be expressed in the most general of terms by Fig. 6.3. The significant factor is that, whilst the cumulative cost of providing additional data increases with each additional unit quantity, the increase in their value to the decision-maker does not. If the cumulative value curve were to lie below the cumulative cost curve from the origin there would be no point at all in collecting the data and using them. As it is, it will usually be found at first that the cumulative value of the data is high in relation to the cost of providing them. As additional data are provided, however, it merely tends to confirm the view obtained from the earlier data and eventually, a surfeit of data may have the effect of adding confusion and uncertainty to a picture that was once somewhat clearer. It is the combination of the two factors, with the maximization of the net value that is important, rather than an attempt to minimize the cost of providing the data themselves.

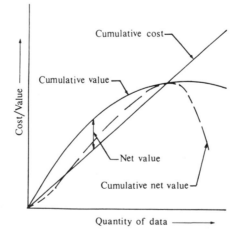

Fig. 6.3   Relationship between cost and value of data.

6.5 The collection of data

Data can be collected both from *internal* and *external* sources. The former concerns data that are collected from sources within one's own organization, whether that organization is a self-contained unit such as a construction site, a branch office, or a particular project within a larger organization. If the boundaries of a system representing a particular project have been defined and drawn round that system then internal sources will all fall within that boundary. Such data may include, for example, the quantities of work produced during a specific period of time, the number of hours and the number of resources devoted to the production of a unit of output, the number of staff and labour employed, the rate of absenteeism amongst staff and labour and the quantities of materials used per unit of output together with any wastage that has taken place.

External sources are those external to the boundaries of the system as defined. They may be, for example, government offices, professional bodies, trade associations and other companies, or, if the unit is contained within another organization, the accounts or purchasing sections of that organization.

Once a decision has been made to gather a certain type of data it is necessary to examine how it can be collected, by what method and from which source. The dangers of concentrating on the collection of certain data simply because they are easily obtained should always be borne in mind. An associated aspect of data collection is that it is *output* or *end results* rather than *activity* which is important. Some groups of resources, for example, may appear to have a high intensity of activity whilst others will appear to be almost casual in their approach to work. At the end of the day it is the productivity and quality that are important and for which a measure against an established standard needs to be made.

Data can be collected in a number of different forms depending on their nature. Some of the forms are:

(a) *Numerical data*: many data will be collected as straight numbers. Such numbers will represent quantities of work completed, hours worked on specific units of production, numbers of operatives employed on certain tasks, etc.

(b) *Frequency counting*: counting the frequency with which a single event occurs. Such counts are relatively simple to make and collect, such as the number of times an item of mechanical plant breaks down, the number of truckloads of excavated material leaving the site, etc. These are usually related to a specific period of time.

(c) *Estimated ratings*: such ratings should be made by experienced personnel who are knowledgeable about the subject of the rating. The ratings will normally be subjective estimates concerned with the quality of a happening. As such they cannot be manipulated in the same way as quantitative measures such as numerical data, and care must be taken in reporting ratings of this kind lest a bias is allowed to creep in due to subjectivity. This category may also include *verbal measures* used to describe performance characteristics. An example of a verbal measure is 'a high quality of supervision'. Clearly the expression is subjective within the standard set by the speaker. The value of the expression rests almost entirely on the consistent interpretation and understanding of the words that are used. On the other hand, the words *subjective* and *objective* are often misinterpreted when it comes to measurement. Objective measures are often taken as being *factual* and hence *true*, whereas subjective measures are assumed to be sheer *opinion* and hence likely to be inaccurate. This, of course, may be so but, on the other hand, there is no reason why it should be.

The bulk of data collection is presently achieved using manual methods. The collector will make a note of the data on a form, normally specifically designed for the purpose, then the data will be summarized and usually, in the process, will be transferred to another form. Finally the data will be analysed, categorized, manipulated, aggregated and so on, before the resulting information is used in order to prepare a report. In the final process, if a computer is to be used to undertake the conversion from data to information and then reporting, the data need to be entered by keying into a computer input system. All of these manual processes of recording, summarizing, inputting and analysing the data are subject to human error of various sorts. Errors of inputting data into a computer by hand, for example, has been established as one of the most significant sources of error in computer processing.

Now under examination and evaluation, both in research and industrial organizations and also in the field, are a number of automated data collection methods in construction and there is little doubt that this aspect of management control systems will become an important feature in the development of future project and construction management control systems. Data collection technologies which are under investigation include the use of *bar coding, optical character recognition, voice recognition* and the use of *radio frequency* methods. Bar coding and voice recognition will be briefly described below; optical character recognition is basically similar in the method of application to bar coding techniques; radio frequency methods refer generally to the transmission of data from source to processor rather than to the data collection process itself.

Bar coding technology has been used in many industries other than construction and perhaps the closest and most obvious example that the majority of individuals are familiar with, is in the retail grocery business where bar codes appear on all packaged foods. In construction, bar coding systems have been used in a limited number of applications.

Bar codes consist of a series of light and dark parallel stripes of varying widths. The height of a bar code, that is the length of the stripes, is not critical to the logic they represent. The graphical striped pattern represents a series of 1s and 0s and the sequence is known as a *cipher* and, as such, can be used to provide fast and error-free entry of data into a computer system. The cipher can be read by means of an electronic input device by passing the device over it providing a much faster data input device than a manually operated keyboard. The device may be either fixed or portable and may be either a *light wand* or a *laser scanner*. The input device senses the ordered and unique arrangement of the light and dark bars and the optical information that is collected in this way is converted into an electronic message. This message is then sent to a *reader* having an internal decoding algorithm. The decoding algorithm is able to identify the message, interpret it and convert it into a machine language that a connected computer can utilize as an instruction. Bar codes can be attached to all manner of materials and their packaging in the form of labels, or they can be printed directly onto sheets of paper. The adhesive labels can be made of metal, polyester film, plastic or paper – all of varying robustness depending on the application situation.

One system of hardware suitable for a bar coding system and described by Rasdorf and Herbert (1990) consists off a desktop *computer* which can, but need not, be connected to a mainframe computer, a *concentrator*, a *printer*, a *remote reader* and a *scanner*. The remote reader (that is remote from the computer) and the scanner/wand are at the point of data acquisition. The printer is optional, but is useful for printing bar code labels or for producing output of the data which are collected. The concentrator acts as a centre piece for the network of the system, connecting the printer, remote reader and scanner(s) and computer together and facilitating communication between them. Software is then required for each particular application in order to accept the collected data and to arrange and organize them in such a form that purposeful output can be created.

In construction, though not exclusively, bar codes have been used for tracking documents such as drawings, purchase orders, change orders and requisitions. The bar code is attached to the relevant piece of paper and is read in conjunction with a central database held in a computer. The database will hold data concerning various projects and will sort documents on the basis of project numbers, sections, etc. Similarly, the bar coding of equipment enables the collection of central data concerning its current location, maintenance and repair history, and with the coding of associated time records can facilitate the storage and retrieval of its operating history. Similarly the ready identification of labour on construction sites can be tracked and when used in conjunction with a central database, will provide managers and supervisors with up-to-date statistics concerning work force disposition. Other applications have been involved with bulk material take-off by using a bar coded menu and using bar codes to input current update data to planning and scheduling software packages.

Voice recognition technology is another method which is being explored and tested as a means of gathering and inputting data for project control into a computer system. Simply put, voice recognition is a means of talking to a computer in such a way that the computer understands what is being said and then acts on it. In practice, a voice recognition system identifies what is being said with words in a stored vocabulary that has been designed for that purpose and then transmits the message in the form of computer recognized characters to a host computer for action using a specially written computer program. A user speaks into a microphone, the speech is then identified by comparing measurable features of it with voice patterns that have been previously stored. A system may be *speaker dependent* in which only one speaker will be recognized or *speaker independent* where more than one speaker is recognized.

Voice recognition can be used in construction to input up-to-date data into control systems stored in a computer and thus save time and effort in establishing error-free data for control purposes. A site cost control engineer, for example, would need only to input the day's schedule and cost data into a desktop computer control system by means of voice recognition technology when summaries and cost reports would be automatically produced.

6.6 Computer processing

Computer data processing was originally introduced to the construction industry as a means by which the tediousness would be removed from many routine clerical processes such as the preparation of wages sheets and that of purchasing and invoicing materials, on the basis of cost reduction, greater accuracy and timeliness. This naturally evolved towards the use of the basic data arising from these processes and generated within a company, for the purposes of information and control as part of a wider system. By the mid-1960s, many engineering and contracting companies had developed simple information systems by using these internal data. For example, data from site wages sheets, after analysis, facilitated the allocation of labour costs to various site activities and operations and hence a comparison of estimated and actual labour costs was possible. However, this type of management information system was inflexible in its structure. It was usually processed by a computer having a limited processor and storage capacity by using a computer program that was specifically written to manipulate data input in a rigidly structured form. Hence a large number of files (sets of related records) each with a fixed data structure, were necessary as part of running the programs and transferring data.

The format of a report output also tended to be limited in the variations it permitted. This rigidity of system did not allow adjustment to meet the different requirements of individual managers by way of information. In addition, as the operations and the organization of a firm varied over time, it was difficult to ensure that the information systems changed with them so as to cater for a different information demand.

With this conception of a computer-based information system built around the internally generated data that were concerned with many of the clerically-orientated transactions forming an essential part of a construction company's business, there were difficulties in maintaining the accuracy and consistency of the basic data. For example, in the preparation of wages sheets, a file would be created containing a large number of records, each of which held the relevant information for one employee. An employee's record would contain certain information about his skills, rate of pay, hours worked (both normal and overtime), incentive scheme payments, contributions to National Insurance and pension schemes, probably details of his family, information concerning his tax situation, etc. As part of the wage-sheet preparation process these records would have to be constantly updated or modified in accordance with the data that were presented. In addition, new records would need to be added to the files and others would have to be deleted from the active list. In these earlier systems, the files were processed by a computer program that was designed to do little more than a clerical assistant had done for the manual system. The computer program ran through the records in the file sequentially, processing each one as required by the demands of the output. The rigid structure of the files did not encourage their use by more than one processing program and, therefore, where data that existed in the wages sheets file were required for another application, an additional separate file had to be created for use with the additional application program.

This arrangement made the process of updating files, modifying and adjusting them, in accordance with the requirements of the system, a tedious, demanding and/or difficult process. There could be little certainty that files intended to contain similar records would have complete consistency from one to the other. Where files were modified to suit the needs of a new applications program, this frequently meant that other programs using similar data, or data from a common source, had to be restructured, rewritten and modified to suit. The outcome of this inconsistency between the data held in different files, ostensibly of the same records, was that the applications programs tended to produce different output that was unfortunately inconsistent with other related processed data.

As time went by, computer equipment became cheaper and processing capacities increased both in storage and speed. The wider availability of direct access storage devices enabled the rapid retrieval of stored data into a central processor. The concept of information systems, integrated throughout a company and transcending barriers between departments and divisions, became accepted. It became practical to store basic data for use in a number of areas throughout a company's activities (though these were not necessarily identified at the outset) in a common database. Data could be retrieved and processed as required and then presented in a report to suit an individual manager or to meet the requirements of a particular department or division of a company. Such a database exists quite independently of the applications programs, several of which may access large sections of the database for different purposes and for producing reports of a different nature. As data are input to the system, their source is identified and the broad relationships between different sections of the data are established. It may be that data from several sources will then be brought together and become the subject of a report. A typical but simplified diagram representing such a system is shown in Fig. 6.4.

In industry and commerce by far the majority of computer applications are concerned with file processing systems. It becomes important, therefore, that access to the files is possible in an as fast and efficient manner as possible. Data files and their structure are usually the key to the effective use of all types of computers, particularly microcomputers. Data files can be created as *sequential* or *random access* files. Sequential data files have records stored in one continuous, ribbon-like stream of information and therefore tend to use

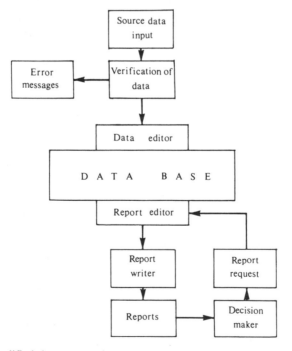

Fig. 6.4 Simplified data processing system.

the disk or tape on which they are stored to capacity. They are space-efficient but suffer from the disadvantage that it is difficult to make changes to the data that are stored in this way. Sequential files are therefore useful for storing information which is static or permanent in the sense that the need to change it is infrequent. Another disadvantage of sequential access files is the relatively long computer time it takes to search through a series of records in order to find the one required. If there are 600 records on a sequential file and it is required to amend the record which is established in the 500th position, it is necessary for the computer to search through 499 records before it finally comes to the one required.

The advantages of random access files tend to be the disadvantages of sequential files. They do suffer, however, from space inefficiency since records have a fixed record length and where only a small amount of information is stored in a record the rest of the available record space remains blank and cannot be used for other purposes. Random access files require to be better planned and to have better designed systems for organizing and using data but, on the other hand, once this work has been carried out they need less effort in use and maintenance than sequential files.

One of the areas of fastest development in data processing which must have a significant effect on management information systems is that dealing with communications. The costs of communication are being reduced very effectively with the establishment of satellites and microwave facilities on a free competition basis. These devices mean more alternative facilities for data processing and the linking more easily and effectively of large computers and their associated facilities.

At one time the purchase and installation of a centralized computer system within a construction company was seen to be the equivalent of buying another large excavator or a specialist item of equipment – except that the purpose and the function of the last two items could be seen to reflect directly on the objectives of the organization and the use to which they could be put was clearly understood. The range of computing systems and technology that was available was limited and permitted few alternative arrangements for the hardware, the software and the means of inputting data and outputting information from the system. In many companies the computer system was seen to be an extension of the office calculating machines and thus the computer personnel, either retrained from within the organization or recruited as specialists from without, were fitted into the organization with little or no disturbance to the existing company structure.

The pace of advancing technology has increased in recent years and no longer can the examination of which computer system to purchase be confined to which manufacturer's machine, and then the specific model mainframe computer, to buy. Networks can be used which link large mainframe computers with mini- and micro-computers, all of which can be used through many terminals; databases can be interrogated from many different locations. The technology of today is such that a manager can select the computer configuration and locations that he requires rather than having to be content with a centralized facility. This means that a computer system can be designed to suit the structure and the policies of the company organization and the information systems must be designed to suit both.

6.7 The value of information

In designing, installing and using a management information system, the importance of the output from it being relevant and timely has been stressed. No consideration has yet been given to the quantitative value of using the system and no quantification of benefits arising from its use has been attempted. It may be possible to estimate the cost of collecting, sorting and analysing data but this is only a part of the complete story. If the information that is generated by the system and communicated to the appropriate manager does not influence that manager in his decision-making processes, then the information has no value. This argument may be taken one stage further in so much as a manager may find that information provided to him only serves to confirm that the decisions which he has already made are the correct ones. Whilst the information may serve the purpose of underpinning the confidence of the manager, it is really serving no effective purpose and thus, again, has little or no value.

At the other end of the spectrum, if a manager were provided with an information report that eliminated all uncertainty from his decision-making then it would have ultimate value and the manager would not be required. In practice a condition between the two ends of this spectrum is most likely to occur although a strong tendency towards the former condition is more likely to prevail than the latter. If the information provided reduced all uncertainty decision-making to zero, then its value may be seen to be the cost of the manager to the organization, including all his related overhead expenses, plus the benefit that accrues from the fact that the decisions made will always be correct. From any calculation of the value of the information derived from the benefits arising out of it, there must, of course be a subtraction of the cost of collecting, classifying and analysing the data that form the basis for the reporting system as illustrated by Fig. 6.4.

Clearly, when in the position of making a decision it is highly desirable that as much of the uncertainty surrounding the decision situation as possible should be eliminated compatible with keeping the cost of data to a minimum. However, when this situation has been reached, such uncertainties as remain must be quantified if at all possible so as to eliminate the difficulties of dealing with qualitative rather than quantitative values. One method of doing this is to express them in the form of probabilities. Having arrived at the probability of various outcomes it is desirable to determine the utility to the decision-maker of those outcomes by the use of the utility curves that are relevant. Subsequently, the expected utility of the various consequences of the decision situation can be calculated and the decision can be made on the basis of maximum expected utility.

Whilst it may be desirable to eliminate or reduce uncertainties wherever possible, it is not always practicable to do so. There are two main reasons for this. Firstly, the cost of doing so may be prohibitive. For example, in producing a manufactured item it may not be feasible to make a detailed examination of every single item produced because of the considerable cost of carrying them out. However, there is no certain way of ensuring that every manufactured item is entirely in accordance with the required specification unless each one is inspected individually. Secondly, nobody can predict the future with absolute certainty. There can be no absolute certainty that a specific event will take place in the future in spite of the strongest prediction that it will do so. An engineer who prepares the design for a building to be constructed in an earthquake zone can never be certain that the maximum severity of earthquake for which he designs will not be exceeded in spite of the fact that up until the date of design it has not been in that area.

Whilst these two practical difficulties of cost and prediction of the future are real deterrents to the removal of uncertainty, it is usually possible to reduce the degree of uncertainty which exists in a decision-making situation by seeking more information which is relevant to it. In the examples given previously, if the manufacturer collects data about the historic quality of his product, particularly noting the frequency with which defects occurred, and the designer examines trends of the severity of earthquake shocks in other areas of the world and the resistance to them of buildings designed to the same principles that he has used for his current design, each will have taken some steps to reduce the uncertainty surrounding the current problem with the aid of additional relevant information.

Even on the assumption that relevant information becomes available to aid the decision-making problem, it is not always used in a completely objective fashion and to the best effect. Situations of which the information-receiver has recent experience, or has been the subject of some prominence, will often assume greater proportions on the whole than is warranted by the facts. For example, a site manager who recently has become disadvantaged by absenteeism of site operatives, when asked to give an estimate of the extent to which absenteeism occurs in site construction, will probably overestimate the frequency with which it actually occurs. An additional aspect of this problem is the tendency when considering estimates of this kind to think in absolute rather than in proportionate terms. On larger sites there are likely to be a greater number of absentees than on small sites with fewer total operatives. It is, therefore, correct to compare the absenteeism on different sites in terms of the ratios of absentees though all too frequently more emphasis is placed on the absolute numbers rather than the proportions, particularly in the case of a manager who has recent personal experiences of the problem.

Many problems in decision-making are evaluated in the first instance by calculating the probability that each alternative outcome will be achieved. When additional relevant information becomes available the probabilities will almost certainly be revised and *Bayes' theorem* can be used for this purpose.

Suppose that a contractor is offered a fixed price contract for repairing a reinforced concrete structure for a sum of £500 000. The contractor has some uncertainty about the material that he should use and this is a very important element of cost in the contract. The use of Material A will cost £150 000; the use of Material B will cost £180 000. The total contractor's costs (including the materials) will be £480 000 and £510 000 for the use of Materials A and B, respectively. The contractor estimates the probability that Material A can be applied successfully is 0.65. If the material proves unsuccessful, he will have to resort to Material B of which he has considerable experience and knows can be used successfully.

The contractor's decision problem in terms of profit outcomes can be tabulated as in Table 6.1.

Table 6.1

		Use Material A	Use Material B
Decision 1:	Accept contract	£20 000	−£10 000
Decision 2:	Reject contract	0	0

If Material A proves satisfactory, Decision 1 would be the correct one; if he has to resort to Material B, Decision 2 would be more appropriate. The contractor's need is for further information concerning the suitability of the materials.

Taking a risk-neutral attitude (that is with a straight line utility curve) the expected cost of the project to the contractor will be:

$$(0.65 \times 480\,000) + (0.35 \times 510\,000) = £490\,500$$

The contractor would therefore expect to make

$$£500\,000 - 490\,500 = £9\,500 \text{ profit}$$

On this basis the contractor would take up the contract since the expected value is positive.

One of the ways in which the contractor could gain additional relevant information would be to arrange for a test to be undertaken on Material A to establish that the probability of it being successful really is 0.65. Such a test would cost money but if it proved successful it would eliminate the need to take into account the expected consequence of resorting to Material B at a higher cost. In other words, if the test failed, the contractor would not take up the contract.

The expected profit of the situation intending to undertake a test is:

$$(0.65 \times 20\,000) + (0.35 \times 0) = £13\,000$$

that is £3500 more than without the further information. The contractor can therefore spend up to £3500 on the test without the expected profit being reduced below the level of that expected without it. The value of the additional information can be said to be £3500.

There are two important aspects to be noted about this problem. Firstly, that there are only two outcomes to the problem – accept or reject the contract and therefore it is something of a specific case. Bayes' theorem can be used in a more general way to deal with problems having a greater number of outcomes. Secondly, that the further information that was received as a result of the test was assumed to be *perfect information*, that is, it confirmed exactly the outcome which obtained. The receipt of perfect information is rare. Additional information will normally have some influence on the way in which a decision-maker will act but it will nevertheless leave some uncertainty as to the course to be taken. Information of this kind is known as *partial information*.

Bayes' theorem, in general terms, (the proof is not offered here) can be expressed as follows:

If $A_i (i = 1, 2, 3, \ldots, n)$ are n mutually exclusive and the only possible occurrences, such that an event B can occur only if one of those n results has occurred, then the probability that A_i occurred when B is known to have occurred is:

$$P(A_i/B) = \frac{P(A_i)P(B/A_i)}{P(A_1)P(B/A_1) + P(A_2)P(B/A_2)\ldots}$$

where $P(A_i)$ represents the prior probability of outcome A; $P(B/A_i)$ is the conditional probability that B occurs, A_i having occurred and, $P(A_i/B)$ represents the revised probability of outcome A_i, given that event B has occurred.

Example 6.1

An engineer is undertaking a feasibility study for the design of some foundations for a heavy plant installation. The estimate of cost has to be prepared quickly and much of the cost depends on the particular design that will be adopted. There is insufficient time to drill test boreholes and only a brief seismic test can be carried out in the time available.

The three foundation conditions are foreseen as strip footings (A_1, heavy raft (A_2) and piles (A_3). At the outset the engineer estimates that the probabilities of each are $P(A_1) = 0.6$, $P(A_2) = 0.3$ and $P(A_3) = 0.1$.

In relation to the brevity of the seismic tests the engineer produces a table of probabilities relating 'good' (G) and 'bad' (B) seismic results to the foundation types that will be required. This is illustrated in Table 6.2.

Table 6.2

	Seismic good	Seismic bad	Total
Strip footings (A_1)	0.7	0.3	1.0
Raft (A_2)	0.5	0.5	1.0
Piles (A_3)	0.2	0.8	1.0

The probabilities in the table are the conditional probabilities $P(G/A_n)$.

The seismic test produces a 'good' result, what are the revised (posterior) probabilities that each type of foundation may be required?

From Bayes' theorem:

$$P(A_1/G) = \frac{P(A_1)P(G/A_1)}{P(A_1)P(G/A_1) + P(A_2)P(G/A_2) + P(A_3)P(G/A_3)}$$

$$P(A_1/G) = \frac{0.6 \times 0.7}{(0.6 \times 0.7) + (0.3 \times 0.5) + (0.1 \times 0.2)}$$

$$= \underline{0.712}$$

and $P(A_2/G) = 0.254$ and $P(A_3/G) = 0.034$

It will noted that these three probabilities add up to 1.000, and as a result of the additional information from the 'good' test result the probability of the simple type of foundation increases from 0.6 to 0.71.

Chapter 7
Cost Fundamentals

7.1 Production costs

The production of goods involves the co-ordination and control of one or more *factors of production*. Factors of production may alternatively be described as resources and they include labour, raw materials, mechanical equipment, buildings and finance – in fact, anything that can make a contribution to the enhancement of a product's value. The total cost of the factors of production must include all the elements of cost attributed to the process even though it may not be immediately obvious that such an item contributes directly to the total production cost. The cost of transport between the production centre and a point of consumption is an example. Since production generally only takes place if the producer expects to recover his costs and probably to make a profit from his activities, it is necessary that the selling price of the product should always be at, or in excess of, the total of all the costs of the factors of production incorporated in the work.

Clearly, because of the large number of different factors of production that are available to a producer, there are many different ways in which combinations of each factor can be used to achieve a required end product. It is the combination of factors which is used, and the method of working to bring about the effective combination, which offers the greatest scope for economy of production costs rather than the cost of the individual factors themselves. This is because of the relatively small variations in the cost of the available factors to those whose business it is to combine them. For example, in designing and constructing a reinforced concrete silo, the unit costs of the factors of production, the cement, aggregates, reinforcement, labour, equipment, etc., are reasonably uniform as far as all designers are concerned. If every designer produces a similar design and every contractor uses a similar construction method, thus overall between them producing a similar combination of the factors of production, then the cost of the completed product must be almost the same in each case. No one contractor, however, purchases a sufficiently large proportion of the output of the cement industry, for instance, that he can influence the price of cement in his favour and against his business competitors (other than to negotiate a small discount for bulk purchase which, even so, is also available to others). The same situation applies to other resources.

On examination it will be found, therefore, that similar work varies in cost (and hence the ultimate price differs) in the eyes of different contractors, or manufacturers, because they calculate costs on the basis of combining the required factors of production in different ways and because they use different construction techniques in the production process. To use, again, the example of the silo, one contractor may use sliding formwork, another may use fixed formwork and yet another may use timber rather than steel formwork. Gang sizes may also vary. One contractor may use a small concrete mixer, another a large one, and so on; the variables are many and the combinations of the variables are almost infinite in the practical situation.

From a designer's point of view, one designer may use more concrete and less steel reinforcement, another may have a short, large diameter silo rather than a tall one of considerably smaller diameter. The available combinations of the factors of production are infinite but it is generally the concern of economics, and engineering economics in particular, to assist in the adoption of an optimum procedure – not only for the initial design and construction, but also with regard to the future life of the product. In some instances the required future life of the product may be the all-important factor which ultimately influences the selection of the alternative in the analysis of investment projects. The nature of some of the costs which form part of the overall picture will now be examined to see how and why they vary and the possible influences that external effects can have on them.

7.2 Fixed, variable and semi-variable costs

When a construction company owns an item of mechanical equipment, such as a dumptruck, it is concerned with two types of cost that arise out of its ownership and use. These can be classified as *ownership* costs and *operating* costs. Ownership costs, as their name implies, represent the cost of possession of the equipment without necessarily using it, or even moving it from the point of unloading after delivery on purchase. Operating costs are those costs which are incurred in addition to the ownership costs during the time that the machine is in production. Broadly speaking, ownership costs are *fixed* costs and operating costs are *variable* costs. It is worthwhile to examine this classification of costs a little more closely. In the case of an excavator, once purchased, it will depreciate regularly whether it is in use or not. The mere fact that it is no longer absolutely brand new but secondhand, is enough to reduce its value in financial terms on the open market (assuming that the particular model can be supplied in the first place as a new machine with no extended delivery period). If new and technically improved models are introduced, then a model which is already secondhand, but not necessarily very much older, will immediately become obsolescent; in addition, some physical deterioration must take place with the passage of time. As well as depreciation and obsolescence creating a cost, if the machine is to be put into use it must be taxed and insured. Both of these latter costs can be regarded as fixed, since they are not necessarily related to the extent of the use to which the excavator is put. Fixed costs can therefore be considered as those which tend to be unaffected by the amount of work carried out, whereas operating costs, and hence variable costs, are all of those costs which tend to be incurred more or less in proportion to the extent to which use is made of the machine. Included under this heading are the costs of fuel, oils, greases, servicing and repairs, and the cost of having an operator with the machine, though the latter may well be regarded as a fixed cost under certain circumstances.

Re-examining the fixed costs, as described above, it is not strictly correct to lump all of these costs together as being fixed. For example, the costs attributed to depreciation and obsolescence, and included in the fixed costs, will be largely dependent on the amount of use to which the excavator is put. Wear and tear will, in this case, reduce the ultimate secondhand value of the equipment. An element of the fixed cost can therefore be extracted from the total and labelled as a *semi-variable* cost. Its proportion of the fixed cost will depend upon the extent of use of the machine during the period under consideration. The cost of using this excavator for one hour is therefore composed of three different elements of cost – fixed, semi-variable and variable. On examination, it will be appreciated that it is not possible to establish, without doubt or free of contention, the actual total cost of using the excavator for any *one particular hour* of its life without making some assumptions or estimates about its use for a period of time in the future. The establishment of cost is necessarily related to a given time-span.

Fixed costs, in this example obsolescence, tax and insurance, are inevitably going to be incurred over a period of time. Whilst it is possible to redeem a licensing fee or to obtain a refund of an insurance premium that has been paid, they are generally paid with the intention that they should remain in force for the various future periods until they expire. Until the time comes to renew the licensing fee and/or the premium, the cost is incurred by prepayment and it must be included in the calculation to establish the cost of the eqiupment per unit of time. As the period of time for which the fixed costs are calculated gets longer, they tend to become more variable and fuzzy by nature. They do so because the longer-term decisions to incur cost will ultimately need to be reconsidered and a decision can then be made to change a commitment if it is so desired.

Semi-variable costs for a particular unit period cannot be predetermined with precision because some assumption must be made about the likely monetary depreciation that is going to occur as a result of the extent to which the machine is put to use. In the cae of an excavator, for example, the actual wear and tear can only be determined historically. If a future hourly cost is estimated prior to the period of use, some assumption must be made about the likely hours to which the machine will be put to use during that period.

Variable costs are those which are actually related to the time the equipment is in use. The cost of fuel is an example. The rate of use of fuel tends not to vary widely, other than for exceptional reasons, and it is more or less directly proportional to the time that the excavator is in use. In the case of variable costs of this nature, they are known as incremental costs because they are added to the fixed costs (and the variable costs already incurred) as and when the machine works. An alternative name for variable or incremental cost is *marginal cost*. This term is explained by defining the margin as being at the total of the costs already incurred and the marginal cost is that cost which is required to extend the period of use, and hence the total cost, by a small increment. For example, assume that all the fixed costs for the excavator have been paid for three months ahead and that what are estimated to be the semi-variable costs for that period are also paid. In practice, if the machine is now worked for one more hour instead of allowing it to stand idle, the marginal cost of doing so is solely the cost of fuel, oil, grease and probably the wage of the machine operator. The rest of the costs are incurred in any event and are therefore fixed for the period under consideration. It would, of course, be unrealistic to substitute marginal cost for the actual cost of one hour's use of the excavator in this example, since this will give an entirely false idea of the actual cost of operation and may ultimately be the result of incorrect forecasts being made in the future.

To be able to distinguish between fixed, semi-variable and variable costs is important to a construction manager in his decision-making. It must be borne in mind that, as time passes, fixed costs can change to variable costs and so on. In the above example, if the operator of the machine is assigned to the machine for the whole of a working week, whether the excavator is productive or not, then his cost is a fixed cost for that period. If subsequently the operator is only available while the machine is actually working, then his cost reverts to that of a variable cost by nature.

7.3 The identification of variable and semi-variable costs

The apparently varying relationship between fixed, variable and semi-variable costs has already been discussed. There is often a need to predict, from historic data, the probable proportions of future fixed to variable and/ or semi-variable costs. One method of doing this is to examine the trends which appear to arise out of the historic data and make an extrapolation for the future. Statistical methods can be used to establish such trends within known degrees of accuracy, and a simple example will be examined to illustrate the fitting of a straight line to given data. The mathematical derivation of the formulae which are used is covered elsewhere. The means to be described here is known as the *Method of Least Squares*. Its basis is the fitting of a curve to collected data by means of minimizing the sum of the squares of the distances from the fitted line to the points obtained from the data.

The method will be illustrated by using a problem which involves depreciation and obsolescence to determine the fixed element in a total cost. If a fleet of similar vehicles is considered, all of which have been run for different distances during the first year of their service, it is possible that a trend can be established in the ratio of the total depreciation in their value to the distance run. In this instance, depreciation is calculated as the difference between the purchase price and the market value at the end of the first year in service.

Using y to represent the depreciation in money terms and x to represent the miles run by that vehicle during the same year, the data in Table 7.1 have been collected and plotted in Fig. 7.1.

Table 7.1

Depreciation (£)	y	600	610	700	600	780	680	800	740	860	740	880	1000
Mileage run	x	6000	8000	8800	9600	10 400	12 800	12 800	14 000	16 000	17 200	19 200	20 000

On examination of the data as plotted in Fig. 7.1, it will be evident that there is a trend through the points in the figure and it would not be too difficult to fit a straight line through them by eye. Such an exercise, however, tends to involve subjective judgement and it is difficult to measure the *goodness* of fit. Recourse to one or other of the statistical curve fitting methods is likely to give a more realistic and objective result. The method of least squares is useful for fitting a curve to a coplanar set of points.

The equation of a straight line takes the form $y = mx + b$, where m represents the *slope* of the curve and b represents the value at the point where the curve intercepts the y-axis. From the series of values for x and y in the data set out in Table 7.1, it is now required to obtain values for the constants, m and b, for a line which gives the best fit to the data. Having established the values of these constants, it will be possible to predict the likely value of y, given any value of x, for situations not expressly occurring amongst the data of Table 7.1. In doing so, it is assumed that the value of x is correct, and that any deviation due to the curve not fitting the data exactly will occur in the value of y so obtained. This deviation is known as the *y-deviation* and if the values of m and b are calculated so as to minimize the sum of the squares of the y-deviations, the line is called the *line of regression of y on x*. On the same basis, there will be a *line of regression of x on y* which can be established and if the value of y is then assumed to be known exactly and the value of x is obtained from this curve, the deviation is assumed to be contained in the x-value. In this second case it is the sum of the squares of the *x-deviations*

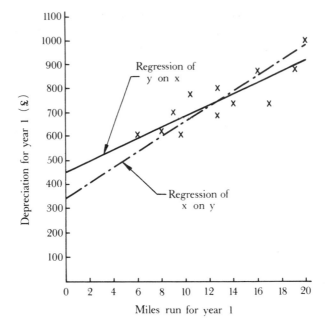

Fig. 7.1   Curve fitting to data.

which is minimized. The required value of x can be calculated from the equation of the line, $x = My + a$ where M and a are constants as before.

In order to obtain first the line of regression of y on x for the data presented, the following two equations must be satisfied:

$$\Sigma y = (\Sigma x)m + nb \tag{7.1}$$
$$\text{and } \Sigma xy = (\Sigma x^2)m + (\Sigma x)b \tag{7.2}$$

where n is the number of readings of data available.

If equation (7.1) is divided through by n it becomes

$$\bar{y} = m\bar{x} + b \tag{7.1a}$$

where $\bar{y}$ and $\bar{x}$ are then values of the arithmetic mean for the data, x and y. This means that the curve of the equation (7.1a) must pass through the average point of the data and also provided a means of both checking the calculation and simplifying the problem of dealing with very large numbers.

It can be assumed (though in some circumstances it may be known to be the case) that the value of $b = 0$ and therefore the curve to be fitted takes the form of $y = mx$. In this case Equation (7.2) can be used in a simpler and special form

$$m = \frac{\Sigma xy}{\Sigma x^2} \tag{7.3}$$

By translating the axes of the diagram to the average point of the data, $(\bar{x}, \bar{y})$, through which the line of best fit passes, the slope of the curve is not affected. If new values are now defined as $x^1 = (x - \bar{x})$ and $y^1 = (y - \bar{y})$, the equation of the curve required is

$$y^1 = mx^1 \quad \text{where} \quad m = \frac{\Sigma x^1 y^1}{\Sigma (x^1)^2}$$

In similar fashion the line of regression of x on y referred to the average point, $(\bar{x}, \bar{y})$, as a new origin can be established by using

$$x^1 = My^1 \quad \text{where} \quad M = \frac{\Sigma x^1 y^1}{\Sigma (y^1)^2}$$

The fitting of a straight line to the data can now best proceed by tabulating the calculations as in Table 7.2. The slope of the line of regression of y upon x,

$$m = \frac{\Sigma x^1 y^1}{\Sigma (x^1)^2} = \frac{5\,228\,999}{225\,000\,000} = 0.023239$$

therefore

$$y^1 = 0.023239 x^1$$

or substituting for $y^1 = y - 749.17$ and $x^1 = x - 12\,900$

$$y - 749.17 = 0.023239(x - 12\,900)$$
$$= 0.0232x - 299.78$$
$$\text{and } y = 0.0232x + 449.39$$

In addition,

$$M = \frac{x^1 y^1}{(y^1)^2} = \frac{5\,228\,999}{167\,091} = 31.29$$

and $x^1 = 31.29 y^1$
or substituting for $y^1 = y - 749.17$ and $x^1 = 12\,900$

$$x - 12\,900 = 31.29(y - 749.17)$$
$$= 31.29y - 23\,442$$
$$\text{or } y = 0.032x + 336.9$$

Both lines will pass through the point $(\bar{x}, \bar{y})$ and, if all the points obtained from plotting the data were exactly on a straight line, both lines would in fact be coincident. However, both lines can only be coincident if the product $mM = 1$ when referring them to the origin $(\bar{x}, \bar{y})$ and using, therefore, the equations $y^1 = mx^1$ and $x^1 = My^1$. The product mM is used to measure the

Table 7.2

x	y	$x^1 = x - \bar{x}$	$y^1 = y - \bar{y}$	x^1y^1	x^{12}	y^{12}
6 000	600	−6900	−149·167	1 029 252	47 610 000	22 251
8 000	610	−4900	−139·167	681 918	24 010 000	19 367
8 800	700	−4100	−49·167	201 585	16 810 000	2 417
9 600	600	−3300	−149·167	492 251	10 890 000	22 251
10 400	780	−2500	30·833	−77 083	6 250 000	951
12 800	680	−100	−69·167	6 917	10 000	4 784
12 800	800	−100	50·833	−5 083	10 000	2 784
14 000	740	1100	−9·167	−10 084	1 210 000	84
16 000	860	3100	110·833	343 582	9 610 000	12 284
17 200	740	4300	−9·167	−39 418	18 490 000	84
19 200	880	6300	130·833	824 248	39 690 000	17 117
20 000	1000	7100	250·833	1 780 914	50 410 000	62 917
154 800 $\bar{x} = 12\,900$	8990 $\bar{y} = 749·17$	0	0	$x^1y^1 =$ 5 228 999	$x^{12} =$ 225 000 000	$y^{12} =$ 167 091

linear relationship between the variables and the *coefficient of correlation, r,* is defined as

$$r^2 = mM$$

It is by using the coefficient of correlation that evidence can be provided as to whether an accurate division of costs on a fixed/semi-variable/variable basis can be achieved using linear regression. The value of *r* which is obtained gives an indication of the reliability of the data and in the above example

$$r = \sqrt{0.0232 \times 31.29} = \sqrt{0.726} = \underline{0.852}$$

r must be on a scale between 0 and 1; with a value of 1, perfect linear correlation is obtained and all the points will lie on one line. Below a value of 0.7 for r^2, the linear relationship is not of significance so that, in the example where $r^2 = 0.726$, only marginal significance can be attached to the results obtained.

In Fig. 7.1 it will be seen that, if the regression line for *y* on *x* is considered, the fixed element of the cost amounts to £430 as calculated and the variable unit element amounts to £23.2 per 1000 miles travelled during the year. In the case of the alternative regression, these figures are £378 and £32 per 1000 miles, respectively. Due to the lack of complete correlation of the original data, there is this divergence between each regression line and some interpolation needs to be made.

There are a number of deficiencies and weaknesses in using least square methods for such purposes. In the first place, the assumption is being made, in the example chosen, that depreciation is dependent only on the variable, 'distance travelled', in the given time. This may not necessarily be true, but may also, for example, be dependent on the weather conditions, the general nature of the road surfaces over which each vehicle travels, the physical nature of the areas through which they travel, the characteristics of the materials carried and the skill and experience of the vehicle drivers. It may well be that each of these variables is quite different for each vehicle and that this point is not brought out from the data that have been collected. The data, therefore, may not have been collected on a uniform basis. The question of considering more than one variable can be dealt with by using a multiple regression technique.

Another factor which affects the validity of the regression technique is consistency amongst the data collected. It is important that data are collected from a population which has similar attributes. They do not, of course, need to be exactly identical, but if one were considering, for example, the breakdown characteristics of trucks used on a construction site, it would be necessary that statistics were collected for similar types of truck of approximately similar sizes, engine, capacity, use, etc.

It may be dangerous to assume that linearity exists in the data, since a curve of another form may be best fitted to the points obtained. The method of linear regression should not therefore be selected automatically as the technique to be used to establish the relationship. Whilst at one time its choice was influenced by the complexity of the calculations involved with other types of regression, the development of electronic digital computation has led to a greater freedom in the use of other techniques. This factor certainly facilitates the use of multiple regression for the consideration of the possible effects of more than one independent variable. The assumption of linearity can lead to the conclusion that the value of *b* obtained as a result of the regression is necessarily the fixed element of cost. This is only a safe assumption if the depreciation, in this example, is known for zero mileage. Care must therefore be taken in the interpolation of the results in order to ensure that misuse does not arise from the regression.

Again, extreme care must be taken with the recording and identification of data to avoid the possible misinterpretation that can take place in comparing data which are collected on different bases. In particular, it is necessary that the data are adjusted so that they take account of their probable collection at different times historically. It is therefore necessary to apply an index to the data in order to bring them in conformity with a particular date and so allow for escalation and other influences. Data that are not adjusted in this way, unless they do not vary with time, will only lead to an incorrect conclusion.

It is also necessary that all data collected should have a wide range since it is impossible to establish a meaningful curve through a cluster of points with practically no spread in either direction. There is danger in extrapolating results from curves beyond the limits and the range of data which is linearly regressed. It is quite possible, with a set of data that has been collected over a limited range, that linear regression will appear to give a satisfactory answer. Extrapolation beyond this range, however, may give incorrect results because of the existence of another form of relationship for data beyond that range, in spite of the fact that a good fit was achieved with the original linear regression.

Finally, it should always be remembered that there is seldom a direct causal link when two variables are correlated, even if the correlation coefficient is at a highly meaningful level such as 0.9. If it is wished to established a causal relationship between two variables then it is a much more complex business than calculating the correlation.

7.4 Break-even analysis

This book is largely concerned with the decision-making process in the broad area of investment in construction projects and the subsequent control of expenditure in their implementation. In general terms, a decision is only required in a situation in which more than one alternative is presented as a possible outcome of that situation. Decision-making is therefore a process of choosing between alternatives. Many ways and means of evaluating alternatives have already been discussed on the basis of what a particular project offers, what is required by way of financial investment in it and what should subsequently result from the investment once it is made.

The question of incremental costs can be developed further to provide information which will assist in the decision-making process. For this purpose they may be defined as being those changes in cost which occur when an operation or an activity changes its tempo or its character.

It is enlightening to consider the relationship between the volume of output which is achieved by a productive unit, the cost of producing that output and thereafter the resulting profit or loss which arises or is expected to arise. The concept of a break-even chart is an attempt to illustrate graphically the relationship between these three factors. As a means of visualizing what is happening under given circumstances the break-even chart is a useful device; its practical value is often limited. It is, however, a useful tool in explaining the relationship and the general nature of those costs in the way that they have been described.

A simple break-even chart is illustrated in Fig. 7.2. From it can be seen the profit or loss that should arise from any given level of sales or turnover. An example will illustrate the principles on which the break-even chart is constructed.

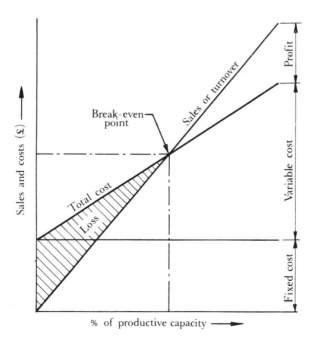

Fig. 7.2   Break-even chart (1).

Example 7.1

If a large batching plant and concrete mixer is erected on a construction site, it is probable that it will have been selected on the basis of the output and the quality of concrete that is required. Assume that, under normal circumstances, it can be expected to have an annual output of 16 000 m³ and that the contractor who is carrying out the work bases his tender and, in particular, his price per cubic metre of concrete, on achieving this figure during the next year's operations. One hundred per cent of productive capacity, as illustrated in Fig. 7.2, is therefore 16 000 m³. Further, assume that this concrete will, in effect, be *sold* to the client at, say, an average of £50.00 per m³.

The total sales or turnover arising from the plant is therefore £800 000 and a straight line graph can be drawn as the first part of the break-even chart.

There will be certain fixed costs associated with the plant. If it is hired, then the hire rate will be the principal factor in these fixed costs. If the plant is bought outright, then the depreciation of the equipment, plus insurance, are likely to be the major components of cost. Since these are constant for the time being, no matter the total production through the plant, they are shown on the break-even chart as a horizontal line, say at £135 000. To these fixed costs must be added the variable and semi-variable costs such as the cost of labour (some may be fixed if they are permanently in attendance on the plant), materials, obsolescence, etc. Assume that these variables and semi-variable costs amount to £533 000 if the full output of 16 000 m³ is achieved during the year under consideration. Thus at full productive capacity, the contractor stands to make a profit of £800 000 − (533 000 + 135 000) = £132 000 in the year. What happens if he only turns out 10 000 m³ in the period under consideration? The position will be as follows:

		£
Income from sales		500 000
less fixed costs		135 000
		365 000
less variable costs, 10 000 × £33.31		333 100
Profit		£31 900

The contractor's profit on mixing concrete, therefore, falls from an expected £8.25 per m³ as a result of the drop in output below that anticipated. In fact, if the production falls below 8745 m³ during the year, what was anticipated to be a profit is likely to turn into a loss. The break-even chart thus shows this relationship between profit/production/cost very clearly and, in addition, it shows the rate at which profit will be made once the turnover has achieved the level at the break-even point.

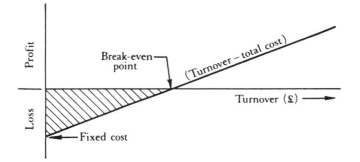

Fig. 7.3 Profit/volume chart.

The same information can be shown in at least two other ways. Figure 7.3 shows a *profit-volume chart*, no explanation for which is required, and Fig. 7.4 shows the break-even chart with the fixed costs plotted *above* the variable costs instead of below them. This latter method does have the advantage that it shows more clearly that it is the fixed cost which is not being recovered below the break-even point, the point which sales must at least reach if the operation is to become profitable. The break-even chart may also be used for certain types of investment analysis such as the problem which follows.

Example 7.2

A water-pipe, approximately 2000 m long, is required to supply a new water tower. A 200 mm diameter pipe will cost £30 000 to install, a 250 mm

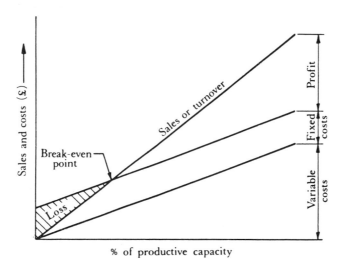

Fig. 7.4 Break-even chart(2).

diameter pipe, £60 000, and a 300 mm diameter pipe will cost £90 000 to install. In each case the total cost per hour of pumping will be £3.50, £2.90 and £2.00, respectively. The life of the pipelines is forecast to be 20 years after which it will have no salvage value and it will be necessary to replace it if the scheme is retained in operation. Determine the most economic size of pipe when 5000 hours pumping per year are expected if the interest rate is 10 per cent on the money invested in the scheme. Determine also what length of time of pumping will be required in every year to make the other sizes of pipe equally acceptable with the 300 mm diameter pipe. Draw the break-even chart for the investigation and determine during what periods each pipe will prove the most economical installation.

The total annual costs of pumping for 5000 hours per year are as follows:

	Diameter of pipe		
	200 mm	250 mm	300 mm
Capital recovery on initial cost of pipeline at 10% interest per year and 20 year life	£3 520	£7 050	£10 570
Cost of 5000 hours of pumping	£17 500	£14 500	£10 000
Total annual cost	£21 020	£21 550	£20 570

From the above table it will be seen that the 300 mm diameter pipe is the most economic for the duty required of 5000 hours of pumping per year.

Let the hours of pumping required for equal cost with the 300 mm pipeline installation equal x.

(a) In the case of the 200 mm diameter pipe:

$$3520 + 3.5x = 10\,570 + 2.0x \text{ from which}$$
$$x = 4700 \text{ hours}$$

(b) In the case of the 250 mm diameter pipe:

$$7050 + 2.9x = 10\,570 + 2.0x \text{ from which}$$
$$x = 3911 \text{ hours}$$

The break-even chart is illustrated in Fig. 7.5.

7.5 Limitations of break-even analysis

The break-even chart is simple in concept and often it is just such a tool or graphical illustration which is required to highlight the major influencing factors in a situation such as the one described above. In particular, it

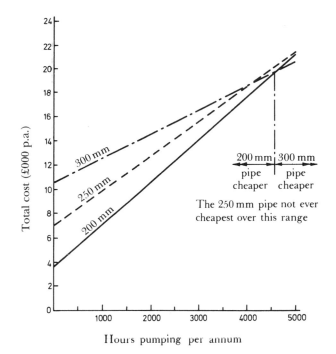

Fig. 7.5 Break-even chart for pipe selection.

demonstrates very clearly the relationship between the three variables, cost, volume and profit, and the effect that changes in each or any of these are likely to have on the others. Unfortunately, it does have some serious limitations which stem mainly from the fact that the chart can only be an illustration of the details of a situation at one specific and instantaneous point in time.

There can also, for example, be difficulty in distinguishing between those costs which are truly fixed by nature and those which are truly semi-variable. The example given of depreciation and obsolescence illustrates that difficulty. As far as a break-even chart is concerned, it is often necessary to assume that costs fall very clearly into either fixed or variable so that they may be represented by straight lines (or near straight lines) on the chart. In addition, it is often assumed that as output rises, fixed costs will remain unchanged and that variable costs will remain unchanged for each unit of production. This is often not the case since an increase in output sometimes necessitates a change in working hours which in turn may require overtime and the inevitable non-productive element of an enhanced wage rate has then to be paid. Variable costs per unit will increase. Some increase in fixed costs may well arise from the necessity to install additional assets, such as plant and equipment, in order to cope with the increased production.

It will be appreciated that a break-even chart may have real deficiencies when it comes to predicting what profits are likely to be if increased production is achieved, and a misleading picture may be obtained. The chart has deficiencies too in situations where the sales or turnover curve is based on the sales of a variety of products, each or some bearing different proportional costs. Whilst the proportions of each individual product remain much the same in the total sales, this presents no great problem. It is when the produce mix alters that some difficulty of portrayal and interpretation may well occur.

A break-even chart should be read and interpreted with these deficiencies in mind. It should not be seen as a precise, determinate answer to the future of a production policy but rather as a general guide to what might well happen in the light of facts known at the present time. Further, that the curves are not always straight lines, as illustrated, but frequently can be of such a shape that profit does not increase at a linear rate once the break-even point is reached and passed.

7.6 The costing process

The *costing* process consists of recording all those expenditures which are related to specific items of work in such a way that the cost of that work, in convenient form, can readily be identified. Costing, therefore, necessitates allocating cost data to specific headings within a predetermined classification or coding system. Costing is a generic term for a process which results in the establishment of the cost of carrying out parts of a larger project or the definition of those costs related to a particular item of production or over-head. It is also concerned with the comparison of these costs with a budget or a predetermined standard of cost.

The costing process provides a service to management. An efficient costing system is essential in order that production may proceed in a properly controlled environment. It must be understood that costing as such, is not a controlling function in itself but that it provides the data and then, after processing, the information on which control decisions can be made. Costing in most aspects of the production or manufacturing industries is a well-developed, well-established and sophisticated technique. The data to be collected are precise and, due to the predominance of repetitive operations in a controlled environment, the processes of collection and allocation become a routine and regular operation. In these industries the data collected allow a firm prediction to be made of likely future costs of processes not generally dissimilar from those giving rise to the cost data. In construction projects of a *one-off* nature, with their ever-changing environment, conditions are not generally conducive to the promotion and use of sophisticated management and control techniques (though this is rarely a legitimate reason for not applying the principles involved).

It is necessary to draw a distinction between *costing* and *accounting*. Accounting is a process which concerns the proper recording of financial information related to the transaction of a firm or business. The process includes book-keeping. The information recorded is of an historical nature and it is classified and summarized along well-tested and well-established lines. Accounting will record all the details of a firm's trading with its clients, sub-contractors, suppliers, associated companies and employees. It is a process necessary in law as well as for the convenience of the parties involved.

Those who may be interested in establishing the financial standing of a company, or in knowing the financial results of its past trading, include the shareholders (the owners), both current and prospective, the Registrar of Companies, the tax authorities and possibly prospective creditors. The accounts of a company therefore facilitate the preparation of a profit and loss statement for the firm as a whole. Normally however, they will not be oriented towards providing information to allow dynamic control at the project level of operations within a company. An accountant is largely concerned with ensuring that the assets of a firm are preserved, or that their value is adjusted to take account of the current economic and financial conditions.

The following example illustrates the different approaches of accounting and costing to the preparation of an interim profit and loss statement for one of a number of contracts operated by a construction company.

Example 7.3

The accounting approach gives relatively little information upon which management can take positive action. Accounting attitude:

Site Account for Contract B4297 – Dormer Road Flats

	£000		£000
Materials	24 000	Payment by client	
Wages	30 000	against valuation of	
Plant	20 000	work completed	89 000
Site on-costs	7 000		
Gross profit	8 000		
	£89 000		£89 000
Head office			
overheads	3 000		
Precast concrete			
department	1 000		
Net profit	4 000	Gross profit	8 000
	£8 000		£8 000

The contract is making a gross profit of £8m on a turnover of £89m (9 per cent). After making a contribution to the Head Office overheads of the firm and the precast concrete department, the net profit amounts to £4m or just over $4^1/_2$ per cent of the turnover (neglecting tax). This financial situation in itself is not unsatisfactory, but even so there is no indication as to what areas of the work are being carried out efficiently and at a profit, perhaps concealing poor performance elsewhere. There is no indication, other than one of an overall nature, that the cost standards for the work are being achieved.

The costing approach, set out in Table 7.3, classifies the cost incurred under appropriate headings and gives a considerable amount of information upon which management can take action. In this instance, piling and concreting are currently financially efficient operations and excavation appears to have been carried out at a loss on the basis of the information presented. If more excavation of a similar nature has to be undertaken then, clearly, the method and efficiency of men and plant, and ultimately perhaps the estimated cost, needs to be investigated in order to attempt to avert a continuing unprofitable situation. A further breakdown of the value of the completed work into materials, wages, plant and on-costs will enable comparison to be made in more detail and to isolate, within trade headings, which are the less efficient items. (One would need to

Table 7.3
Cost Account for Contract B4297 – Dormer Road Flats

	Excavation £000	Piling £000	Concreting £000	Total £000
Materials	—	10 000	14 000	24 000
Wages	6 000	5 000	19 000	30 000
Plant	12 000	6 000	2 000	20 000
Site on-costs	1 000	2 000	4 000	7 000
Total	19 000	23 000	39 000	81 000
Head office overheads	1 000	1 000	1 000	3 000
Precast concrete department	—	—	1 000	1 000
Total cost	20 000	24 000	41 000	85 000
Value of work done	18 000	25 000	46 000	89 000
Profit	—	1 000 (4%)	5 000 (10·9%)	4 000 (4·5%)
Loss	(2 000)	—	—	—

know, in addition, whether the estimated cost of the work has been unbalanced in favour of piling and concreting as part of the tendering policy, thus concealing true expected performance standards.) Managers, therefore, need to take action in some areas of operation, in spite of the overall financial position being satisfactory – the action needs to be focused on one operation in this case and the action centres are highlighted by the costing report.

In summary, costing for a project needs to be carried out for the following purposes:

(a) To provide information that enables periodic statements of profit and loss to be drawn up for each project at regular intervals of time, probably at monthly intervals for large projects but almost certainly weekly intervals on smaller or fast-moving or more finely balanced projects. In turn, the project statements will be broken down for individual operations or sections of the work. The project statement of costs will itself facilitate the production of a composite statement showing the overall company position on all its projects.

(b) To highlight particular activities or operations within a project which are being pursued in an uneconomic fashion, and to allow reasons for being so to be assigned. For example, excessive wastage of materials, inefficient utilization of mechanical equipment or labour and the incorrect balancing of resources as a whole.

(c) To provide cost data on which management can act within a satisfactory timescale. In the case of activities which are currently being pursued uneconomically, costing information must be available at least within the duration of the activity. This enables a decision to be taken with the real prospect of an improvement being effected before the activity duration expires.

(d) To enable the staff who have estimated the cost of the work to gather feedback information (properly and adequately described) in order to check the efficiency of their original assumptions in preparing the estimate of cost. Also, to provide this information in such a way that it will facilitate the preparation of future estimates.

(e) To facilitate the preparation of budgets for the future operations of a company, particularly in respect of the contribution to be made to the company's overheads and its general administrative costs.

(f) To provide operational information on equipment and plant in order to assist in the economic selection of equipment for future work and the selection of plant for purchase.

(g) To provide the necessary information in order to guide a company into the areas of work which it is best equipped to carry out and which it finds most profitable. Hence it forms one part of the basis for a long-term corporate plan.

7.7 Contract costing

Contract costing is a system of costing which is used in industries where a business has a number of individual contracts which are undertaken to a customer's special requirements and will take some considerable time to complete. Building, civil engineering and shipbuilding are three such industries. A separate contract account is allocated to each contract and that account bears a contract number. Because of the detail and size of the individual contracts, it is usual to break the work down into a number of sub-accounts against which contract costs are allocated. Each contract thus has its own profit and loss account. Because the individual contracts tend to be long, it is necessary to assess the profit/loss which occurs on each as the work progresses for each accounting period. SSAP 9, previously referred to in Chapter 2, provides guidance on this subject:

> 'The profit, if any, taken up needs to reflect the proportion of the work carried out at the accounting date and to take into account any known inequalities of profitability in the various stages of a contract. Many businesses, however, carry out contracts where the outcome cannot reasonably be assessed before the conclusion of the contract and in such cases it is prudent not to take up any profit. Where the business carries out contracts and it is considered that their outcome can be assessed with reasonable certainty before their conclusion, then the attributable profit should be taken up, but the judgement involved should be exercised with prudence'.

Example 7.4

A building contractor currently has three contracts in operation, C123, C124 and C126. Data collected in the costing system during the last accounting year are summarized in Fig. 7.6.

The first section of the contract accounts is set out in Fig. 7.7 and is, in effect, a statement that compares what has been incurred by way of cost since the commencement of the contract and the value of the work returned to meet those costs. In the first half of this section are the costs which have been brought forward from a previous year(s) together with those costs that have been incurred in the year for which the cost accounts have been drawn up. Against these costs are listed and totalled several of the costs which are to be carried forward to the next year commencing 1 January 19x5. *Materials on site* is one example. *Cost of uncertified work* is not yet in the system though the work has been carried out and it will be paid for during 19x5. An important balance which comes out of the cost accounts for each of the three contracts is that of *Cost of work certified*. This is a vital component for the calculation of profit.

	Contract C123 £000	Contract C124 £000	Contract C126 £000
Contract price	578	946	323
Balances brought forward at 1 January 19x4			
Cost of work to date	450	320	—
Materials on site	10	16	—
Plant and equipment – written down values	24	98	—
Profit previously transferred to profit and loss account	3	—	—
Balances carried forward at 31 December 19x4			
Prepayments to sub-contractors	25	13	—
Materials on site	10	44	8
Plant and equipment – written-down values	50	134	—
Value of certified work at end of year	497	836	43
Cost of uncertified work at end of year	48	15	—
Costs recorded during year 19x4			
Materials delivered to site	28	270	—
Direct wages	31	300	42
Payments to sub-contractors	10	32	—
Overheads	20	80	8
Written-down value of plant sent to sites	40	63	—
Design consultants' charges	6	10	1

Fig. 7.6 Costing data.

The reconciliation of *cost of work certified* with the *value of work certified* is shown in Fig. 7.8. Until the value of work has been certified there is no agreement as to the quantities of work undertaken for payment purposes and therefore the profit which has been made cannot be confirmed at this stage. Some of the figures for profit/loss that appear in Fig. 7.8 are the result of a calculation shown in Fig. 7.9. The calculation of Fig. 7.9 requires an evaluation of the *cost of work completed but not yet certified* and also an estimate of the *cost of work to completion*. Having arrived at these figures and assuming that the contract price has not altered, the profit/loss can be established (for the time being in the case of contracts not yet completed).

SSAP 9 makes recommendations and sets out principles for the amount of profit/loss which should be taken in the year under consideration. In many companies a very conservative approach to taking profits on contracts which are still in progress has been adopted. Often no profit has been declared in the profit and loss account until a project has been completed. This means that only those contracts which are completed in a year for which financial

	Contract C123 £000	Contract C124 £000	Contract C126 £000
Balances at 31 December 19x4			
as a result of year's operations			
Cost of contract to date	450	320	—
Materials on site B/F	10	16	—
Plant on site B/F	24	98	—
Materials delivered to site	28	270	—
Direct wages paid	31	300	42
Sub-contractors	10	32	—
Written-down value of plant issued	40	63	—
Overheads	20	80	8
Design consultants' charges	6	10	1
	619	1189	51
Balances carried forward			
Materials on site C/F	10	44	8
Plant on site C/F	50	134	—
Prepayments C/F	25	13	—
Cost of work not yet certified C/F	48	15	—
Cost of work certified	486	983	43
	619	1189	51

Fig. 7.7   Year end cost balances and balances carried forward.

	Contract C123 £000	Contract C124 £000	Contract C126 £000
Cost of work certified B/F	486	983	43
Profit taken this period	8	—	—
Profit taken previously	3	—	—
Profit not taken	—	—	—
	497	983	43
Value of work certified	497	836	43
Loss taken	—	147	—
	497	983	43

Fig. 7.8   Reconciliation of costs and values.

	Contract C123 £000	Contract C124 £000	Contract C126 £000
Cost of work certified	486	983	43
Cost of work not yet certified	48	15	—
Estimated cost to completion	33	95	280
	567	1093	323
Contract price	578	946	323
Anticipated profit/(loss)	11	(147)	0

Fig. 7.9 Profit calculation.

accounts are being prepared feature in the balance sheet and the profit and loss account. Such an attitude may well preclude a fair and reasonable view of the company's activities being presented in the accounts. SSAP 9 recommends the application of the *prudence concept* where losses are expected to be incurred on projects. This results in provision being made for the loss as soon as it is seen that such a loss will be incurred. In Example 7.4, therefore, the anticipated loss of £147 000 on Contract C124 is written off immediately.

SSAP 9 contains recommendations as to how the anticipated profit for work still in progress should be treated. Profit that is anticipated and included in the value of work in progress, and can be anticipated with *reasonable certainty*, can be taken in proportion to the total value of work certified to date against the total contract price. For profit to be anticipated with reasonable certainty, a contract must be nearing completion. Also due allowance must be made for costs arising out of the possibility of rectification of unsatisfactory work at a later stage together with any costs that may arise in the defects liability period. In Example 7.4, Contract C126 is only approximately 15 per cent completed and, therefore, it is too early in the contract to view *profit* as having reasonable certainty. The contract should be something like one third complete before profit is taken and even then only if it is reasonably certain. As far as Contract C123 is concerned, this is 86 per cent completed based on the value of work certified and the contract price and, therefore, some profit can be seen with reasonable certainty. In this case the whole of the anticipated profit has been taken in view of the conservative way in which it has been estimated. Where the completion of the work is less than the 86 per cent for Contract C123, but more than about one third, it would be prudent to take no more than two thirds of the anticipated profit.

	Contract C123 £000	Contract C124 £000	Contract C126 £000
Cost of work completed to date	534	998	43
Profit/(loss) taken to date	11	(147)	0
	523	851	43
Progress payments received	447	752	39
Work in progress	76	99	4

Fig. 7.10 Work in progress calculation.

The costing exercise of Example 7.4 also illustrates the way in which data are generated to transfer into the balance sheet. Work in progress can be calculated as in Fig. 7.10.

The cost of work completed is the sum of the costs for certified and not yet certified work. To this is added the total profit which has been taken or, if a loss is incurred, as in the case of Contract C124, it is subtracted. From the sum so obtained must be subtracted the payment that has been received so far. Such payments are usually subject to the terms of the contract agreed at the outset and it is usual for between 3 and 10 per cent of each payment to be deducted and retained until a later date. For example, half of the accumulated retention may be released on completion of the work and the balance at the end of the maintenance period. However, for the purposes of illustration, in Fig. 7.10 the progress payments received have been arrived at by deducting a uniform 5 per cent from the value of work certified in each case.

Other items that will be required for the make-up of the balance sheet are given in Fig. 7.11.

	Contract C123 £000	Contract C124 £000	Contract C126 £000
Cost of work not certified B/F	48	15	0
Materials on site B/F	10	44	8
Plant on site B/F	50	134	—
Prepayment B/F	25	13	—

Fig. 7.11 Balance sheet items from cost accounts.

7.8 The need for cost control

One of the principal objectives of most organizations is that of achievement at minimum cost. This contributes to the one essential, though not exclusively so, of all business enterprises, that of profit. Profit in simple terms is what is left after all the income accruing to the enterprise has been received and all of the costs or outgoings have been paid. For all construction projects, costs must be monitored and controlled, whether from the point of view of owner, designer or a contractor. Some of the techniques for so doing can be used by each party equally well, whilst others, because of the different nature of the business, cannot readily have universal use. When it comes to a construction contractor, for example, due to the nature of some of the contractual aspects and standard practices in the civil engineering and building industries, a system is required which adequately reflects those practices. Such a system is not readily applied within other industries without major modifications being made to it. On the other hand, an understanding of the problems which occur in each phase of industry enables better co-ordination of the activities of the many participants in a single project. Cost control, whatever the industry, may be defined as the regulation, by executive action, of the cost of carrying out the various activities which go to make up a project or a contract.

It is necessary at all stages of activity with which costs are associated to distinguish between cost control and analysis, and cost optimization. The former has been defined above and cost analysis and costing have already been discussed. They enable the detection of favourable and unfavourable variances against standards which are set for the measurement of performance and expenditure. Cost optimization, on the other hand, has an entirely different purpose. It is concerned with the establishment, usually by analysis, of the methods, resources and the programme of work which should be adopted so that a project will be carried out at minimum total cost. It can involve many tedious and costly calculations, the pay-off from which is sometimes doubtful or which cannot be established with accuracy. As such, it will be appreciated that cost optimization, or time-cost trade-off, by which name it is alternatively known, is a tool concerned largely with planning, since it is desirable that it be used prior to the commencement of a project.

There are some significant problems which arise when considering the particular form of project cost control to adopt. Perhaps the most significant of these concerns the cost of installing and operating the system itself. The cost of providing a means of collecting basic cost data, usually in the form of the hours worked by individual men on a daily basis, can be considerable, and this is only the first of a series of costly steps. The hours which have been collected must then be allocated, perhaps broken down into smaller groups as a result of investigation, converted into money terms, compared with the established standards, and then sundry reports to meet the needs of the management must be prepared. The resulting demands which are placed on individuals concerned with supervision or management at various stages in the process are high, and frequently – there is no better example than in the construction industry – little attempt is made to train those who are involved in the process as to how best to conduct their business affairs and meet the requirements of the system. Often there is a premise that the collection and analysis of cost or its associated data is not the function of a technologist. Where technologists fill the principal management roles, there is no outlook more certain to undermine a cost system from its very inception.

One further area which requires attention is that concerned with motivation. This is necessarily linked to the foregoing point regarding managerial support. The collection of the data at their source must be accurately carried out if a meaningful control is to be set up. Accuracy at this stage depends largely on the man who is actually making the collection being interested in his job and being motivated by the need to produce a useful end result.

In designing and using a cost control system, it is as well to keep in mind the main purposes which such a system serves. These are:

(a) to provide immediate warning of uneconomic operations, both in the long and the short term;
(b) to provide the relevant feedback, carefully qualified in detail by all the conditions under which the work has been carried out, to the estimator who is responsible for establishing the standards both historical and future;
(c) to provide data to assist in the valuation of those variations that will arise during the course of the work;
(d) to promote cost consciousness;
(e) to summarize progress.

It is as well to keep in mind, too, the extent to which the initial contractual commitment of the parties concerned in a project, and the professional role which they are playing, determine the means and the detail of control which is necessary and will be most effective. An engineer or an architect who is responsible for a project is interested in cost from a feedback point of view and not necessarily from the point of view of detailed day-to-day control of resources. The majority of the cost data in which he is interested will be established prior to the commencement of the work and at the contract negotiation stage. The remainder will be available after the work is completed. A client clearly needs to have not only the value of completed work to a contractor, which he has to pay as the work proceeds, but also the probable future costs and their phasing. The owner's control system is likely to be designed so that it produces information for this purpose. The client exerts considerable influence over his ultimate degree of control by the form of contractual relationships and organization that he chooses at the outset of the project. For example, if he elects to have the work done on a *cost plus a fixed fee* basis, then he has to be aware of, and to approve, through his representatives, every single item of expenditure for which he has to reimburse cost. His knowledge and control must therefore be very detailed, not only as far as the costs of the work are concerned, but also with regard to the exact detail and scope of the work. In this case he has freedom to vary the quantity of work carried out as it proceeds. On the other hand, if a client chooses to agree a fixed sum for a well-defined unit of construction, then he can justifiably neglect the detailed control of the cost of the work, simply paying attention to cash flow. The third party who is involved in a project, a contractor in one form or another, needs to control the cost of the work in some detail, since it is he who has a precise standard towards which he must aim if he is to operate with a surplus. It is he and he alone who can control his own resources in detail from day to day so as to meet these standards.

7.9 Budgets and budgetary control

In order for a firm to operate its business in a satisfactory manner, it is necessary for its senior management to establish objectives, towards which the endeavours of the firm's staff can be directed. The most important and, to many, the most meaningful objective, will be that concerned with the amount of profit to be made. This may be stated in absolute terms such as a straight sum of money either before or after tax, or as a ratio such as the percentage return of profits on the capital employed in the business. In addition, there will almost certainly be other objectives, probably concerned with some or all of the geographical and product areas in which the firm will trade, its relationship with its employees, the location of its production centres, the service it will provide to its customers, its general relationship with the public and many other relevant topics.

Where possible, it is highly desirable that such objectives as these should be set in quantitative terms and that they should be of such magnitude that their achievement requires real effort and some extension of capability on the part of the company's staff. On the other hand, the targets which are set by way of objectives must be realistic in the light of the overall resources which are to be made available to the company's management. The set objectives will be meaningless unless they are related to a specific period of time for achievement. Having set the objectives, plans can be prepared for the course of action to be taken in order to achieve them. This planning needs to be expressed in such a form that it can be used as a means of co-ordinating the many activities which necessarily must take place as a result. A plan expressed in quantitative (usually financial) terms to achieve a given objective(s) in a specific time is known as a *budget*. Like all plans of this nature, it must be drawn up prior to the action taking place and in sufficient time to enable full discussion and adjustment, if necessary, by those people who are to use it.

A budget provides one means by which control of an undertaking can be effected whether at the company, departmental, factory or project level. It is a device through which planning and control can be integrated. One of the more important aspects of a budget is that it defines the objectives to be achieved by all levels of management and presents those managers with the limits of restraint within which they are required to work. From senior management's point of view, it enables the delegation of authority within specific and quantitative terms. A comparison of what is actually achieved in practice against the budget figures enables action to be taken, either to adjust the emphasis in resource employment in order to achieve the original objectives, or to revise the policy originally established to a more realistic end in the light of what has been experienced so far during the budget period. A budget, therefore, must be prepared with the co-operation and participation of a manager to whom the authority is to be delegated. He must at all times fully understand the implications of the plan which is being set out and to which he is expected to work.

A commonly accepted and widely used definition of budgetary control is provided by the then Institute of Cost and Works Accountants:

> *'The establishment of departmental budgets relating the responsibilities of executives to the requirements of a policy, and the continuous comparison of actual with budgeted results, either to secure by individual action the objective of that policy or to provide a basis for its revision.'*

This definition assists in defining the requirements of a budget to be determined and the ways in which it can be used in order to facilitate control of a business, a department, a section within a business or a one-off project undertaken by a client or contractor.

The aims which a system of budgetary control should attempt to achieve are as follows:

(a) to set out, in quantitative terms, the objectives of a firm, a department, a project, or some other sub division of a business;
(b) to enable comparison to be made between alternative plans for achieving differing objectives, the subsequent analysis resulting in the choice of the most expedient and satisfactory plan making the best use of the resources which are available;
(c) to facilitate the co-ordination of the activities of a number of different sections making up the whole of an organization, in order to make the most efficient use of limited resources, for example working capital. Subsequently, to allow measurement of the achievement against the standard set;

(d) to provide a guide as to the level of achievement which is feasible for the organization when actual performance is measured against the targets set;

(e) to provide a plan against which the effects of action to correct adverse trends, or to take advantage of beneficial ones, can be measured both in performance and cost.

It will be appreciated that from these aims of a budget, budgetary control cannot be effective in an organization unless:

(a) senior management has a clear and precise idea of what a company's objectives are and it is strongly committed to achieving them;

(b) the organization structure and responsibilities for each section of the organization are clearly defined;

(c) a continuing record of performance and achievement is maintained and is available for the preparation of future budgets;

(d) a clear statement of the total resources which will be committed to a budget area is available;

(e) budgets are treated as flexible rather than fixed plans, capable of adjustment when the appropriate conditions warrant such action;

(f) managers and supporting staff are educated to use budgets properly and effectively, and have a hand in their preparation.

Before the detailed preparation of a budget can be commenced it has to be decided for what period the budget will be drawn up. It is not possible to lay down hard and fast rules as to the best or proper budget period, since this will vary considerably with the nature of the business being undertaken and the subject of the budget. In capital intensive industries where large investments are made, in themselves perhaps taking many years to fulfil, and the benefits from which take a large number of years to mature, a budget period of 25 years would not be too long. For example, in the electricity generation industry, the investment in a power station is considerable and when complete its working life probably continues for at least 30–50 years afterwards. Budgeting and planning in a context such as this need to be very long term. Such budgets do not need updating at monthly intervals, but rather on an annual basis, because of the coarseness of control which can be exercised. On the other hand, a project investment of, say six months' duration needs a much closer and finer control and the comparison of achievement with the standard may need to be on at least a weekly basis in most cases.

In many businesses, on a company basis, there is a tendency to accept one year as a basic period for detailed budget purposes. This is largely due to the fact that company accounts are normally published on an annual basis and forward planning therefore tends to follow this period. The selection of such a period does not preclude the use of longer term budgets over a period such as five years, or shorter term budgets on, say, a quarterly basis, at different levels of operation. Budgets having different purposes may well have different budget periods. For example, the capital expenditure budget for a company will necessarily be long-term with a probable minimum period of three years. A sales budget may be set for a period of six months. When considering a budget as a means of project control, then the budget period will need to coincide with the duration of the project. If this exceeds twelve months, then consideration needs to be given to the use of short-term budgets within that period. The usual period for the review of a budget, which is used to control a project, is one month, though clearly this is also subject to change as the extent of control demands.

7.10 Fixed and flexible budgets

It has already been shown that what are often assumed to be fixed costs can frequently be divided into fixed and semi-variable costs. In like manner, the fixed costs or overheads that an engineer or a contractor incurs in operating a construction site may have a semi-variable element in addition to the fixed element. For example, the number of supervisory staff which are needed to ensure effective and proper site working will almost certainly depend on the rate at which the site construction work is progressing. If the level of activity is high, then a larger number will be required than if the level is somewhat lower. The rate of production, therefore, can influence the total cost of achievement in a particular cost centre because of this effect of overheads. If an actual cost for, say, overheads for supervisory staff is collected, it must be related to an actual rate of production. There is a danger, however, if care is not taken, that it may be compared with a budgeted cost which was originally planned for a different rate of production. It is very necessary to be aware of the effects of such changes on the overall outcome of a project.

A *fixed* budget is one that takes no account of varying rates of production; a *flexible* one is such that certain elements of the budget are established for different levels and/or rates of production. A flexible budget can be assumed to be fixed, but only for the level or rate of production for which it is calculated. An example will illustrate the main features of the two types of budget.

Example 7.5

This example is a very much simplified case for demonstration purposes. Taking the case of a single fixed budget first, the 100 per cent capacity column of Table 7.4 provides the budget for a selection of the overheads for a site contract having, at this stage, a budgeted turnover of £100 000 per month. The individual amounts in this column are the standards against which actual costs are to be measured. If there is only this single budget, any

Table 7.4

	Flexible budget				
Monthly turnover (£)	80 000	90 000	100 000	110 000	120 000
Output (%)	80	90	100	110	120
Costs (£):					
Supervisory staff	8 000	8 000	10 000	10 000	12 000
Office staff	2 000	2 000	2 000	2 800	2 800
Office accommodation	500	500	500	500	500
Power and light	150	180	200	240	280
Printing, etc.	100	120	150	180	220
Totals	10 750	10 800	12 850	13 720	15 800

variation in turnover will then necessitate a *pro rata* adjustment of it to give revised and more realistic standards. For example, given only a single fixed budget at £100 000 turnover, the total overhead for a monthly turnover of £110 000 calculated from this will be

$$\frac{(12\,850)}{(100)} \times 110 = \underline{£14\,135}$$

It can be seen that this is not realistic since by setting out a flexible budget in detail and considering the effect of increased turnover on each individual item, the answer is shown to be £13 720. The flexible budget recognizes the fact that certain overheads will rise by fairly large discrete steps, so reflecting the addition, for instance, of discrete numbers of staff, rather than by a linear increase on a uniform *pro rata* basis over a period of time.

In the example, if the total actual costs recorded against the items shown amounted to £14 000, using the fixed budget calculated above as the basis for calculation, the variation appears to be favourable to the extent of £135. Using the flexible budget, the variation will be unfavourable or adverse to the extent of £280. A fixed budget is therefore of limited value only, and where changes in the rate of output are expected, a flexible budget taking account of fixed, semi-variable and variable costs should be prepared over the envisaged range of outputs.

7.11 Cost coding systems

The success of a cost control system will depend to a large extent on an ability to develop a sound system of identification coding for the basic cost data.

This is particularly true of a system which is to be computer based because of the need for rapid interchange and handling of data and information from one point of the system to another. An adequate coding system will simplify the data-handling facility and also provide economy of storage in the case of a computer. A good and adequate system of coding is one which simplifies the task of referring to the items to be coded. For example, brevity of description frequently provides the need for coding, but clearly any code reference for a subject which is longer than the original description, or more difficult to file in a systematic fashion, cannot reasonably meet the requirements of a coding system. Two reasons for coding have already been referred to – brevity and ease of systematic filing. A third reason of importance is the necessity to provide, under a relatively simple code reference, a wide range of characteristics or attributes which can be used to define the subject to be coded. A four-digit code name, for example, enables four levels of the characteristic definition of the subject to be represented.

All systems of coding should be designed so as to provide adequate hospitality. *Hospitality* is the facility which defines the extent to which future additions can be made to the list of items to be coded without the need to alter or adjust the basic structure of the coding system. Such hospitality needs to be available at the appropriate level or position in the coding system, that is at each digit position, rather than at a single position in the total range of hospitality of the system.

The coding process consists of assigning a symbol or a group of symbols to each item in a list of items, so that any item being coded can be identified conclusively from all the other items which appear on the list. Perhaps the list of items to be coded has been put together in a completely random fashion, or alternatively it may be classified so that the order of the items in themselves is meaningful. The coding system to be adopted may follow either of these two patterns and may, in itself, render a form of classification to the items. The generally accepted business coding systems use either alphabetical or numerical symbols, or a combination of both, for coding purposes. Of these alternatives there is a preference for purely numerical systems.

The use of purely numerical symbols for code numbers has the limitation that there are only 10 symbols, 0 to 9. This means that only ten variations are available at each position in the code number. On the other hand, numbers are better understood and are more meaningful when it comes to rating precedence and order. The evidence of the facility of numbers as a code symbol is provided in everyday life by such codings as Boeing 767, 31 on the wine list, No. 9 iron, etc. Because of the ten numerals that are available, a two digit code name can accommodate 100 items, three digits can accommodate 1000 items, four digits, 10 000 items and so on.

Purely alphabetic systems of coding have the distinct advantage that there are 26 symbols and hence there are that number of choices for any one digit position in a code word. In practice, not all of these symbols can be used, since it is usual to omit the letter *I* because of the danger of confusing it with 1, the letter *O* because of the numeral zero, *Z* because of the numeral 2 and *J* and *Q* for similar reasons. This leaves 21 letters available for use and a two-digit code name can therefore encompass 441 items, a three-digit code name can encompass 9261 items and so on. Another advantage of using alphabetic systems rests on the fact that it is often possible to represent a characteristic or an attribute at the initial level of coding by the initial letter of that attribute. An example in construction work where *E* can be used to represent *excavation*, *C* for *concreting*, etc. Sooner or later, however, the need arises to code two items, both having the same initial letter, for example, *pumping* and *painting*, and an exception must then be made to the system. Once this is done, the logic of the coding is destroyed and the value of the symbol in this respect is greatly reduced. Disadvantages are that alphabetic symbols suffer from the fact that there tend to be more errors of transcription when using them and they are certainly more difficult to speak and write when combined to form a long code word which is not easily pronounceable as a whole.

A combination of numerals and alphabetic symbols in code words is not of great value in a business coding system and generally suffers from most of the combined disadvantages outlined above, whilst having few of the advantages. There are common instances of these combinations in current use. The system for allocating registration numbers for motor vehicles in the United Kingdom is an obvious example, although the system is not universally adopted in other countries. Where it is found either desirable or necessary to combine the two types of symbol, it is found, as in the case of the motor car registration number, that each type of symbol is better grouped together, rather than allowing a free mixture of the two. The problem of differentiating between numbers and letters is less acute in this instance.

Having made a decision about the type of symbol that will be used to represent the attributes of the subjects to be coded, it is then necessary to establish the rules by which the code words will be constructed, and what the meaning of each digit in the word will be. In general, the shortest possible code word should be the aim, without skimping the hospitality or the sensitivity of the code system within the requirements of the storage or retrieval of the information and data which are to be coded. Brevity of code name assists in eliminating errors of transcription and it will also assist in making the code names for items in common use more easily remembered. As to the length of code word, on balance it is more useful to have all the words of uniform length as a precaution against losing some of them in transcription or recording without a ready checking aid. This is particularly true for work processed by a computer, where automatic checking systems need to be provided in order to ensure, as far as is possible, that no errors have been made in the input data to the system. Another quality of a code name is that it should, if possible, provide the means of sorting and selecting information under all types of classification headings. Using a code for construction work as a further example, it may be required to pull out all the information relating to *excavation* or *bricklaying* for a particular contract; it may be desirable to review all costs of labour for one week, all the costs incurred in batching concrete or all the costs related to a particular excavation. These aspects can be covered by ensuring the facility of sorting one or other digit or groups of digits within the code name.

Having discussed the general requirements for a suitable coding and the way in which the code names can be built up, it is then necessary to decide on a method of allocating the code words. One simple method is to assign code names to the items to be coded in a completely random fashion. For example, a list of the items is drawn up and, if a numerical code is to be used, then numbers are assigned to each of the items in turn. As a new item is added to the list so the next number in sequence is assigned to it. Such a method of applying code names simply provides the means of building up a register of items in numerical sequence. The code name is not in itself significant and means nothing to a user by way of describing the attributes of the item until the master register is consulted, unless it happens to belong to an item in common use. This is a popular form of coding because, very often, the future requirement of the coding system is not known, nor can it be known, at the time that it is created. It is simply allowed to grow with the list of items. It is not particularly demanding on the part of the originator and can claim to have infinite hospitality.

Another mode of constructing code names is that which incorporates some form of mnemonic. A mnemonic code name is one which has an inbuilt reminder of the meaning or significance of the code word without reference to a register or index explaining each detail. A suitable code for a large number of items is exceedingly difficult, probably impossible, to construct with mnemonic qualities. Examples of mnemonics are common in computer programs written in high-level languages such as FORTRAN, where a short word in alphabetic symbols is constructed to represent a variable. The variable will adopt a wide range of values as the result of input to the program of calculations which are carried out when the program is processed. For example, a code word such as LCOST might be used as a variable to represent a series of values of the labour cost for a particular operation. The labour cost may be derived as a result of a calculation involving hours allocated to that operation and further input data in the form of the wages that have been distributed to the labour involved. The scope of mnemonic codes is very limited and in general they must be confined to very select and small lists of items.

The third and probably the most important and useful method of coding, as far as accounting and costing procedures are concerned, is a form of code system based on a standard classification. The basis of such coding is to divide the list of items to be coded into classes at various levels of detail. Firstly, the items are divided into broad and general categories, bearing in mind that if an all-numeric code is being used there cannot be more than ten of these at each level, and then the first level is further broken down in accordance with some attribute of the items. The process of breaking down the previous level continues until the coding system enables a comprehensive description of each item to be obtained in the degree of detail that is required. This form of coding for a large number of items has several advantages. It is a system of coding which quickly lends itself to memorization at the different levels of detail. In the first place, the more general levels of the system are remembered, gradually spreading to the levels where the attributes represent more and more detail of the items concerned. There is rarely any difficulty in coding new additions to the list of items to be coded if the structure of the code has been well thought out and adequate hospitality has been provided in the first instance. Thirdly, it is a systematic method which is based on a form of logic and it will generally facilitate the smooth operation of the business in which it is applied. It will also tend to highlight anomalies in the business systems currently in use in that environment, because of the questioning and searching attitude that will be applied during the coding operations.

An example of a code which can be used for construction work in the field is illustrated by Figs 7.12 and 7.13. This is an example of a coding system which is based on systematic classification and an all-numeric form of code. Figure 7.12 shows the overall structure of the code in the form of a family tree. Level 1 of the code represents the contract, level 2 represents the general type of trade that is being undertaken, level 3 is a breakdown into structural elements within the trade shown in level 2, and level 4 is a breakdown into the resource on which the cost was incurred. It will be seen that the level of breakdown could well be taken into as much detail as is required, bearing in mind that there is little point in having a coding system which breaks the work down into greater detail than that in which the data, cost in this example, can be collected prior to coding. For example, there is little point in

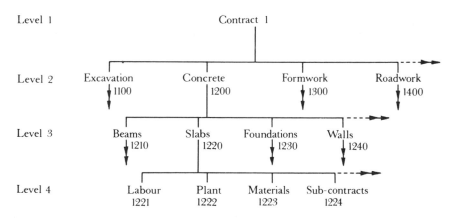

Fig. 7.12 Cost code breakdown.

Second digit

		1	2	3	4	5	6	
Excavation	1	Bulk	Topsoil		Pit	Base–ment		
Concrete	2	Beams	Slabs	Found–ations	Walls	Columns		
Formwork	3		Vertical		Sus–pended			
Roadwork	4		Base	Sub–base				
Bricklayer	5		Facing	9″ Solid	11″ Cavity			

First digit

Fig. 7.13 Cost code matrix.

being able to code, for a cost control system, the cost of labour in placing concrete in a combined suspended beam and floor slab, if it is not possible to collect the times spent by the concrete-placing gang on each of those activities separately. If the beam and the slab are integral, and the concrete is placed in one operation, the time of the concrete-placing gang can be sensibly allocated only to the whole operation. It then becomes very much a matter of guess-work as to the proportions of its total time that the gang spends on each part of the work.

In Fig. 7.12 the tree has been extended through concrete at level 2, slabs at level 3 and then on to level 4 which, in this case, will apply to every one of the items at level 3. To complete the tree it is necessary to extend each operation at level 2 in a similar fashion. The nature of the classification used for this example is purely random and its only function is to illustrate the general principles involved. Figure 7.13 is an example based on the same classifica-tion, but showing a very much simpler and more convenient means of portraying the detail of the various levels. It is a means of building up the cost code numbers and it applies to two digits at a time. For example, 22 represents concrete in slabs and 43 represents the sub-base to a road. The numbers are labelled as first and second digits, but this method can be applied to digits anywhere in the code number. If the code numbering system is broken down to another two levels of detail, then an additional matrix can be built up for every entry in the one that is shown in Fig. 7.13. It will be noted that spare spaces for digits are left in Fig. 7.13 to provide the necessary hospitality for the code.

The coding system illustrated above is based on the operational nature of activities for a contract and the lowest level represents the resources that are

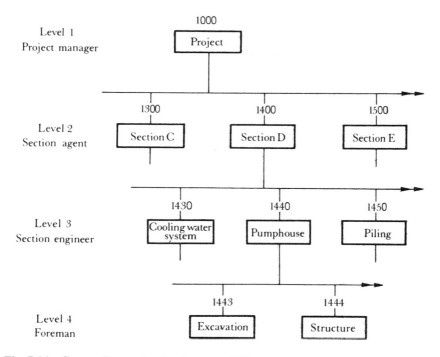

Fig. 7.14 Cost coding on levels of responsibility.

used. Cost codes can be based on other classifications, and Fig. 7.14 illustrates one that is based on the levels of responsibility for the management of a construction contract. Such a code system enables the reporting of a project to be varied in the amount of detail required by managers at different levels of the organization and to some extent it delineates the responsibilities of each manager at those levels.

Problems

7.1 A chemical processing plant produces a compound which, at full production, can be sold for a total sum of £1 200 000 per year. The fixed overhead for the plant amounts to £450 000 per year and the total variable costs when at full production are £650 000 per year. By making some minor modifications to the plant the production can be increased by 25 per cent with only a 10 per cent increase in fixed overhead.

Plot both situations on a break-even chart and determine to what extent the enlarged plant needs to be utilized in order to achieve a similar profit to that achieved before at 100 per cent production.

7.2 A precast concrete manufacturing company produces 10 000 beams per year when working at 50 per cent of the factory's capacity. Each beam is sold for £15. The costs involved in manufacture at this rate of production are £30 000 for materials, £60 000 for wages, £30 000 on fixed overhead and £12 000 on variable overhead. Variable overhead is constantly proportionate to the number of beams produced.

A contractor places an order with the company for 10 000 additional beams at the reduced price of £11.75 each. Due to the additional quantity of materials that are required, a 2¹/₂ per cent discount on previous prices is obtained but only for the additional work. On the other hand, the additional labour that is required has to be drawn from a reservoir of untrained labour and the overall efficiency of labour in production throughout the factory falls by 5 per cent on all production.

Illustrate the variation in profit between the prior and the new situation.

7.3 An analysis of the trading results of a small works department for two consecutive years is tabulated below:

		Year 1 £	Year 2 £
Wages		66 000	73 000
Materials		110 000	146 000
Overhead	variable	12 000	17 000
	fixed	20 000	27 000
Profits		10 000	15 000
		£218 000	£278 000

Explain how the changes in the material price and wage rates have influenced the profit figures in view of the fact that these have been increased by 8 per cent and 10 per cent, respectively. Sales prices have been increased by 10 per cent.

7.4 Find the lines of regression and the coefficient of correlation for the following data:

x	2	3	4	5	6	7
y	2	3	5	7	8	10

7.5 Find the lines of regression and the coefficient of correlation for the following data:

x	0	1	2	3	4	5	6	7
y	3	4	4	5	6	8	9	11

7.6 The following table lists the number of private houses which have been completed in a certain area over a period of ten years. Also in the table are listed the additional number of families owning two motor cars over the same period. Is there any correlation between the two sets of statistics? Comment on your answer.

Year	Houses completed (000's)	2 car families (00's)
1	3	2
2	4	5
3	5	7
4	7	8
5	7	10
6	8	13
7	8	15
8	9	18
9	10	16
10	12	19

7.7 A small product used in the plumbing systems of houses has a break-even point with a sales income of £120 000. The normal sales income is £180 000 with fixed costs of £100 000. A new product of better design involves additional costs of £20 000 which results in sales income increased to £240 000 with the profit/volume ratio improved to 20 per cent. What net profit does the improved design yield?

7.8 A building contractor currently has two contracts in progress, A23 and D14. The costing system shows the following figures for the last accounting year:

	Contract A23 £000	Contract D14 £000
Contract price	215	423
Balances brought forward *at 1 January 19x3*		
Cost of work to date	50	20
Materials on site	3	2
Plant and equipment – written-down values	12	15
Profit previously transferred to profit and loss account	1	—
Balances carried forward *at 31 December 19x3*		
Prepayments to sub-contractors	2	3
Materials on site	4	1
Plant and equipment – written-down values	14	56
Value of certified work at end of year	123	249
Cost of uncertified work at end of year	15	21
Costs recorded during year 19x3		
Materials delivered to site	19	70
Direct wages	31	130
Payments to sub-contractors	35	40
Overheads	13	24
Written-down value of plant sent to sites	18	63

By preparing a cost account for each of the two contracts for the year 19x3 establish the cost of the work carried out during that year, recommend how much profit/loss should be taken/written off for the year and write a brief account of the financial state of each contract.

Chapter 8
The Owner

8.1 Feasibility studies

When a private, industrial or commercial undertaking conceives a plan for an addition to the capital assets of its organization, the undertaking is embarking on a project which should, of necessity, pass through a number of well-defined stages. While the stages may be well-defined, the emphasis on each may differ from project to project as each project almost invariably differs from the last. The four stages of a construction project are those of:

(a) Feasibility and design;
(b) Purchasing/ordering;
(c) Construction;
(d) Commissioning and operating.

Each of the four stages is not necessarily separate and distinct; there will be merging and overlapping. The total process can be brought to a halt at the completion of any one of the stages or part-way through it.

When embarking on a project, other than one of insignificant cost in relation to the scale of operations of the undertaking, the common practice is for the owner to appoint a project manager to co-ordinate the activities of all those individuals and organizations which will be required to make a contribution to the work and to bring it to a successful conclusion. The project manager will normally be responsible to the client who is making the investment and will look after his interests. He may be drawn from the owner's staff, and be supported by additional resources from the owner's organization, or the owner may appoint an external firm or individual to provide a project management service. Whatever method is used to provide the project management service, it is essential, for the best chance of success, that the chosen project manager should have an intimate knowledge of the abilities and facilities of all the participants and of modern methods of management and control. The possible alternative organizational and contractual details of an owner's relationship with the other parties to the project are many and varied and will require thorough investigation before work on the project commences.

Before the cost of a project is sanctioned, a general appraisal of its merits within the overall structure of the owner's undertaking will be needed. The proposed project may, at this stage, be simply a promising idea that a particular avenue for investment will prove profitable. It will almost certainly be competing with other promising ideas and more routine but necessary investments and hence it will be competing with other projects for what is always a limited amount of capital for investment. It will be necessary, therefore, to develop the basis of the idea to a stage where preliminary estimates of total cost can be made and the proposed project can be thoroughly examined for both technical and commercial viability. During the whole of this process it must be remembered that developing ideas by means of a feasibility study can be an expensive procedure demanding the allocation of high-priced resources, the latter perhaps temporarily withdrawn from other important activities. As a result of the feasibility studies, some proposed projects will have their early promise confirmed and there will be little contention about their acceptance. In any one group of proposals, however, there are likely to be a few of approximately equal merit and distinction between them may be difficult.

One essential part of a feasibility study is an estimate of the capital costs for providing the facility, such working capital that is required to get it into use and the pattern of the various cash flows that will subsequently be associated with the asset while under construction and then when in use. Other aspects to be investigated will include the technological feasibility. It is at the stage when the feasibility study is completed that the owner reaches a point where an extremely important decision must be made, namely, whether to proceed with the project or not. During the feasibility study the project manager will seek the help of experienced professionals in the fields of architecture, building, civil engineering, quantity surveying and/or structural engineering as required, depending on the nature of the project and the design and construction processes involved. These professional staff may be engaged from outside the organization or they may already be appointed to the owner's staff.

From expertise within his own business the owner will probably already have an approximate idea of the range of costs that he can reasonably consider for the project and will have advised the study team of what he believes these should be. Studies will have been made, in parallel, of all the other factors to test the financial viability of the scheme and once the capital estimate is complete for the constructed facility the potential rate of return for the project can be assessed, using one or more of the various methods described in Chapters 3 and 4.

When the feasibility study is complete and the estimated total investment in the project is established, even if the rate of return is in excess of the company's cost of capital or the minimum rate of return it requires to achieve for other reasons, the decision to proceed or not may well depend on what other projects are available for investment and the state of their feasibility. Capital rationing has already been discussed in Chapter 3.

8.2 Project capital finance

The completion of the preliminary economic and financial analysis of an investment project marks the end of one stage in a sequence of examinations and judgements which will ultimately lead to the implementation or rejection of a project as a viable investment. As a result of an economic appraisal, a decision can be made, in the light of predetermined parameters, as to whether to proceed or not. Having decided that it is worthwhile to proceed, there is a need to examine in detail how the finance can be provided in order to be available at the appropriate time and in the most economic way. There is a need to examine in depth the timing of the cash flows (both in and out) that will result from the investment and to determine the optimum available form of finance that should be employed. If the money is to come from the internal sources of a company that wishes to make the investment, then it must necessarily result in a potential loss of earnings from its employment else-where. Alternatively the finance will have to be obtained from sources outside the company. In this case it can usually be identified more closely with a specific cost of doing so. The cost of finance is an important element in setting out a financial plan.

In establishing a financial plan for a project requiring capital investment, it is necessary to determine the sources of capital that are currently available to an investor. The sources used for this finance may include lenders of many descriptions, and shareholders. It is important to know what financial and legal responsibilities the user of capital has to its providers by way of interest and/or dividend payments and voting rights, together with the necessary and sometimes inevitable repayments of the capital sum. The latter must be defined both in quantitative terms and in their timing. By far the larger proportion of capital used to finance projects is borrowed in one form or another.

Borrowing can be of two kinds. Money is often borrowed by enterprises on a more or less permanent basis in the form of *debentures*. In other cases, money is borrowed on the basis of the repayment of a lump sum at a date fixed at the time of the loan, or by a system of phased repayments over a predetermined period.

In the case of a debenture issued by a company to raise funds, it is likely to be either a *mortgage* or a *floating* debenture, though these are not the only types of debenture available for this purpose. A mortgage debenture is one which is secured on specific assets, such as land or buildings, belonging to the company; a floating debenture is secured by no particular asset of the company but rather by all those assets not specifically assigned to mortgage debentures that have already been raised. Lenders who purchase debentures do not become owners-in-part of the business, as do shareholders. They merely lend money and in return receive a fixed annual rate of interest which is quite independent of the profitability of the borrower's activities. Debenture holders amongst the providers of capital do, however, retain a prior call on the company's assets in the event of business failure. If a company fails to pay the interest on debentures, then the debenture holders, by concerted action, have a right to put the company into liquidation, taking from the sale of the company's assets sufficient money to recover the principal of their loan plus any outstanding interest. Because of this, a company which is considering the issue of debentures must make sure that its assets will retain their value in the future, even if times become so hard that the company may fail. The company must also be reasonably confident that it will make a steady profit over the years, so that it can be sure of covering the debenture interest.

In the case of the other type of loan, there are a number of forms which it can take, depending mainly on the characteristics of repayment. Local government loans, for example, are required for many of the investments made by a local authority, and these loans are subjected to stringent controls. Since such investments on the part of the borrowing authority rarely give rise to a significant income, because on the whole they take the form of such projects as schools, sewage works, roads, etc., repayment must come from such sources as the local authority rates, council tax or government subsidies and yet further loans.

There are three principal methods of repayment of such loans and each of these will be illustrated by means of a simple example in order to contrast the implications in repayment terms of each.

The first method, and probably the most common, is that known as the *annuity method*. It derives its name from the fact that repayment of the loan and interest is made by instalments of equal value each year (or other agreed period). Each repayment instalment covers both repayment of capital and the interest on the outstanding loan. This type of situation is described in Chapter 4 when the use of the uniform series capital recovery factor was established.

The second method is to pay back each year equal instalments of capital together with the interest on the outstanding balance.

The third and final method is the situation in which a sinking fund is set up. In it is accumulated, throughout the period of the loan, sufficient moneys to repay the loan in a lump sum at the end of the loan period. The investment in the sinking fund is most frequently a constant annual amount which is related to the sum which is borrowed. To this has to be added the total interest on the whole of the loan which will usually be paid on an annual or other agreed basis. Investment of moneys placed in a sinking fund usually take place at low interest rates which in turn reflect the security requirement of such a fund.

Example 8.1

A loan of £10 000 is required for a period of seven years. It can be negotiated at an interest rate of 8 per cent per year. Compare three methods of repayment of the loan as described above, assuming that if a sinking fund is required it can be set up bearing an interest rate of 4 per cent on the invested moneys.

(*Note*: In Tables 8.1, 8.2 and 8.3 the rounding of numbers leads to some very small inaccuracies in totalling.)

(a) Method 1 – annuity method

Table 8.1

Year	Capital outstanding at start of year (£)	Interest for year (£)	Total funds outstanding at year end (£)	Annual repayment of capital + interest (£)	Repayment of capital only (£)
0	—	—	10 000	—	—
1	10 000	800	10 800	1 920	1 120
2	8 880	709	9 589	1 920	1 211
3	7 669	612	8 281	1 920	1 308
4	6 361	508	6 869	1 920	1 412
5	4 949	395	5 344	1 920	1 525
6	3 424	274	3 698	1 920	1 646
7	1 778	142	1 920	1 920	1 778
Totals		3440		13 440	10 000

(b) Method 2 – equal instalments of capital plus interest on the outstanding balance

Table 8.2

Year	Capital outstanding at start of year (£)	Interest for year (£)	Capital repayment (£)	Total annual repayment (£)
1	10 000	800	1 429	2 229
2	8 572	686	1 429	2 115
3	7 143	571	1 429	2 000
4	5 715	457	1 429	1 886
5	4 286	343	1 429	1 772
6	2 858	229	1 429	1 658
7	1 429	114	1 429	1 543
Totals		3200	10 003	13 203

(c) Method 3 – single repayment at end of loan period plus sinking fund

Table 8.3

Year	Sinking fund balance at year commencement (£)	Interest on fund balance (£)	Investment in sinking fund at year end (£)	Interest on loan (£)	Total annual outgoing (£)
1	—	—	1266	800	2 066
2	1 266	51	1266	800	2 066
3	2 583	103	1266	800	2 066
4	3 952	158	1266	800	2 066
5	5 376	215	1266	800	2 066
6	6 857	274	1266	800	2 066
7	8 397	337	1266	800	2 066
8	10 000	—	—	—	—
Totals		1138	8862	5600	14 462

In Method 3 the need is to calculate the constant sum of money which must be invested in a sinking fund at the end of each of seven years in order to accumulate, with the interest which it earns, £10 000.

The method of determining the amount has previously been discussed in Chapter 4 and use can be made of the uniform series sinking fund factor, Equation (4.2). In this case the annual amount to be invested

$$= 10\,000(A/F, 4\%, 7) = 10\,000 \times 0.1266 = £1266$$

In addition to the investment in the sinking fund, interest has to be paid annually on the total amount of the loan. This amounts to £800 per year. The tabulation of the details is set out in Table 8.3.

If each of the three cases is examined, it will be seen that, for the given figures, Method 2 results in the lowest *total* repayments over the loan period. The major characteristic of Method 1 is that the annual repayment is a constant sum made up of a steadily increasing capital repayment together with a steadily decreasing interest component as the capital outstanding decreases. With Method 2, the declining interest repayment as the capital outstanding decreases, combined with a constant capital repayment, leads to a steadily decreasing total annual payment. There is some virtue, from the point of view of budgeting, in having a constant repayment year by year. On the other hand, Method 1 does mean that the repayment is made later on the whole. This, of course, is reflected in the total paid for the loan because of the time-value effect of money.

This total cost of the loan using Method 3 is far in excess of the other two methods. This is due to the fact that the interest rate for the sinking fund is different from that for the loan itself. If the two rates are similar, then the repayment details will be the same as for the annuity method. With a constant annual outgoing of £2066 in order to finance a loan of £10 000, it will be noted that the uniform series capital recovery factor is

$$\frac{2066}{10\,000} = 0.2066$$

This represents a true rate of interest of slightly over 10 per cent.

8.3 Investment estimates

When an owner anticipates the investment of capital in constructed facilities, a series of estimates need to be prepared, each with its own particular purpose, so that well-defined stages of project development can be undertaken. It is the nature of estimates that inevitably they must be wrong. In spite of this an estimate of capital cost forms an essential basis for the management control process. There is a need, having established the plan, which in this case is the estimate, to monitor the progress and performance against it. Should there be a significant deviation when measured against the plan, then some corrective action must be taken so as to bring the project back on course, or alternatively, the plan needs to be adjusted in the light of current experience. In many aspects, the estimated cost of a project must form the major part of the project management's plan. The estimate, if it is prepared in some detail, not only sets an end objective in terms of a financial commitment, but also will provide the considerable detail of the physical achievements that are being made on a construction site. It will become a major indication of the quantities of work that are required of the contractors and will be broken down so as to give some detail to the various facilities that are to be constructed.

As the preparation of an estimate passes through each of the ensuing stages it becomes further refined and the likely degrees of error when compared with actual cost, if the project goes ahead, will decrease. These estimates will range from a preliminary estimate of cost, having no great amount of detail, to a very detailed estimate and hence the foundation for the preparation of a budget from which the project can be controlled through design and construction. Estimating is often referred to as an art; it clearly is not an exact science. However, what is required to be precise is the subdivision and presentation of an estimate that is going to be used for control purposes. It must be presented in such a way that is meaningful to its user when control begins and it can be readily followed and most easily understood. The estimate is a measure of the scope of a project. When detailed design begins to take place then it becomes a check upon the scope of work included in a design and enables a project manager to have an early indication of any changes in scope that are necessary or trends in directions away from the definitive estimate. It provides an opportunity for definitive estimates to be adjusted so that, prior to construction, due allowance can be made through all the processes for an increased scope of work.

One of the major uses of an estimate is to form a basis by which a forecast of the remaining cost to completion can sensibly be made. As the work proceeds many of the physical quantities of construction will have been completed and provisional quantities will have become firm. The scope for making major design changes will have become limited and the estimate of the amount of work to completion will have become better defined as the time proceeds. In addition, there will be experience of cost of the work which has been undertaken and it will be possible for this experience to produce an estimated completion with ever increasing accuracy as well as to prepare an accurate cash flow forecast of future commitments.

The actual amount of time and detail which is put into the preparation of an estimate for capital investment purposes is often reflected by the urgency with which the investment is to be made. In the case of an investment which is required, perhaps to comply with the law, the commitment of funds may need to be made so as to prevent the interference with the use of a facility. In this case, the accuracy of the estimate to within narrow degrees of error is not a first priority or an essential feature of it. It may well be that an initial cost appraisal in a very approximate form may be the one that is used to set the project under way.

In another case a company may wish to invest in a building which is one of a series that is currently being built. A hotel chain, for example, may be investing in a series of five similar hotels in one particular area. When one or two have been built an accurate idea of what the next one will cost will be reasonably well established and, with a modest adjustment to the previous costs to take account of a different site, perhaps increased costs and different local conditions in the construction industry, the company will have an estimate of cost which is probably within plus or minus 10 per cent or less of the probable actual outcome. In circumstances such as this it will clearly not be necessary for the enterprise to go through a series of formal detailed estimating processes, consuming large amounts of both time and human resources, before they can appraise the investment and make arrangements for the project work to begin.

Having made a preliminary appraisal of the estimated cost of the investment, and interest in the development is still maintained, an *order of magnitude* cost must be established that will reflect a greater detail of the facility that is to be constructed with the cost apportioned to major elements in the total scheme. At this stage there is unlikely to be any detailed design work carried out though there may be broad brush layout plans in sufficient detail for a prospective owner to appreciate the overall magnitude of the facility that is being constructed and its physical concept. The basis of such an estimate is likely to be obtained from a data bank of cost data relating to similar construction elsewhere. It may, at this stage, be based upon unit costs for floor areas, cubic capacities or various rates of output, such as are used in *parametric estimating*, and there will be some measure of the risk that something untoward might happen, or that some particular item of construction which cannot be foreseen at this early stage, will prove to be exceptionally complicated and expensive. A contingency will be included in the total sum.

A typical estimate at this stage of the work, if a project which included several buildings were being considered, would be broken down into units of buildings. It would include for external works, such as roads, external lighting, drainage and so on, and there would be a sum added to this for the preliminaries and contingencies which have already been mentioned. Depending upon the nature of the work, and particularly whether it is a simple form of construction or very complicated, whether it is work of which the estimator has a great deal of experience, or whether it is something entirely novel, will depend upon the accuracy with which an order of magnitude estimate can be prepared. For simple straightforward and familiar work, the estimate at this stage may well be within plus or minus 10 per cent or less. For complicated, unusual work then the limits vary and may be as high as plus or minus 25 per cent.

From the order of magnitude estimate, progress is made to the next stage, that of a *preliminary* or *predesign estimate*. If the order of magnitude estimate is acceptable to the owner, then the next stage in the sequence of design can proceed. Buildings will be defined in more detail with regard to their size, number of floors, facilities provided, some detail about foundations, plant for heating will be established and so on. At the end of the preliminary estimate stage, it should be possible to have an estimate which, in the case of each building, is likely to be broken down into the detail of sub-structure, its superstructure broken down into the various elements such as roof, floors, internal finishes, and building services together with any external works, all of which will be in greater detail. If the investment includes plant or machinery, then the size and duty rating of this plant should be estimated approximately and other services like instruments, piping, and electrical supply can then be determined in sufficient design detail in order to obtain an approximate cost. It is likely that the degrees of error in the preliminary estimate ($\pm 15\%$) will be reduced over those of the order of magnitude estimate but, again, this is likely to depend very much on the nature and complexity of the work involved. It should, however, give the owner a degree of confidence in going forward with his investment unless some untoward factor is exposed by this further stage of estimating.

The third stage of estimate that should be prepared is usually called a *definitive estimate*. This will be prepared as a result of progressing the detailed design work for the various constructed facilities and a more detailed indication of the quantities of materials that will be required can then, at this stage, be obtained. It will be determined from this detailed examination whether the broad plans of previous estimates will be achieved. Sufficient work should, however, have been undertaken in previous stages to ensure that no major upsets in the estimated cost of the works should be revealed. The estimate produced at this stage will become the basis for control by the project management team and should be certainly well within the limits of error of plus or minus 10 per cent of the eventual cost. Whilst, in most instances, a blend of these three definitions of estimating stages is used, the pre-construction estimate for an investment project should include a reasonable contingency allowance in order to provide for unforeseen conditions and the necessary design changes during the project life.

In many instances, owners may change their minds about the scope of the work that they require as construction proceeds. This can result in major variations in both the design and cost. It is not the purpose of a normal contingency fund to cope with the cost of major changes of this nature but rather the small variations that occur due to unforeseen circumstances in the construction of the foundations, some change in detail of finishings, etc. Where major changes are brought about, it is sometimes necessary that a toal revision of the estimate is undertaken and reappraisal of the likely return on the investment is made for re-sanctioning before the changes are put in hand.

8.4 The significance of cash flows

There are two distinct and opposite points of view regarding the distribution and movement of the capital finance used to undertake a capital project. This is particularly true of a construction project. On the one hand, the owner or promoter of the project wishes to minimize his total investment, to delay the rate at which he pays out the capital sums in order to minimize his interest charges, and to begin to receive revenue from the project as a result of it becoming operational as soon as possible. As a result, the owner is interested in an economic project design in the overall sense, as great a delay as is possible in paying the contractors who are working for him, and as short as possible a construction programme with a minimum operational period before maximum production is reached. In addition, an owner obviously wishes to borrow money, if necessary, on the most advantageous terms. On the other hand, a contractor who is constructing and/or supplying a capital asset is interested in prompt payment by his client for work completed and a construction programme which allows him to minimize the working capital he will need to provide from his own resources. Exceptionally, in some instances, a contractor may misguidedly believe that it is to his advantage to maximize the capital investment of his client.

It is necessary to consider the availability of *working capital*. The availability of working capital in a company (whether owner or contractor) is vital to the smooth and efficient working of that company. Overtrading can be equally as dangerous and disastrous as undertrading. The temptation to overtrade is great for a relatively new, rapidly expanding and dynamic company and it results in a shortage of liquid assets because of the need to provide for increasing larger stocks of materials, work-in-progress and more unpaid accounts. A similar financial strain can be placed on a company by a shortage of work, when the inward cash flows reduce to such proportions that insufficient surplus is available to provide a fund of liquid capital.

Simple examples will illustrate a number of these points.

Example 8.2

A client establishes the economic viability of a project which has an initial estimated cost of £5 000 000. The £5 000 000 to finance the project can be borrowed at an interest rate of 14 per cent per year together with a commitment charge, on the whole sum, of half of 1 per cent per year. (This commitment charge is payable half-yearly in advance from the initial reservation of the funds until the whole of the sum is borrowed and is a normal procedure where funds are required to be reserved for future use.) Any excess over £5 000 000 will be provided from the company's own funds at a cost of 14 per cent.

In order to have some idea of the total cost of the project, a tentative construction programme over four years is prepared, together with a cash flow statement. Expenditure for any one year is assumed to take place at the middle of that year. A construction period of four years is assumed in the first case and the outlays are as shown in Table 8.4 (Year 0 is the time at which construction finishes and production should begin).

Table 8.4

Year	Capital outlay (£)	Commitment charge at ½% p.a. (£)	Interest at 14% pa on cumulative outlay (£)	Cumulative outlay (£)
4	–	12 500	–	12 500
-3½	500 000	12 500	875	525 875
-3	–	12 500	36 811	575 186
-2½	1 000 000	12 500	40 263	1 627 949
-2	–	12 500	113 956	1 754 405
-1½	1 500 000	12 500	122 808	3 389 713
-1	–	12 500	237 280	3 639 493
-½	2 000 000	–	254 765	5 894 258
0	–	–	412 598	6 306 856
Totals	5 000 000	87 500	1 219 356	6 306 856

Table 8.5

Year	Capital outlay (£)	Commitment charge at ½% p.a. (£)	Interest at 14% p.a. on cumulative outlay (£)	Cumulative outlay (£)
−3	−	12 500	−	12 500
−2½	1 000 000	12 500	875	1 025 875
−2	−	12 500	71 811	1 110 186
−1½	2 000 000	12 500	77 713	3 200 399
−1	−	12 500	224 028	3 436 927
−½	2 000 000	−	240 585	5 677 512
0	−	−	397 426	6 074 938
Totals	5 000 000	62 500	1 012 438	6 074 938

It would seem, therefore, that nearly £1¼ million accumulates as an interest and commitment charge on this size of a project under the conditions stated. The importance of having the best possible terms of borrowing is therefore underlined.

The desirability of as short as possible a construction programme is emphasized in Table 8.5 with a different phasing of expenditure.

It will be seen that by reducing the contract period by one year, with a slight rearrangement of the expenditure programme, a saving of £231 918 can be effected. If now the programme is further adjusted so as to bring as much expenditure as possible towards the commencement of production, both by delaying payment to the contractors and by re-phasing the work, Table 8.6 illustrates the effect.

A table based on the second construction programme, but using an interest of 8 per cent, is set out in Table 8.7. The drastic effect of an increase in interest rate on the cumulative outlay should be noted.

Table 8.6

Year	Capital outlay (£)	Commitment charge at ½% p.a. (£)	Interest on 14% p.a. on cumulative outlay (£)	Cumulative outlay (£)
−3	−	12 500	−	12 500
−2½	500 000	12 500	875	525 875
−2	−	12 500	36 811	575 186
−1½	500 000	12 500	40 263	1 127 949
−1	−	12 500	78 956	1 219 405
−½	1 500 000	12 500	84 308	2 816 213
0	2 500 000	−	196 085	5 512 298
Totals	5 000 000	75 000	437 298	5 512 298

Table 8.7

Year	Capital outlay (£)	Commitment charge at ½% p.a. (£)	Interest at 18% p.a. on cumulative outlay (£)	Cumulative outlay (£)
−3	–	12 500	–	12 500
−2½	1 000 000	12 500	1 125	1 026 125
−2	–	12 500	92 351	1 130 976
−1½	2 000 000	12 500	101 788	3 245 264
−1	–	12 500	292 074	3 549 838
−½	2 000 000	–	319 485	5 869 323
0	–	–	528 239	6 397 562
Totals	5 000 000	62 500	1 335 062	6 397 562

The total charges have risen from £1 074 938 to £1 397 562 further emphasizing the need for the best possible terms when financing projects.

Up to this point, only the finance required during the construction period has been considered. Interest has been calculated on the capital outlay as it has accumulated with other charges and no income has as yet been received to offset the expenses involved in construction. Let it be assumed that production starts as expected at Year 0 and in Year 1 a net profit of £400 000 is available to set off against the investment capital and charges. Let it further be assumed that in each subsequent year £1 200 000 is available for this purpose. The outcome of the decision to invest can well be illustrated graphically and Fig. 8.1 shows one form which such an illustration might take.

The outlay and recovery of finance for two alternative construction programmes are illustrated in Fig 8.1; the programmes are shown in Tables 8.4 and 8.5, respectively. This figure illustrates the length of time, under the given financial conditions, that it takes to recover capital from income on the basis that interest is paid only on the capital which remains outstanding at the end of each year. The shorter, more intense, expenditure programme has already been shown to lead to a smaller peak capital outlay and therefore, given similar net incomes as in this example, the capital recovery time will be shorter. Whilst expenditure during the construction period is at a greater rate in the case of Table 8.5, the start of the project can either be deferred for 12 months over that illustrated in Table 8.4, leading to the same calendar start-up date, or production can be put in hand 12 months earlier. In the case of Table 8.5, complete capital recovery takes place one year sooner than in the alternative. The net income of the first year is insufficient to pay-off the interest on the peak capital debt incurred, hence this results in a continuation of the 'expenditure' trend of the diagram for one to two years. It should further be noted that if an interest rate of 18 per cent is paid on the capital, the capital expenditure will never be recovered because the available annual

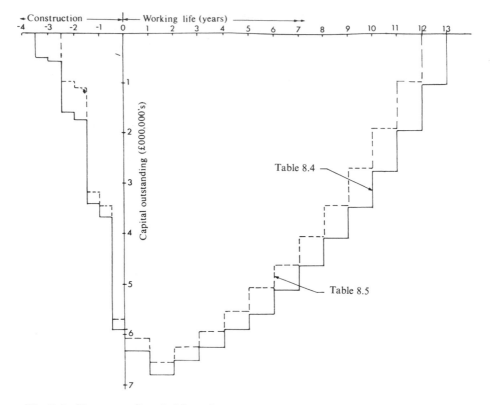

Fig. 8.1 Recovery of capital from income.

income will never be more than the interest which accrues on capital outstanding at the end of each year. The NPV of the project is negative.

A diagram such as that shown in Fig. 8.1 lends itself to an appreciation of
the risk involved in making any of the alternative patterns of investment. If
the two examples are plotted, it will be seen that, with the shorter
construction programme, the capital can be recovered two years before that
of the other programme, if they are both assumed to have the same calendar
starting date. In addition, at no point in the working life of the investment is
the capital outstanding greater for the short construction programme. Since
risk is almost invariably directly proportional to both the time taken for
capital recovery and the amount of an investment which is outstanding at
any one time, the shorter construction programme offers a less risky
investment when measured in these terms. On the other hand it is not always
easy, or possible, to obtain a shorter construction programme without
incurring premium costs such as those of overtime and bonus or incentive
payments. There is also a greater risk that the shorter programme with a
higher work rate will not be achieved, and therefore, because the investment

is at a higher rate, it is very necessary that the design of such a project is finalized and made firm at an earlier date. The period over which withdrawal from the project, once started, is economically possible is much shorter.

8.5 Design costs

The more detailed design work that is undertaken, the more definitive becomes the estimate for the project simply because the work content for the capital investment becomes better defined with each successive stage. The general strategy of a capital investment is set by the owner at the completion of the investment appraisal stage. When detailed design commences, the owner having taken a decision to proceed, there is a need to implement a cost control process. This has the long-term view of ensuring that the best interests of the owner are reflected in the design and that the work is designed with as good a chance as is possible of its capital cost falling within the overall budget prepared for the owner at the end of the appraisal stage. The managerial process in controlling cost at the design stage of the work is one that continues throughout the detailed design process and also into the construction process in so much as variations in design are frequently necessary once construction work is underway.

A design cost control system needs to meet the following objectives:

(a) achievement of the owner's financial budget in relation to the design and construction costs resulting from his investment;
(b) achievement of the owner's objective to have a facility which will have the capacity that was specified at the appraisal stage;
(c) provides a basis for accurate and effective cost control by the owner throughout the whole of the project's design stage;
(d) provides a means by which the owner can predict with some reliability and considerable accuracy his financial commitment to the end of the project from any particular point in time throughout its duration;
(e) can have linked to the cost control process some indication of the physical progress of the work in relation to the plans and schedules which form part of the budget.

With the owner preparing a budget from his capital estimate at the appraisal stage, he is able to direct the designer towards the quality of project that he requires during the design and construction stages. He is able to set a series of guidelines which clearly indicate the solution that is required.

During the design phase every opportunity must be taken to keep the cost of the designed facility to a minimum – compatible with satisfying the requirements of the owner. The following example demonstrates one of a number of analytical techniques which are available to assist in the optimization process for design to a minimum cost and to eliminate some of the alternatives that are available.

Example 8.3

A bridge is required across a river estuary. The total length of the bridge is 740 m and there are no reasons, other than considerations of cost, why the piers supporting the bridge superstructure should not be sited entirely to suit the design. The total weight of structural steel in the superstructure for each span of the bridge may be found from the formula

$$W = 150L^2 + 900L$$

where W is the weight in kilogrammes and L is the length of each span in metres.

If the structural steel in the bridge superstructure costs £400 per 1000 kg including fabrication and erection, and the total cost of one pier or abutment is £400 000, find the best practical length of each span, assuming that all spans are to be of equal length (see Table 8.8 and Fig. 8.2).

The above example illustrates a problem in economic choice, wherein the time-value of money does not play a part. It is a study in present economy. If the bridge were one for which it is proposed to charge a toll for passing over it, then the study would be taken further to examine the likely rate of return for a series of changing schedules. However, from the point of view of the design the objective is to establish minimum cost only. This example represents a situation in which the total estimated cost of the bridge varies with a variable in the design – in this case, the length of a span of the bridge. It will be noted that as a span of the bridge goes up in length, so the cost of the structural steelwork rises for that span but the cost of the associated support piers goes down per unit length of the span. Allowing for the fact that there

Table 8.8

No. of spans	Length (m)	Superstructure cost (£)	Substructure cost (£)	Total cost (£)
5	148.0	6 837 600	2 400 000	9 237 600
6	123.3	5 742 400	2 800 000	8 542 400
7	105.7	4 960 114	3 200 000	8 160 114
8	92.5	4 373 400	3 600 000	7 973 400
9	82.2	3 917 067	4 000 000	7 917 067
10	74.0	3 552 000	4 400 000	7 952 000
11	67.3	3 253 309	4 800 000	8 053 309
12	61.7	3 004 400	5 200 000	8 204 400
13	56.9	2 793 785	5 600 000	8 393 785
14	52.9	2 613 257	6 000 000	8 613 257
15	49.3	2 456 800	6 400 000	8 856 800

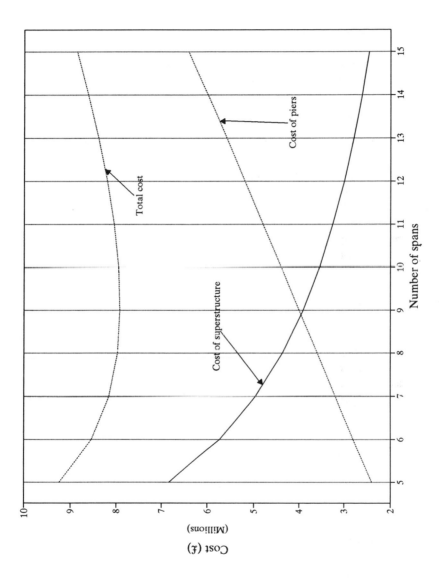

Fig. 8.2 Analysis of bridge designs

will be a small number of piers, which can only change in discrete intervals, there will be a value of the variable, L the span, for which the estimated total cost of the bridge will be a minimum.

The economic relationships can be illustrated in such cases, by using graphs from which the underlying economic influences can be readily interpreted. The display of information in this way enables a clear picture of the whole problem to be obtained and it enables the need for further analysis to be determined. Figure 8.2 illustrates the bridge design problem.

It should be noted that, in general, whilst the above examination of the initial design of the bridge locates the general arrangement for which the minimum total cost should be achieved, this study is not directly a measure of the economic efficiency of the project. Not always will the minimum cost design necessarily coincide with the design for maximum operating efficiency.

8.6 Expenditure forecasting

Having decided to proceed with a capital investment programme, at the earliest possible moment an owner will require to have an indication of the phasing of the payments that will have to be made over the duration of the construction period. There will be fees for designers and other professional advisers who are engaged for the preparation of designs, of contractual matters and supervision of construction. These fees will generally be a small proportion of the total costs and the largest and most frequent payments will generally be made to contractors engaged in the construction of the facilities.

A useful method for forecasting the cash requirements of a project is that which makes use of an *S-curve*. This curve draws its name from the fact that the cumulative expenditure for a project typically takes its shape as that of a letter *S*. Because of this it is possible, as a result of observation and use of historic records, to draw up a curve on an empirical basis which predicts the way in which expenditure will occur. One such alternative is to assume that one quarter of the expenditure on the work will be incurred in the first third of the contract programme, one half in the middle third and the remainder in the last third of the programme.

Figure 8.3 illustrates the principles underlying the construction of an *S*-curve for the cost of carrying out a project. It is an example of a curve that might be used by an owner to forecast the pattern of future expenditure or might also be prepared by a contractor to determine the pattern of his cumulative direct cost expenditure and as a basis for other controls. The curve itself is a forecast, or budget, within which progress can be controlled. Once a time-based programme has been drawn up for each operation or activity, an allocation of the cost for each unit of time can be made. In Fig. 8.3 the time-base has been drawn up in months and the programme is set out

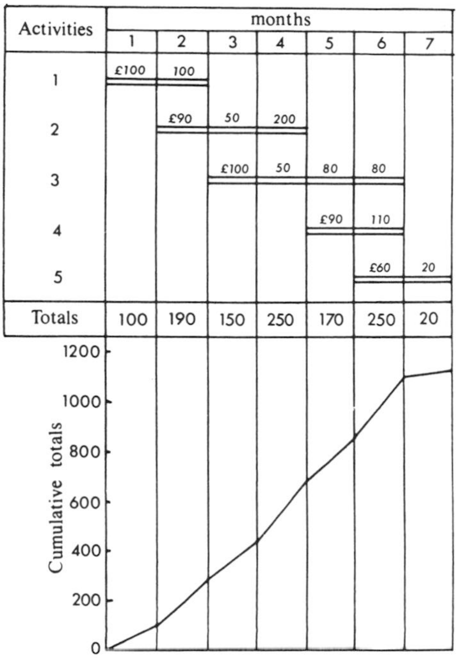

Activities	months						
	1	2	3	4	5	6	7
1	£100	100					
2		£90	50	200			
3			£100	50	80	80	
4					£90	110	
5						£60	20
Totals	100	190	150	250	170	250	20

Fig. 8.3 Expenditure S-curve.

in the form of the familiar bar-chart. The pattern of expenditure for each time unit and on a cumulative basis can be examined and the S-curve in this example gives a clear picture of this pattern and the various rates of working in terms of cost.

The owner, at this stage, will not have very detailed information about the construction programme and the expenditure related to it. However, such curves can be constructed on the basis of approximately estimated programmes of expenditure and answers can be obtained well within the limits of error required for cash forecasting of this type. If a greater degree of sophistication is required, then the S-curve can be assumed to be that of value of work and a stepped curve of actual payments to a contractor at monthly intervals, and making due allowance for delays and retentions, can be superimposed.

Where an owner has a series of similar construction projects, such as power stations, hotels, office blocks, blocks of flats and so on, it is sometimes possible to derive an equation that will approximate to a cumulative expenditure curve for future projects of a similar type. Drake produced such a formulation for hospital construction at the Department of Health and Social Security (DHSS).

The equation for an S-curve is:

$$y = S[x + Cx^2 - Cx - \frac{1}{K}(6x^3 - 9x^2 + 3x)]$$

where y = cumulative monthly value of work executed before deduction of
        retention monies or addition of fluctuations
   x = the ratio of the month, m, in which expenditure y occurs to the
        contract period, P (measured in months)
   S = contract sum
   C and K are parameters.

Drake has published a series of values for the parameters C and K (for hospital construction). The parameters have been established for typical S-curves within a classification of contract value by first establishing the co-ordinates of two points on a series of historic curves and then substituting each pair of values of the co-ordinates in the equation. The two simultaneous equations so formed are then solved for C and K. For example, for a hospital costing approximately £3 million, $C = 0.110$ and $K = 3.980$; for one costing £6 million, $C = 0.192$ and $K = 3.458$.

The use of a formulation for an expenditure curve, such as in this example, goes beyond that of predicting the need to provide finance at specific intervals of time in order to pay for work carried out. If the parameters C and K, the contract sum and the cumulative expenditure to any month of the contract duration are known, then it is possible to forecast, by use of the formula, the likely duration of the project. Corrective action can be taken if the duration as calculated appears to be greatly in excess of that planned.

In determining future cash flows, by the use of a suitable index, future inflation costs can also be allowed for over the future period of planning.

Peer produced a similar model having similar functions when developing a cash flow planning system for housing construction. He analysed the detailed construction schedules for four typical projects of different size and which were constructed at different rates. The cumulative cost curves (excluding general overheads) were brought to a common denominator by plotting the cost and duration data to percentages on the horizontal and vertical axes and hence plotting a percentage cost expended to its related percentage duration expired. Three formulae were obtained, each producing similar curves. The curve obtained by polynomial regression of the fourth degree yielded the most accurate estimate, whilst those obtained with the aid of the tanh and error functions were said to be easier for manual manipulation.

The curve obtained by polynomial regression is:

$$y = [0.009 + 0.2731(W/T) - 1.0584(W/T)^2 + 5.4643(W/T)^3 \\ -3.6778(W/T)^4]C$$

where y = cumulative cost
        C = total cost

W = number of time units passed

T = total number of time units, i.e. total construction duration.

Therefore (W/T) = relative fraction of total time expended.

The application of the Peer curve is similar to that of the DHSS curve above.

A further, simpler model used the US Corps of Engineers (Perry) when forecasting the earnings of a contractor against the expenditure of project times is, $y = sin^2(90x)$ where y is the decimal fraction of the contractor's earnings and x equals the decimal fraction of the total project time completed.

8.7 Owner's optimum construction programme

Premium payments, such as overtime and bonus, are sometimes involved in shortening a construction programme. Reliable information about reducing a construction programme, thus enabling a decision to be made by an owner as to whether it is worthwhile or not, must necessarily come from the contractor who may be undertaking the work. Only the contractor will be aware of the detailed resources that are required to carry out the work and what is entailed in either increasing their quantity beyond that which he considers to be an optimum economic level to meet the contractual terms under which he was appointed or working them longer hours. The latter frequently results in an increase in the unit cost of production and usually, a falling-off in productivity towards the end of, or during, an extended work period. It will often be desirable, before orders are placed and particularly for projects which entail a very large investment, for an owner to obtain a number of tenders from an interested contractor for various durations of the construction work (these need only be stated in overall construction duration). Additional expenditure to reduce project construction duration from the normal can then be set against savings in interest as calculated above. Figure 8.4 shows diagrammatically the relationship of cost and time in the selection of the optimum programme.

Figure 8.4 is based on what has been called the *normal construction cost* and *normal construction period*. In obtaining a contractor's tender for carrying out construction work, an owner will normally either set an overall programme requirement in terms of duration, or he will expect a contractor to state a time period within which he, the contractor, believes he can complete the work, together with a cost associated with that duration. In either case, the value of the tender may be considered to be the normal construction cost and the programme to be the normal construction period. This point for a hypothetical project is indicated in Fig. 8.4. If that construction programme can be accelerated, the various examples detailed

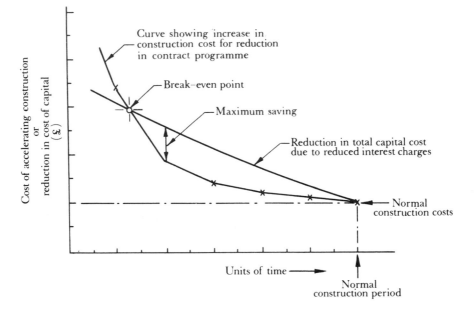

Fig. 8.4   Optimizing client's requirement for construction programme.

earlier (Tables 8.4 and 8.7) show how the total interest incurred on finance
can be diminished, and thus the total capital cost of the project can be
reduced. The upper curve in Fig. 8.4 illustrates the relationship between
capital cost and project duration. As the construction programme is reduced,
so the total cost of capital is reduced. Hence the curve has a negative slope
from left to right.

The cost of accelerating the construction will generally increase with the
extent of the reduction. The lower and more sharply curved graph of Fig. 8.4
shows a relationship between these two factors. Where the two curves cross is
the break-even point, at which the cost of accelerating the programme just
balances the savings in interest payments. This point indicates the minimum
programme duration that can be traded-off for savings in interest on finance.
The graphs show another programme duration which may be of more
interest to the owner than that shown by the break-even point. This is the
point at which his maximum saving will occur; it is that point at which there
is the maximum cut-off on an ordinate between the two curves and it is
indicated in Fig. 8.4.

Having settled on an optimum programme and established the programme
of capital commitment for an owner, it is important that the construction
targets are kept so as to avoid the continual reassessment of the viability of
the investment project due to any changes. The importance can be illustrated
by the use of an *S*-curve.

One particular mechanism, though it is not always available to an owner, is the advantage created by advancing and lengthening the design period allowed for the project. One of the reasons why the progress of construction work sometimes falters, particularly in the early stages of the work, is the non-availability or late supply of finalized design drawings. Given that the design work is not being carried out in-house by an owner, he has little control over the design resources assigned to a project. The designers will normally be in a situation wherein they have to satisfy a number of other clients who are actively pressing for progress on their own particular projects. In normal circumstances it will be cheaper to increase the resources applied to the design process prior to the commencement of construction, than it will be to increase the resources used in construction if lost time on the overall project programme needs to be made up. However, because of this lack of specific control by an owner, and the need for a designer to maintain a reasonably level demand over time for resources within his own organization, the lengthening and bringing forward of the design period may be the only useful and effective device to which he can resort.

Figure 8.5 is a diagram representing the typical situation in which the design period is lengthened and the commencement date of the design is brought forward prior to the construction commencing on site.

It will be seen that at the left-hand end of the S-curve in doing this, and assuming that the total cost of design does not increase because of it, there

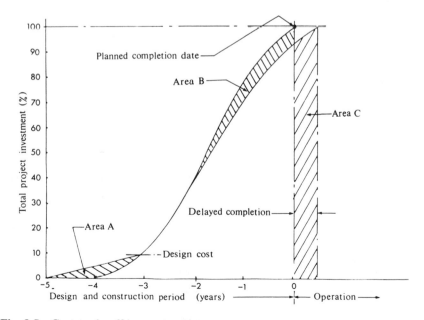

Fig. 8.5 Cost trade-off in construction programme.

will be a different rate at which payments for this service are made. This will increase the interest on capital invested in the process which will be proportional to the shaded area between the two curves. If this premium is paid then the chances that the commencement of the operation of the asset will actually occur at the original target date will be very much increased. The additional interest payment in this case must be set off against the savings by an owner of additional interest on the capital that he has so far invested in the project and labelled Area C on Fig. 8.5.

There is also a slight compensating feature in having a longer programme, that is a delay, which needs to be taken into account in the overall calculation. This is Area B. As well as any additional interest payments on capital investment due to the delay in the start up of commercial operation, the owner will, of course, have to make allowance for the lack of income from his investment arising from later operation than anticipated.

8.8 Project budgetary control

Detailed analysis, such as that which is an established part of standard costing, is not usually applicable to the type of project cost control that is required by an owner's organization. An owner rarely requires to exercise direct control in detail over the resources such as labour, materials and plant, that are to be used in the construction project. Project control for this purpose rarely requires reporting and feedback as promptly as with a contracting organization. The latter needs to have truly current costs promptly if strict and effective control is to be exercised. In the light of this, a variance analysis that can be used effectively in the client's situation should be used. Reporting on the cost progress for a project as a whole is normally required, rather than on smaller elements of it, though there will be occasions when certain sections of the project need to be highlighted.

The owner will usually have an interest in a more detailed elemental breakdown of the costs of the project for other purposes but the timing for the receipt of these will not be of the same order as information for management control.

Before a project is initiated, it is necessary for a budget that links expenditure and time to be agreed by the project manager and the senior management of the company for whom the project is being undertaken. This initial budget will reflect the base estimate of cost for the work to be carried out during the project construction period and that used in the initial investment appraisal together with subsequent changes as design has proceeded. Inevitably, as the work on the project progresses, it will be necessary to revise certain of the estimated costs. These revisions may be necessary because of inflationary effects on the cost of resources, changes in design and/or the scope of work involved (perhaps due to a better knowledge of the site conditions once work commences), changes to process or production requirements, and many other reasons. Thus the first variance which needs to be calculated, the *project revision variance*, is, in effect, an umbrella variance encompassing all cost changes for the project for whatever reason. The effect of this variance, when added to the initial budget, is to give the current estimated total cost of the project, having made due allowance for all known changes, increased prices, etc.

Project revision variance = revised budget – original estimated budget

The project revision variance takes account of, in absolute terms, any changes in overall budgeted cost. It does not enable the variance analyst to assign a reason for the change as is the case with some of the other variances. At each change of budget from the original estimated cost, approval for the change must be sought from the owner's authority appropriate to the size and scope of the revision. Table 8.9 illustrates a typical report on project variance analysis demonstrating the use of such analysis for a project. Such a report would probably be prepared on a monthly basis. The figures in this analysis will be used to illustrate the explanation of the other variances.

The analysis in Table 8.9 is divided into two parts. The eight columns ('End of month' plus numbers 1 to 7) all include figures which result from the collection of data as a result of work done, or are original budget plans derived from the definitive project estimate or are calculated as a result of the progress made up to the reporting point in time. The remaining five columns are the result of the variance analysis.

It will be seen that the original planned budget for the project is £100 000 (Column 4). At the end of the first month's work, work costing £9000 has been completed (Column 5) and the current estimate for the cost of work to complete the project amounts to £93 000 (Column 6). In this respect, it is essential that an estimate for the cost of the work necessary to complete the project be prepared at each update in the light of the experience gained, since it is only in this way that a revised estimate for the total cost of the work can be prepared. In this example, it is seen that the *revised total estimated cost of the project* is now £102 000 (Column 7) at the end of Month 1, leading to a project revision variance of £2000. A similar calculation for Month 2 produces a *revision variance* of £6000; a further £4000 plus the £2000 revision in Month 1. Such information is essential for the proper control of the owner's finances, for the preparation of a budget for future capital expenditure throughout an enterprise and, perhaps, a check as to whether the project is still viable as an investment if large increases occur. Large increases in the revision variance certainly need close investigation.

The project *revision variance* shows the change in the total estimated cost between establishing the original budget and the budget which is current at the report update. In addition, it is necessary to record the change in the estimated cost of the total project since the last review date which, in this

Table 8.9

End of month	Budgeted value of work to be completed	Budget value of work actually completed	Budget value of work to completion	Total budget of work as planned	Actual cost of work completed	Revised cost estimate of work to be completed	Revised total estimated cost of project	Variances				
								Revision	Budget performance	Volume	Efficiency	Total cost review
	1	2	3	4	5	6	7	8	9	10	11	12
				$(1+3)$			$(5+6)$	$7-$ init. est.	$2-1$	$5-1$	$2-5$	$7-4$
	£	£	£	£	£	£	£	£	£	£	£	£
0			100 000	100 000								
1	10 000	8 000	90 000	100 000	9 000	93 000	102 000	2000	(2000)	(1000)	(1000)	2000
2	25 000	18 000	77 000	102 000	25 000	81 000	106 000	6000	(7000)	Nil	7000	4000
3	50 000	45 000	56 000	106 000	47 000	67 000	114 000	14 000	(5000)	(3000)	(2000)	8000

example, is for a period of one month. In this case, for Month 1, it is the same as the project *revision variance* since only one update has been made. The £4000 revision for Month 2, however, is displayed in Column 12. This variance is known as the *total cost review variance*. When added to the total estimated cost of the project at the previous update, it results in the current estimated cost of the project. The *total cost review variance* is obtained by subtracting the figure in Column 4 from that in Column 7 at each update process.

From the month-to-month control point of view, it is the current performance on the project that is important. The *budget performance variance* appearing in Column 9 of the variance analysis consists of two components. It gives the measure of progress up to date and may be calculated as:

Budget performance variance = budget value of work actually completed – budget value of work to be completed.

In Table 8.9 this may be calculated by subtracting the figure in Column 1 from that in Column 2. In the case of the update at the end of Month 1:

Budget performance variance = £8000 – £10 000 = – £2000

This is a negative variance and is therefore unfavourable. In budget terms, £2000 worth of work less than that budgeted for completion in the same period has actually been carried out. However, this variance must be used with caution because it is possible that its two components may have a self-compensating effect and thus conceal the real reasons for its unfavourable nature.

The budget performance variance is a combination of the *project volume variance* and the *project efficiency variance*. The volume variance is calculated as:

Volume variance = actual cost of work completed – budgeted value of work to be completed.

The actual cost of the work completed is the price paid to a contractor (or whoever carries out the work) or, in effect, the actual cost incurred by the owner for the work carried out in the respective month. The *volume variance* therefore shows whether the progress of the work is either ahead (+) or behind (–) the schedule set by the budget and in terms of money.

The *efficiency variance* is calculated as:

Efficiency variance = budget value of work actually completed – actual cost of work completed.

The *efficiency variance* is, therefore, a measure of whether value for money is being obtained when measured against the standard set by the budget. Money has been saved if the variance is negative and overspending has taken place if it is positive. If the volume variance for a project were – £10 000, indicating that the project was behind schedule, and the efficiency variance were + £12 000, indicating that money had been saved in the execution of the work, the budget performance variance would be + £2000. This favourable performance variance indicates that, overall, the project is progressing satisfactorily. However, the two components indicate that the progress is unsatisfactory but that the value for money is the reverse. Management action is needed in respect of the former, with the possibility of achieving an even better budget performance variance. The almost self-compensating effect of the two components conceals the real situation.

In summary, budget performance variance, therefore, indicates whether the value of work completed is up to budget or not, a negative variance indicating that the work is behind budget. It does not, however, indicate whether delays are related to the programme, overspending or production efficiency. Volume variance indicates whether the rate of spending is up to programme as budgeted or not. The efficiency variance indicates whether value for money is being achieved. A negative efficiency variance indicates more has been spent than was originally budgeted for a given amount of work.

8.9 Network analysis as a basis for cost control

When a total cost for a project is established, it must be established on the basis that a programme has been drawn up for the time performance of the work involved. The cost must inevitably be linked with carrying out the project in the terms of that programme. Cost and time are inseparably linked and if, for example, the duration of a particular part of the project, or the programme as a whole changes, then in the majority of cases the cost of doing the work must also necessarily change. It is, therefore, almost impossible to control the cost of carrying out work unless the time performance is also closely linked to it and controlled at the same time. There may be isolated instances wherein a similar cost is derived for moderately different programmes of work, due to an adjustment of the method of working which is employed or the disposition of the resources. If the case of a labour-only item is considered, such as excavating a pit for a foundation by hand, it may happen that five men can do it in a time of ten hours or that ten men would do it in five hours' duration, given that there is sufficient room for them to work. It is not difficult to see the interaction between direct and fixed costs in such a situation. However, the general relationship between time and direct cost for a specific activity is in the form of the curve shown in Fig. 8.6. In broad terms, it costs money to accelerate the progress of work above the optimum duration because this frequently involves the introduction of overtime working, decreased efficiency because more resources are used in the same area of working, reduced efficiency of supervision and reduced

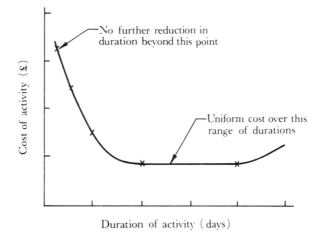

Fig. 8.6 Variation in activity cost with duration.

productivity because of longer hours. At the other end of the scale, the resources can be reduced to such an extent that they are not able to work efficiently as a balanced team for reasons such as waiting for plant, inability to carry out operations requiring a gang of minimum size and so on. Therefore, the cost of work carried out at the two ends of the scale will be greater than the optimum, although the point has already been made that the optimum may be over a range of resource combinations rather than a single point on the scale.

If it is the optimization of total project cost within a specific programme that is required, rather than the control of cost once the project is in progress, network analysis provides the basis of a useful, if sophisticated and at times complicated, method of doing so. It is a method which is based on the interaction of direct and fixed costs when considering various projects and their durations, known as *time-cost trade-off* or *time-cost optimization*. The application of such methods tended to be neglected in the early days of network analysis because of the restrictions placed on their use by lack of computer capacity; this restriction has now been removed. The point is emphasized, however, that cost cannot be controlled satisfactorily in isolation; there is a need to control it in conjunction with a control of time.

In the context of the cost control of projects, it is also necessary to be aware of another influence on the cost efficiency and progress of the work. Control is required over performance in terms of compliance with the specification for the work. While the standards set by the specification do affect the duration and cost of a project directly, they are not the subject of control in the same diversity of ways as are time and cost. However, a severe specification will almost certainly increase the length of time required to carry out the work.

Traditionally, the time control of the duration of construction projects has been carried out, in those instances where formal control methods have been exercised at all, by the aid of a variety of methods; most frequently by means of a Gantt or bar-chart. To a lesser extent, cost and budgetary control has been monitored by the use of a financial status chart taking the form of an *S*-curve showing the expected budget performance compared with another *S*-curve which shows expenditure to the date of the review. Neither of these aids to planning and control can be said to be truly effective for construction projects, though they do go some way towards providing a more effective view of operating conditions. Not least of the advantages that their use brings is the fact that they encourage the user to think in quantitative terms about the work content, the planning and the progress of a project.

The bar-chart is well known to planners and programmers. It displays the operations making up a total project. They are listed vertically, one beneath the other, on the left-hand side of a graph, with the time scale reading horizontally from left to right along the top or bottom axis. A horizontal bar against each activity description indicates on the timescale, the period of calendar time over which the operation is scheduled to be carried out. Such a means of planning has the advantages of simplicity, cheapness in use, familiarity over a large area of use as a means of communication, graphical display and a means of reporting progress, and it provides a base for simple resource aggregations.

On the other hand, it has the disadvantages that planning and scheduling must be carried out at one and the same time as the bars are constructed; that dependencies of one operation or activity on others are not shown; that control is restricted to that of duration, and it is not easy to connect the physical quantities of work involved in each operation with a specific period of time. Therefore, time progress does not necessarily give an accurate indication of the actual physical progress of the work. There are various stages of sophistication in the preparation and display of a bar-chart, largely depending on the amount of information which it displays.

The traditional form of the financial status chart in use is that of the budget *S*-curve, the construction of which has already been discussed. The *S*-curve can be drawn from experience of previous similar operations in the company or, at the estimating stage, it can be prepared on an arbitary basis. The chart is frequently used as a means of control by plotting the actual expenditure curve against the budget curve as shown in Fig. 8.7. It is sometimes wrongly assumed that whenever the *actual expenditure* curve keeps below the budget curve for value, the cost of the work is being controlled effectively and that the work is being carried out efficiently and profitably. In fact, from the control point of view, this combination of information is almost meaningless. As with the work schedule chart, the actual expenditure curve is not related to the physical quantity of the work which has been carried out, nor to the efficiency of the operations.

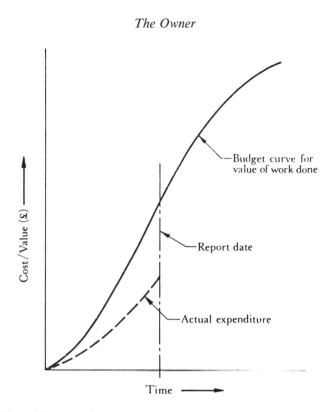

Fig. 8.7 Financial status chart.

The inadequacy of such a chart may be demonstrated by the use of a simple example. Figure 8.8 illustrates the budget curve established from the simple outline bar-chart for a project estimated to have a total cost of £629 000.

In Table 8.10 is set out the budgeted monthly expenditures (in £000s) from which the budget curve is constructed. A uniform linear rate of expenditure over each activity is assumed. If the activities are appropriately chosen this assumption need not be too wide of the mark. The bottom line of the table shows the actual expenditures for each of the first six months of the project.

The dotted line of *actual expenditure* superimposed on the *budget curve* in Fig. 8.8 shows that, after two months, *cumulative work completed* amounts to £50 000. The budget curve shows that £71 000 worth of work should have been completed and the difference of £21 000 may have one or both of two causes. Firstly, the work may have been completed more cheaply than budgeted and secondly, not as much work has been completed as was planned for this period. Figure 8.8 gives no indication of what the true position is. It may, for example, be that £45 000 is the cost of *excavation* and only £5000 the cost of the foundations. Hence there is an overrun of £5000 for *excavation* (up-to-date, because it may not be finished) and an under-

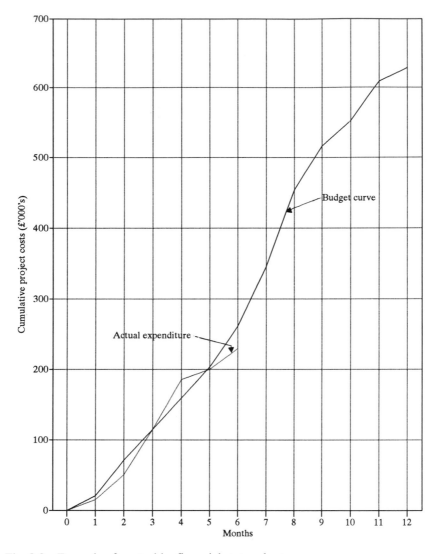

Fig. 8.8   Example of control by financial status chart.

spending of £26 000 on *foundations* indicating that this work is well behind programme. These aspects of the progress of the work are concealed by the curves of Fig. 8.8.

The use of an *S*-curve is very valuable for the purpose of estimating the amount of working capital required for a project and hence calculating the interest on finance outstanding. In addition, in other circumstances, it is useful for cost control providing that, to the basic information contained on Fig. 8.8, is added the curve indicating the value of work done to date together

Table 8.10

Activity	0	1	2	3	4	5	6	7	8	9	10	11	12	Totals
								Months						
(a) Excavation		20	20											40
(b) Foundation			31	31	31	31								124
(c) Brickwork to first floor				13	13	13	13	13						65
(d) Concrete to first floor							45	45	45					135
(e) Brickwork to eaves								27	27	27				81
(f) Concrete to roof									36	36	36	36		144
(g) Waterproofing												20	20	40
Totals	0	20	51	44	44	44	58	85	108	63	36	56	20	629
Cumulative totals	0	20	71	115	159	203	261	346	454	517	553	609	629	
Cumulative work completed	0	15	50	115	185	200	230							

Note: All figures in £000

with a projection of the actual expenditure curve to the completion of the project.

One method of planning and programming which overcomes many of the disadvantages of the bar-chart is that of *network analysis*. Network analysis covers a range of techniques many of which are known by different names, the three most common of which are *critical path method (CPM)*, *PERT* and *precedence networks*. All are based on the use of similar principles. These methods have been explained in great detail, more than adequately, elsewhere. The CPM for planning and programming relies on the graphical representation of an operation or an *activity*, which is part of a larger project, by an arrow. The length of the arrow and the direction in which it points are not significant as far as the activity is concerned – the only feature about the arrow is that work on the activity is assumed to progress from the tail of the arrow towards its head. Completion of the activity is assumed to occur at the point of the arrow head. Arrows are then arranged so as to represent the logic and dependencies of all the activities making up the project. Some activities can proceed concurrently with others, but many cannot be started until a number of others are completed. The extent of work which is represented by a single arrow depends largely on the judgement of the planner who is constructing the network. Figure 8.9 illustrates a very simple network plan for placing concrete to a foundation.

Having constructed a network, a duration is then assigned to every activity, though some arrows known as *dummies* (as often as not shown by a dotted or broken line) are of zero duration and appear in the diagram only to maintain the logic; the length of all the possible continuous paths through the network are added up to give the total duration for each path. The longest path through the network represents the minimum overall duration in which the work can be completed using the plan represented by the network and is known as the *critical path*. All activities on the critical path must be carried

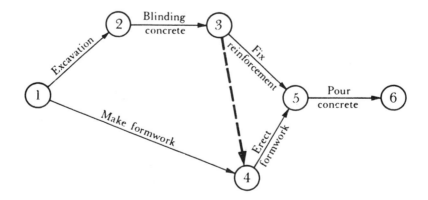

Fig. 8.9   The arrow diagram.

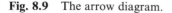

out within their allowed durations if the overall programme for the project is to be achieved. Any programme slippage on a critical activity must be made up on one of the other critical activities if progress is to be as planned. Other activities have some leeway when measured against the critical path (of which there may be more than one) – such leeway is known as *float*. Clearly there is some option, though limited, as to when these latter activities can be carried out. Figure 8.10 extends the analysis of Fig. 8.9 and shows the critical path through the sample network.

Network analysis methods, when used for controlling projects, may be said to have the following advantages:

(a) the systematic approach to planning encourages a logical discipline in the overall process;
(b) it allows the practice of management by exception, inasmuch as it concentrates attention on critical activities rather than those which have an abundance of float;
(c) it allows the considered adjustment of scarce resources by taking into account the float of each activity;
(d) it provides a model of the project which can be tested in many ways before the commencement of the work;
(e) it allows the true effect of variations in the work to be calculated, particularly insofar as the variations affect other activities, perhaps remotely and not obviously connected to the ones being varied;
(f) it provides a standard method of planning and control which in itself leads to standard documentation. As a result, the communication of the plan within an organization becomes easier and more effective;

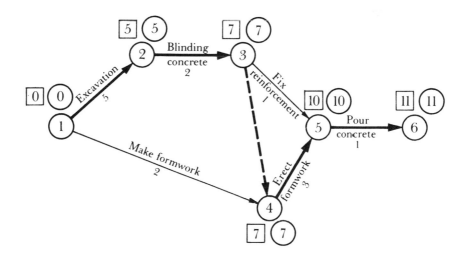

Fig. 8.10 The critical path.

(g) it facilitates the preparation of a precise materials delivery schedule;
(h) it portrays clearly the interrelationships between one activity and the other;
(i) it facilitates a change of supervision in the work during its progress;
(j) it allows the separate consideration of the functions of planning and scheduling;
(k) it pinpoints the responsibilities of individuals within the organization;
(l) it harnesses the facility (if required) of a computer to make quick and accurate calculations.

From these many advantages it will appear that network analysis methods of programming are likely to be more effective in use for many projects than the traditional methods already briefly described.

The use of network analysis is not, however, without its drawbacks. They tend to arise because the programming of projects is largely concerned with the direction and control of human beings. They are drawbacks which can be overcome to a large extent with continuing experience in the use of the method.

The drawbacks may be summarized as follows:

(a) there is the temptation to create larger organizational empires and to over-produce paperwork;
(b) attempts are made to use the system as a decision-taker rather than as an aid to decision-making;
(c) the technique tends to cut across existing systems of organization;
(d) there is a marked reluctance, through unsuitability, unfamiliarity and the fear of exposure, to move from the static traditional methods to the dynamic methods of network analysis;
(e) the method cannot be exploited to the full unless it is integrated with other management systems;
(f) in certain circumstances the method may be too inflexible. The setting-up of one particular, rigid network of activity sequence may obscure, for the time being, the existence of other possible methods and sequences;
(g) there may be some difficulty in arriving at the appropriate degree of detail required for the initial network;
(h) if network analysis is used on a time-only basis, uneconomic use of resources may result;
(i) the establishment of float may result in the premature slackening of effort;
(j) a widely-based training programme must be undertaken in the use of method;
(k) greater planning effort (and therefore greater cost) is necessitated by the use of the tool.

It will be appreciated that a network is an excellent means of displaying and co-ordinating the programme of work for a project and that it enables progress to be monitored, and the programme to be adjusted in considerable detail during the course of the work. Since cost and time are inseparably bound together, it is convenient to control both of these elements through the same medium – the network readily provides this medium. If it is decided to control cost and time by means of a network, this point should be borne in mind at the time that the original network is constructed. The activities should be arranged so that they are convenient for the control of both elements. A schedule of work can then be drawn up showing which work will be carried out during each activity and the appropriate costs or values can be assigned thereto.

There are two major problems associated with the use of cost control and network analysis methods. The first is an organizational problem which arises as a result of the conflict of a project-oriented system with an existing company accounts structure. Company accounts are usually formulated in such a way that the accountability of the company over a large number of projects is established, and the single project is the lowest level at which the accounts are significant. The second problem area arises in connection with the efficiency of the cost control system.

There must be some balance point established whereby the cost of collecting data does not exceed the benefits which are likely to accrue as a result of operating the system. Should the system demand the use of too much detailed data as input, then it is highly probable that the system will rapidly fall into disuse. Another problem arises from the contractual conditions applying to the work. Often the method of applying for payment for completed work differs from the way in which the costs, in controllable form, can be collected. Existing systems are frequently directed to obtaining payment rather than control. Further, considerable precision is required in the planning and scheduling of work if cost is to be linked to it, since discrepancies in the one can readily be carried over to the other if slackness is permitted.

In general the additional work involved in using such methods will only be warranted on very large projects, since the cost of executing such a system lies largely in setting up the means in the first instance. Networks themselves usually pay-off for time control to a greater extent on projects which are complex by nature and have many concurrent paths of activities running through them. This is also true for projects which are best suited to control of cost by these methods.

8.10 Relating costs to a network

When using a network diagram as a base for cost control, the lowest level of cost breakdown will depend on the following:

(a) the number of management levels to which the reports are circulated. If, for example, distribution is to be made down to a supervisory level and, therefore, a wide-ranging level of an organization, there will need to be a corresponding degree of detail in both input and output;

(b) projects of short duration and low total cost will tend towards the lowest unit of cost being formed for an activity expressed as a single arrow. Larger projects will necessitate the use of a group of arrows as the lowest subdivision of cost group. It follows, therefore, that a higher degree of activity and cost detail will probably obtain for a small project, whereas less detail can usefully be employed in a large project;

(c) for a project involving repetitive processes, such as laying a pipeline, there will be less detail arising from the work, for example, than in a highly complex project such as building and equipping a research laboratory;

(d) the nature of the individual operations will influence the amount of detail required, since some operations involve large quantities of material movement (excavation is one), which can be carried out at relatively low cost. Other operations involve the use of high priced materials (permanently incorporated in the work) and labour, but take a very short time to carry out.

In practice, when using a network as a means of controlling both time and cost for a project, it is not easy to obtain a breakdown of the work into activities which represent both the most convenient time and cost quantities for control purposes.

Because of the inefficiencies and lack of quality which are inherent in the general flows of information for purposes of control on construction work, it becomes difficult to integrate the control of both the time and cost control functions as is necessary for effectiveness. As a rule, the detail which is required for adequate time control provides too much detail for cost allocation, so that it is not possible to allocate the cost accurately, or vice versa. It is necessary to group activities, each of which is represented by a single arrow on the network diagram, in order to maintain uniformity of activity cost and time. Such groups of activities (or in the event, single activities, as required) are known as *work packages*. Work packages, where possible, are formed of activities which each require control of a similar nature and where cost expenditure is likely to be near linear through the activity. The smaller the work packages, the more flexible and accurate will be the control. It should be noted that work packages are usually expressed in terms of direct costs since these are the only costs which can be directly associated with single arrows, or work packages consisting of more than one activity. It is difficult to relate indirect costs to single arrows or small parts of a project and some special provision must be made for these, where applicable. *Hammock activities* can be used in the network to cope with indirect costs and they are introduced for that purpose alone. Hammock activities start and finish in a network at the events which mark the span of the activities to which the indirect costs refer. Initially they bear no duration as such, since this is calculated as a result of an analysis of the other direct cost activities in the network. It is often the case that a hammock activity will span the first and last events in a network and can, therefore, be assumed to adopt the duration of the whole project on completion of the analysis. The indirect costs are then applied to the hammock activity as a *rate per day* item or as a lump sum uniformly spread over the duration eventually assigned to the hammock activity.

A simple example of cost control based on a network will readily illustrate the principles involved. Figure 8.11 shows a network for a small project which consists of the design and construction of the installation of a small overhead structural steel hopper on two parallel reinforced concrete walls, 3 metres high, one on each side of a new rail track diversion. The installation is for loading small coal into rail trucks. The overall duration of the work is 22 weeks. The durations and the costs of each activity are shown against the relevant arrows in the network of Fig. 8.11. Control is assumed to be on behalf of the owner who has let various contracts for the work and, therefore, the sums of money shown are not broken down into labour, plant and materials components. Overheads and supervision are shown on a hammock activity spanning the total duration of the network at 22 weeks.

Clearly, on such a limited project as in this example, the scope for amalgamating activities into work packages, which reflect similar requirements from the operational point of view, is limited. However Table 8.11 lists six work packages that can be formed to provide the basis of control for the project. Each work package is coded and its total estimated cost is calculated by summing the estimated cost for the activities contained within it. The total estimated cost for the project enables a check to be made to ensure that every item included in the project has been included in the network and has been assigned to an arrow representing an activity. An assumption is often made – and it is usually more or less correct – that the expenditure over a work package duration is linear, at a constant rate throughout the package duration. Clearly, if care is not taken, this can lead to some discrepancies in calculating the pattern of cash flows. Figure 8.12 illustrates the phasing of the payments on a simple bar-chart at earliest start and Fig. 8.13 shows the *S*-curve which can be constructed once the phasing is known for the expenditure, over the duration of the whole project. In cases where there is a wide departure from the linear form of expenditure on an activity, it is desirable to add arrows in the form of dummies with duration but no expenditure assigned to them.

The pattern of expenditure within a work package is further illustrated in Fig. 8.14. The work package is that coded 1003 for the concrete work. It will be seen that the pattern of expenditure is slightly irregular over the six-week duration of the work package. Much of the irregularity will be seen to occur

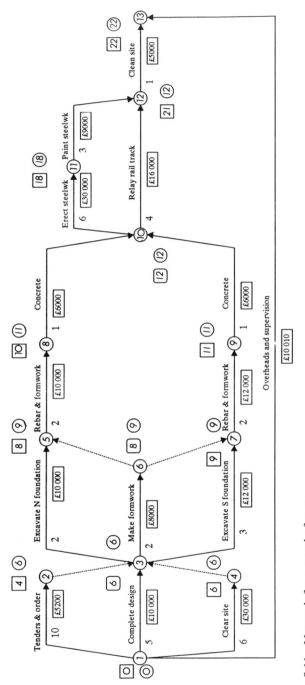

Fig. 8.11   Network for control of cost.

Table 8.11

Work package code number	Work package description	Estimated total cost (£)	Activities in work package i j		Estimated activity cost (£)
1001	Tenders, design and order	15 200	1 2 1 3	Tenders and order Complete design	5 200 10 000
1002	Clear site and excavation	52 000	1 4 3 5 3 7	Clear site Excavate N foundation Excavate S foundation	30 000 10 000 12 000
1003	Concrete	42 000	3 6 5 8 7 9 8 10 9 10	Make formwork Fix rebar and formwork to north wall Fix rebar and formwork to south wall Concrete north wall Concrete south wall	8 000 10 000 12 000 6 000 6 000
1004	Structural steel	39 000	10 11 11 12	Erect steelwork Paint steelwork	30 000 9 000
1005	Railtrack	21 000	10 12 12 13	Lay railtrack Clear site	15 000 5 000
1011	Overheads and supervision	10 010	1 13	Overheads and supervision	10 010
	Total estimated cost	179 210			179 210

as a result of the sequence of activities not being continuous. However, there would not be too much error involved if the expenditure is assumed to be linear over the duration, and this is demonstrated by the broken line on the diagram in Fig. 8.14. Experience in choosing the activities to make up each work package will enable the appropriate proportions between time and expenditure to be obtained. There is no reason why the overheads and supervision should not be spread in linear fashion over the duration of the project since this is usually the nature of such activities in an example such as this.

The use of network-based cost control systems can be advantageous where close control of cost and time progress is necessary. Clearly, the manager using such methods must have a valid reason for doing so; in other words he must have an interest in *controlling* the work progress in both these aspects,

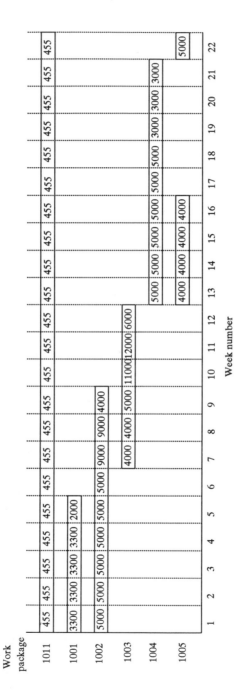

Work package	1	2	3	4	5	6	7	8	9	10	11	12	13	14	15	16	17	18	19	20	21	22
1011	455	455	455	455	455	455	455	455	455	455	455	455	455	455	455	455	455	455	455	455	455	455
1001	3300	3300	3300	3300	2000																	
1002	5000	5000	5000	5000	5000	5000	9000	9000	4000													
1003							4000	4000	5000	5000	11000	12000	6000									
1004													5000	5000	5000	5000	5000	5000	3000	3000	3000	
1005													4000	4000	4000	4000						5000

Week number

Fig. 8.12 Project programme at earliest start.

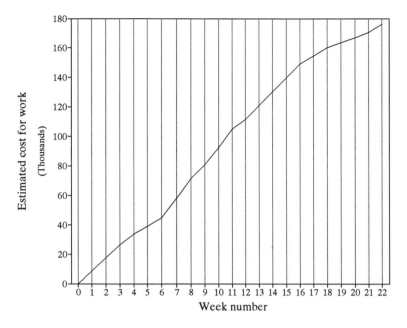

Fig. 8.13 Project *S*-curve at earliest start.

rather than using such a system for *recording* progress and costs. It is also desirable that there should be an incentive to finish a project in a case where this method of control is used. Completion to both time *and* cost must be critical and of approximately equal importance. In some projects either one or other feature, that is either time or cost, is very much in the ascendancy and the other takes a minor role in second place. For example, in relaying rail tracks, the time element is the most important of the two if the work is being carried out on a main line track so that the track can be brought into use as soon as possible and to meet a deadline. The cost is very much of secondary importance since it will be outweighed by the cost of failure to complete the work at the appropriate time. On the other hand, if an owner has only a limited sum of money to spend and cannot under any circumstances exceed this, cost becomes the most important criterion and time can then be of relatively little importance.

There is a further development of the application in the use of *S*-curves allied to network analysis as the tool by which a project is planned and programmed. It is often not fully taken into account that a network diagram, with activity durations added to it, represents not one, but a wide range of possible and feasible programmes. One extreme is represented by each activity in the network commencing at its early-start time; the other extreme by the commencement of all activities at their late-start times. The position of the chosen programme between these two extremes depends on the extent to

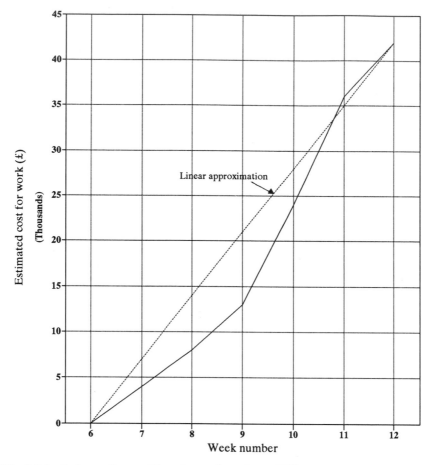

Fig. 8.14 Pattern of expenditure for work package 1003.

which the available float is taken up on individual activities. Since the rate of expenditure on a project depends, for direct cost, on the rate at which the work is carried out, it is possible to establish the two *S*-curves representing both extremes of progress. Figure 8.15 shows an envelope formed of two such cost curves for a project, the left-hand curve representing the early-start situation for all activities and the right-hand one, the late-start curve for all activities. Over some periods of time, usually at the commencement or the completion of a project, only one activity will be in progress and this will be one that is critical. In such cases, the envelope will be reduced to a single line. The horizontal distance on the timescale cut-off between the two curves is related to the float over that particular period in time in the network. It therefore gives an indication of the strictness of the control which is required over that phase of the project in order to maintain the expected progress.

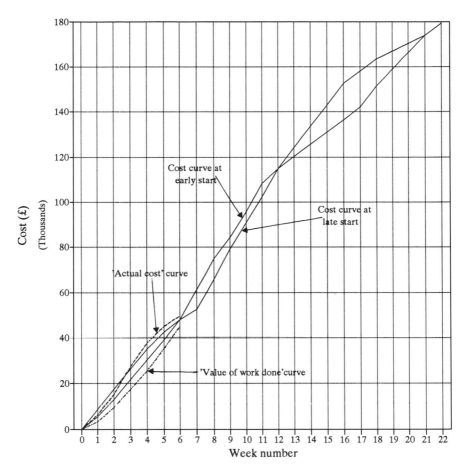

Fig. 8.15 Cost curve envelope.

To this envelope of curves can be added the curves of *actual cost of work done* and *value of work done*. These are represented by the broken line and the chain-dotted line respectively in Fig. 8.15. If the progress of work at estimated cost is to be maintained in accordance with the targets set by the budget and the network analysis, then the curve of 'value of work done' must pass between the early-and-late-start curves, the greater margin of safety as far as achieving the finish date being achieved if the curve is kept as close as possible to the early-start curve. If the 'value of work done' curve keeps within the envelope, but its trend is towards the late-start curve, this is an indication that some of the available float in the programme is being used up, and that the project progress as a whole is slackening with a tendency for future activities to become critical. Once the 'value of work done' curve pierces the envelope on the late-curve side, then some special action must be taken if the original programme target is to be achieved. In such a case, to carry out the remaining activities at the time and cost indicated in the original network will not be sufficient to ensure that the original objectives will be achieved. Conversely, if the value progress curve pierces the upper side of the envelope, productivity must be higher than estimated and too many resources (money might be one) are being applied in order to carry out the work faster than is necessary to achieve the original objectives. The curve of 'actual cost of work done' when compared with the curve of 'value of work done' indicates whether the work is being carried out at the same profitability as was estimated at the time of the original estimate.

8.11 Risk for an owner

Wherever an attempt is made to look into the future and forecast the outcome of activities which take place some weeks, months or years hence, there will exist considerable uncertainty about the definition of those outcomes. The parties that are involved with these forecasting processes are then at risk that the outcomes will not be achieved to the extent to which they were perceived. In carrying out a construction project the two parties who bear the bulk of the risk are the owner and the contractor. There may be other influences on the extent to which the risk is carried by each party if professional advisers in the form of architects, structural engineers, civil engineers, quantity surveyors, and so on, are engaged as part of the construction team to perform certain roles in the works usually predominantly concerned with design.

Risk and cost are directly proportional one to the other. In general, the greater the risk that exists, the greater the premium that must be paid to insure against it and the greater the possibility that larger costs will be incurred than were originally foreseen and not anticipated. On the other hand, in some situations, the possibility that larger gains may be made in a situation of considerable risk is greater than in a less risky but better defined situation. The assessment of risk is therefore important to an owner. It can be undertaken either intuitively or on a basis of statistics. It should also be noted that if due attention is paid by an owner to the likely areas of contractual risk at the early stages of the creation of a capital investment facility, the cost of avoiding unnecessary risk is likely to be very much less than it will be at later stages of the construction process. A small premium paid early is likely to avoid high and extensive costs later.

An owner who invests in a capital project, such as a new factory, is embarking upon a risky venture from the point of view of the capital investment since there is usually no guarantee that there will be a return of a specific predetermined magnitude when the factory becomes operational and the product is to be sold. In addition an owner has no guarantee that he can purchase the investment facility for the capital cost that has been estimated for the purposes of investment appraisal. However, when an owner starts to make arrangements for a construction project to go ahead, he is probably then in his strongest position to establish control of the risks in which he is to be involved. Before an owner begins to make the arrangements for the work, decisions need to be taken about the contractual conditions that will exist between the parties involved, the total organization that is to be used in order to deliver the facility, what the roles of the various participants in the organization will be, what is the timescale within which they need to act, to what extent can the risks involved in the construction be passed on to other participants and how the responsibility for risk can be devolved amongst these parties in a reasonable and a legally enforceable manner.

One of the first acts of an owner in instigating a construction project must be to specify, in very clear and unambiguous language, exactly what it is that is required to be available at the end of the construction process. The size of the facility that he requires, the quality of the workmanship, the general arrangement and the location of the various parts of the whole, the sizes and the quantity of the works involved and the conditions under which it will be carried out, such as any arrangements that may be required to blend the work in with the existing activities on adjacent sites, the facilities that will be available to contractors engaged in the work, the schedule of when the activities will need to be undertaken and any conditions which relate to poor performance by the contractor must be defined in precise terms. One of the features that an owner should try to avoid at this stage is the imposition of exculpatory conditions on the other parties to the project. An example might be an attempt to make a contractor bear the cost of all alterations to the design and construction that may result from the ground conditions, whatever they may be, being different from those that might reasonably have been expected. This would clearly be an attempt to place the total risk on a contractor for the state of the foundation conditions that are ultimately established and is unreasonable, thus placing the contractor at considerable risk if he accepts the condition without qualification. Many exculpatory conditions are unenforceable in law. They can only lead to very large contingencies being inserted by contractors in tendered sums to insure against the risk. Alternatively they may lead to unreasonably low tenders from contractors who are either completely unaware of the risks they run or are prepared to carry the risks at a considerable gamble in order to obtain the work in what may be a difficult time for them. The question of exculpatory conditions must always be at the back of the mind of the owner in establishing the contractual conditions for the design and construction phases of a project.

The next phase of the owner's activity will be that of deciding on the general organization which will be needed in order to undertake the design and construction of the work. Different hierarchical organizations are likely to place different measures of risk on the parties. With regard to the design of the project, the owner may be able to use professionals on his own in-house staff or he may have to appoint professional groups on a consulting basis to carry out the work. On very large projects the owner must bear in mind that if he uses a segmented approach to the appointment of professionals, using different groups to undertake the design of various sections of the work, that the difficulties of organizing these groups to provide a cohesive design approach to the whole project will place a considerable burden and risk on his own organization. The contractual relationships between these professional organizations must be very carefully designed and an in-house representative must be appointed who can monitor their output and the quality of the work that comes from them.

One significant area in which the owner has the opportunity of adjusting the degree of risk which he is prepared to bear himself and that which will be carried by a contractor is in the determination of the nature of the contract between an owner and a contractor. If the owner decides to place the work with a contractor on a *fixed price lump sum contract* he is then devolving a large part of the risk for escalation of cost on to the contractor. If the owner decides that it would be better for the work to be undertaken on a *cost plus basis*, then much of the risk for cost escalation will fall on the owner and will depend upon the extent of the control that the owner can exercise over the work as it progresses. If an *admeasurement contract* is to be used wherein the contractor prices a bill of quantities with unit rates, then the risk is divided between the two parties; the owner takes the risk that the quantities in the bill are correct and the contractor takes the risk that the unit prices which he inserts in the bill are correct.

The owner also has an opportunity to influence the degree of risk which he will bear when making the decision as to how a contractor will be appointed in relation to the nature of the contract. If the owner decides that the cost of the work will be negotiated with a single contractor, thus eliminating competition, the owner must face the difficulty of choosing a contractor who will offer him a competitive performance even though the contractor is not under competition for the work. The owner must make sure that he arranges the contractual conditions in such a way that he will attract good and reputable contractors to give a competitive price rather than those with relatively little experience of the work to be constructed and hence little experience of the risks that are involved.

Another class of risk in a construction project arises out of the nature of the work itself and will include such matters as delays caused by the weather or by conditions that were unforeseen at the time of the original design of the works; risks of inflation and in the cost of resources to be used in the work; errors made in the preparation of estimates of cost by a contractor and in the preparation of method statements describing the way in which the work will be carried out; difficult labour relations problems; shortages of specific trades in the labour requirements.

These risks arising out of the nature of the work are generally those that a contractor attempts to foresee when, in possession of the detailed contract documents, a tender needs to be prepared. Depending on the conditions which the owner has specified in the contract documents will be the extent to which a contractor needs to allow in his tendered price for the discounting of such risks. Given a competent and reputable contractor preparing a competitive tender, an owner may assume that the cost of insuring against the majority of the construction risks will be passed to the owner. However, the added risk for the owner is that a low tender will be submitted by a contractor who either does not fully recognize the construction risks or sees fit to gamble that they will not arise and if they do that he can pass the cost onto the owner. In this event, if the contractor is of low financial stability the owner may eventually incur much larger cost penalties than were originally thought feasible.

8.12 Evaluation of owner/contractor risk shares

Expected monetary values and utility theory have been referred to in Chapter 5. An owner and a contractor are likely to have different attitudes to risk reflected by their own individual utility curves. These differences are likely because of the different relative financial situations in which each will find himself, thus explaining their differing attitudes to large or small profits/ losses relative to their investments and/or the size of the contracts at stake. A very large owner with a multi-national business is likely to be indifferent to specific risks on a comparatively low value construction contract, whereas a small contractor may well conceive that those same risks are too great to be assumed by his own organization without including a substantial premium in the form of a contingency sum in his tender. These differences in attitude provide a basis by which an owner can decide on the most appropriate and desirable division of the risks between a contractor and his own organization. Reference should be made to the published work by Carr, Willenbrock and Erikson, Boyer and O'Connor who have all contributed to this area of work on construction-oriented applications of the use of utility theory.

Example 8.4

An owner proposes to place a contract for the construction of the foundations for a chemical processing plant. The plant will be supported on concrete bored pile foundations. The piles are expected to be founded on a layer of dense sand with an average pile length of 10 metres. There is a risk, however, that the piles may need to be longer. The owner's engineer assesses the probabilities that the piles will average 10, 12, 14, 16 or 18 metres long as 0.60, 0.10, 0.15, 0.10 and 0.05, respectively. The costs of using piles 10 metres long is calculated to be £500 000 with added contingency costs of £50 000, £90 000, £130 000 and £170 000 for piles of average length of 12, 14, 16 and 18 metres, respectively. Given that the contractor requires a mark-up of 20% on cost, has a utility as shown in Fig. 8.16 and that the owner is large enough to take a risk neutral attitude, how should the owner attempt to apportion the risk concerning pile length in order to minimize his probable cost of the foundations?

On the assumption that the contractor arrives at the risk-free estimate of £500 000 for piles at an average length of 10 metres (that is with no risk that he will be required to bear the cost of the additional lengths) he will add to this cost 20% for his mark-up, i.e. £100 000. The contractor then needs to assess the mark-up that he will add to take account of the possibility of having to bear the cost of the additional length of piles if he is required to carry the whole of that risk. This will need to be carried out taking into account his risk-averse attitude as demonstrated by his utility function illustrated in Fig. 8.16.

From the utility function a mark-up of £100 000 for the risk-free situation is shown to have a utility of 2301 utility units (utiles). The contractor will

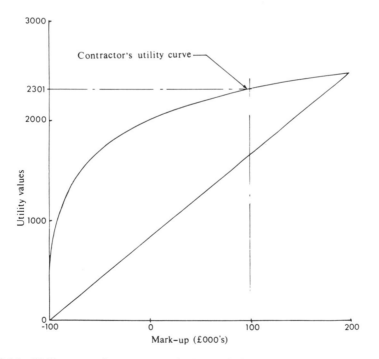

Fig. 8.16   Utility curve for contractor in Example 8.4.

therefore require to adjust his mark-up for a situation in which there are considerable risks of his costs increasing so as to maintain a similar utility value (i.e. 2301 utiles) to that of a risk-free contract. The contractor would calculate his expected utility value as follows:

$$EUV = U(M)0.60 + U(M - 50)0.10 + U(M - 90)0.15 +$$
$$U(M - 130)0.10 + U(M - 170)0.05 = 2301$$

where M = mark-up

The expression $U(M - 130)$ represents the utility of the mark-up less the contingency cost for additional pile lengths in each case.
Such values are in thousands of pounds.
The decimal fractions are the probabilities of each contingency occurring.

Solving the above expression for EUV gives a value of M = £149 000. The contractor would therefore require an additional £49 000 added to his mark-up in order to bear the risk of an increase in pile lengths.

The above equation can be solved by trial and error using the utility function for this purpose. This however would be very tedious and take, on occasions, some considerable time. If the function of the utility curve is known, or can be found by regression analysis, this is likely to be a more satisfactory means of solution. In the above situation the curve function is U(utility value) = 1000 log M where M is the mark-up.

The owner is expected to behave in a risk-neutral manner and will therefore decide on the basis of EMV:

$$EMV = (0)0.60 + (50)0.10 + (90)0.15 + (130)0.10 + (170)0.05$$
$$= \underline{£40\,000}$$

This EMV is to be compared with the additional £49 000 that would be required by the contractor as a risk premium for the additional lengths of piles. The owner would, therefore, pay an additional £9000 for the contractor to bear the risks.

8.13 Inflation and escalation in the construction industry

Inflation has been a serious problem in national economies in recent years. Fluctuations in the rate of inflation have caused serious problems in the economic processes of industry, none more so than in the construction industry. With the difficulties inherent in construction contracting and, in particular, with the difficulty of being unable in many instances to calculate a true rate of return for work undertaken on a construction project, the effects of inflation can cause serious difficulties for contractors. Whilst many contractors consider that they are as well protected as they can be by using some form of indexing in order to recompense them for future inflationary costs, others take a more naive and simple view and choose to ignore the possible effects of inflation altogether. In neither instance does a contractor protect his organization in the best possible way by feeling that a thorough review of inflation and its effects on a project during its construction period should be made, particularly by looking at the inflation forecasts using the best possible information which is at hand at the time of the planning process.

At the outset of a discussion on the increases in costs which take place throughout a construction project it is necessary to be clear as to the definition in use of two commonly occurring terms that are often used interchangeably and with confusion. In this context the term *inflation* is used to mean the sudden increase in the amount of money and/or credit which becomes available for use in relation to the quantity of goods and services which are on the market for sale resulting in a consequent increase in the price of goods and services. As well as causing this increase in price levels, another effect of inflation is to cause a gradual decrease in the purchasing power of a unit of money over time. Both of these effects of inflation, that is the general increase in price levels and a depreciation of the currency, are proportional to the rate of inflation.

The term *escalation* usually carries a less precise definition. The Study Group which was set up by the British Government in 1975 in order to investigate the escalation of costs of offshore oil and gas operations on the United Kingdom continental shelf, and which produced a series of recommendations in a report published by Her Majesty's Stationery Office, found considerable confusion over the definition of the term. Four different ways in which the term was used were identified and these have been discussed further by Likierman. The following definition and explanation of escalation follows broadly that set out in Likierman's paper based upon the work of the Study Group. Firstly, the Study Group's definition of escalation where project work was concerned was as follows:

'The difference between the original estimate of final cost of a project and the final out-turn cost or latest estimate of the final cost.'

Escalation in this context includes unanticipated costs that were not included in an original estimate. It may well be that in compiling an estimate for a project, an estimator includes a forecast of likely escalation throughout the course of the work. This may not be sufficient, however (or it may be too much) and other increases of cost which increase either the final estimated or actual cost of a project would form the escalation for the project cost. It also needs to be borne in mind when considering such escalation that there may well be inaccuracies in the preparation of an estimate which will vary the total escalation and very possibly be self-compensating when combined with the costs of inflation.

Table 8.12 sets out the estimated expenditure for a hypothetical project

Table 8.12

Project year	Expenditure £m	8% input cost increase p.a.	Further 10% input cost increase p.a.	Unanticipated resource increase 6% Years 2–4
1	25	25 × 100.00% = 25.00	25 × 100.00% = 25.00	25 × 100.00% = 25.00
2	25	25 × 108.00% = 27.00	25 × 118.00% = 29.50	26.5 × 118.00% = 31.27
3	25	25 × 116.63% = 29.16	25 × 139.23% = 34.81	26.5 × 139.23% = 36.90
4	25	25 × 125.97% = 31.49	25 × 164.30% = 41.08	26.5 × 164.30% = 43.54
	£100m	£112.65m	£130.39m	£136.71m

which is to take four years in construction. The expenditure is uniform over the four years at £25 million per year. These expenditure figures are at current prices and an adjustment is made in Column 3 to allow, at the initial stage, for an 8 per cent per year cost increase. These are referred to as the anticipated increase in input costs (AIIC). Column 4 of Table 8.12 adjusts those figures on the basis that the figures, including the anticipated 8 per cent increase for each year, are subjected to an additional 10 per cent per year cost increase that was not anticipated at the initial estimating stage. These are referred to as the unanticipated increase in input costs (UICC). The initial estimated cost including anticipated increases therefore amounts to £112.65 million and, because of the unanticipated increases in costs of 10 per cent per year, this equivalent figure is now increased to £130.39 million per year. The final column of Table 8.12 illustrates the effect of underestimating the resources that are required to complete the project in years two, three and four of the construction period. The underestimate amounts to 6 per cent and may be called the unanticipated increase in resource input (UIRI). Thus the estimate is again revised to take account of this and the total and final estimated cost is therefore £136.71 million.

The analysis of these totals of cost can be made as follows:

$$\text{AIIC} = \frac{(112.65 - 100.00)}{100.00} \times 100 = 12.65\%$$

$$\text{UICC} = \frac{(130.39 - 112.65)}{112.65} \times 100 = 15.75\%$$

$$\text{UIRI} = \frac{(136.71 - 130.39)}{130.39} \times 100 = 4.85\%$$

The combined project inflation and escalation elements can therefore be calculated as:

$$\text{Total increased cost} = 1.1265 \times 1.1575 \times 1.0485 = 1.3671\,(36.72\%)$$
$$\text{Project inflation} = 1.1265 \times 1.1575 = 1.3039\,(30.39\%)$$
$$\text{Project escalation} = 1.1575 \times 1.0485 = 1.2136\,(21.36\%)$$

8.14 Valuations, depreciation and replacement

Once the construction project has been completed it passes to the owner who then commissions and uses it for its designed operations. Sooner or later the owner will need to value the plant and buildings as an economic enterprise and perhaps to check periodically on the return being received for the initial investment plus working capital. There are a number of principles that can be used in order to establish value.

Previously, a decision has been made as to whether a project is viable or not by calculating a rate of return and comparing it with the minimum rate that is acceptable. Alternatively, the future cash flows arising from a project have been discounted at the required rate of return and then compared with the initial capital investment. In both instances an attempt has been made to establish the *value* of the project under certain given conditions. Given, for example, a positive net present value as a result of the latter calculation above, and a reasonable degree of risk, a decision-maker might well be inclined to the view that it will be better to invest in the project rather than keep the finance on deposit in the bank. The value of the project to a decision-maker, in such a case, is said to be greater than that of the money when taking everything into account.

By examining this concept of decision-making, it is possible to establish a definition of the value of an asset, in an economic sense, as being *the present value, at the owner's cost of capital, of the likely future net cash flows which will arise from possessing it*. If, for example, it is required to value a revenue-producing property, an estimate of all the expenses and income arising from it will need to be made, a minimum acceptable rate of return will have to be established and a calculation of net value made using these data.

Example 8.5

What is the estimated economic value of a block of flats which produces a net revenue of £250 000 per year, has an expected further life of 20 years and a residual value, at the end of that time, of £500 000 (for the site)? A likely purchaser's minimum acceptable rate of return after tax is 8 per cent.

$$\text{Present value of net revenue} = (P/A, 8\%, 20) \times 250\,000$$
$$= 9.82 \times 250\,00 = £2\,455\,000$$
$$\text{Present value of residual value} = (P/F, 8\%, 20) \times 500\,000$$
$$= 0.215 \times 500\,000 = \underline{\quad £107\,500}$$
$$\text{Value of flats} \quad \underline{\underline{£2\,562\,500}}$$

If the purchaser needs to spend £100 000 on structural repairs and decoration at the time of purchase and before putting the block into revenue-producing use, the value will be reduced to £2 462 500.

An alternative way of viewing the value of an asset is to establish the cost of its replacement, making due allowance for any delay that may arise in obtaining a suitable replacement and then putting it to work. Such a value will, of course, be a relatively high one since the replacement, if new, will be in better condition and, it is assumed, will have a longer potential life than the existing asset.

Value may also be established by selling the asset, though if its nature is such that it has only limited use on the general market, the price received may not represent the *use-value* to the present owner. The *market value* of very specialized equipment, for example, may be quite different from the value placed on it by the owner who can, one presumes, utilize it to a high degree. This is another way of confirming value as being the present value of an asset's associated future cash flows, since these will be quite different for different owners depending on the use to which they can put it. The price which an owner can obtain on the open market for one of his assets may be thought of as the lower limit of its value. This is a sum of money which can be obtained now, regardless of the future cash flows that might arise from owning it. Clearly, if the present value of future net cash flows falls below this value, the asset should be sold.

To establish a reasonable value for an asset which has already given some service to its owner and is now secondhand, it is necessary to compare the cash flows which will arise from it with those that should arise if the owner suddenly finds himself without it. Such a situation tends to lend itself to a demonstration of financial principles rather than to a practical situation since, of necessity, a considerable amount of estimation and assumption must be undertaken. Neither is a replacement situation always free from many other associated problems. The following example illustrates the application of discounting principles to a valuation situation.

Example 8.6

A company owns a mobile crane which it purchased four years ago. Previous experience with this type of crane leads the company to believe that it will have a life of ten years when it will then have a resale value of £2000. If the company were to purchase a replacement crane today to provide a similar duty, it would buy a model having an initial cost of £10 000, an estimated life of eight years and an estimated resale value after that life of £2500. The operating costs of the existing machine are £1000 per year, but in the case of the replacement machine are expected to drop to £800 per year. If the cost of capital to the company is 12 per cent, calculate the value of the existing crane.

The net cash flow for both machines, assuming that the existing crane will be replaced by the newer alternative at the end of its present life and that both will be replaced indefinitely, are listed in Table 8.13.

Converting the operating cost of the new crane to a single payment at the commencement of its working life,

Table 8.13

Year	Existing crane	New crane
0	—	£10 000
1	£1000	800
2	1000	800
3	1000	800
4	1000	800
5	1000	800
6	10 000 + 1000 − 2000	800
7	800	800
8	800	10 000 + 800 − 2500
9	800	800
10	800	800
11	800	800
∞		

$$P_1 = (P/A, 12\%, 8) \times 800 = 4.968 \times 800 = \underline{£3974}$$

The present value of its resale value

$$P_2 = (P/A, 12\%, 8) \times 2500 = 0.404 \times 2500 = \underline{£1010}$$

For the new crane, therefore, the total costs can be represented by a lump sum of £10 000 + 3974 − 1010 = £12 964 every eight years including Year 0.

If a payment, A, is made every t years, the present value of such a series in perpetuity, P, commencing in t years' time amounts to

$$P = \frac{A}{(1+i)^t} + \frac{A}{(1+i)^{2t}} + \frac{A}{(1+i)^{3t}} + \dots + \frac{A}{(1+i)^\infty} \tag{8.1}$$

Multiplying both sides of the equation by $(1 + i)^t$,

$$P(1+i)^t = A + \frac{A}{(1+i)^t} + \frac{A}{(1+i)^{2t}} + \dots + \frac{A}{(1+i)^\infty} \tag{8.2}$$

Subtracting Equation (8.1) from (8.2),

$$P(1+i)t - P = A = P[(1+i)^t - 1]$$

or

$$P = \frac{A}{(1+i)^t - 1} \tag{8.3}$$

In this example, the present value of a payment of £12 964 every eight years, plus an initial payment of that sum at Year 0,

$$= \frac{12\,964}{(1 + 0.12)^8 - 1} + 12\,964 = £21\,748$$

For the existing crane the present value of the costs from Year 6 onwards (with the exception of the operating cost and resale value for the existing crane in that year) will be £21 748 as above at Year 6. At Year 0 they will be

$$\frac{21\,748}{(1+0.12)^6} = \underline{£11\,018}$$

The present value of the operating costs of the existing asset from 0–6 years are,

$$1000 \times (P/A, 12\%, 6) = 1000 \times 4.111 = \underline{£4111}$$

Present value of resale value, six years hence,

$$= \frac{2000}{(1+0.12)^6} = \underline{£1013}$$

Therefore, present value of net cash flows of existing asset =
$$£11\,018 + 4111 - 1013 = \underline{£14\,116}$$

$$\text{Value of existing asset} = £21\,748 - 14\,116 = \underline{£7632}$$

This example has been worked through assuming that the company will always require the service of such a crane and whether they have to replace it now, because of its sudden 'contrived disappearance', or later, because of wear and tear, they will do so with a crane having known associated cash flows.

Sometimes an asset is valued in a more rigorously formal way than by discounting the future cash flows which are expected to arise from possessing it. Accounting conventions, and those imposed by tax laws are used in such a way that the *book value* of an asset is decreased as time goes by, so as to reflect the reduction in its value as a result of wear and tear. The accounting function of depreciation allowances is to prevent the flow of excessive funds to the shareholders or other outside bodies by way of dividends and other returns. Before profit is distributed, therefore, a deduction for depreciation is made from it so that sufficient funds are retained inside a company to offset the reduction in value of the related assets. The method and extent of annual depreciation as shown in company accounts are the result of a decision by the management of the company and in no way represent a cash flow since they are purely book transactions. Depreciation represents one part of the capital that is ploughed back into a business and its purpose is no more than to show the cost of an asset as an expense of the business in the long run. Different types of assets do have, in the true sense, different rates of depreciation, that is rates of obsolescence and wear and tear. Assets tend to drawn together into fairly large groups, each containing a wide variety of plant, buildings and equipment, and therefore rates of depreciation as used in accounts tend to be somewhat arbitrary and average in application.

In the case of tax laws, capital allowances can be set against profits before the tax payment to be made are calculated. Such allowances are usually called initial allowances and writing-down allowances. An initial allowance frequently takes the form of an accelerated writing-down allowance, which is allowed in the first year of the use of an asset or when it is initially purchased.

There are several different methods of calculating both depreciation for accounting purposes and allowances for tax purposes. Two common methods will be demonstrated by the use of examples.

Example 8.7

Calculate, using the *straight line method*, the annual depreciation for a motor car having an initial cost of £10 000 and a secondhand value of £2000 after a five-year life. What is its book value after three years?

Straight line depreciation

$$= \frac{\text{initial cost} - \text{estimated resale value}}{\text{estimated life}}$$

$$= \frac{£10\,000 - 2000}{5}$$

$$= \underline{£1600 \text{ per year}}$$

Book value = initial cost − depreciation to date
$$= £10\,000 - (3 \times 1600)$$
$$= \underline{\underline{£5200}}$$

Example 8.8

Calculate, by the *diminishing balance method*, the depreciation for each of the first three years of the life of a truck having an initial cost of £20 000 and a diminishing balance percentage of 20 per cent. (This method is also known as reducing balance or declining balance.) Calculate in two ways the book value of the truck at the end of three years as a result of applying the diminishing balance method.

Initial cost	£20 000
Year 1 depreciation	4 000, i.e. 20% of £20 000
Book value at end of Year 1	£16 000
Year 2 depreciation	£3 200, i.e. 20% of £16 000
Book value at end of Year 2	£12 800
Year 3 depreciation	2 560, i.e. 20% of £12 800
Book value at end of Year 3	£10 240

An alternative means of calculating the book value is by using the formula

$$\text{Book value, } B = P(1 - d)^k \tag{8.4}$$

where

P = initial cost of asset
d = declining balance percentage as a decimal
k = expired life of asset

therefore in this case,

$$B = 20\,000(1 - 0.20)^3$$
$$= \text{£}10\,240$$

Straight line depreciation has the advantage of simplicity and the fact that it is possible to write-off or depreciate an asset to zero value, an advantage not shared by the reducing balance method. The rate of depreciation, however, is constant – an equal amount each year – a pattern which is rarely followed in the values of assets. The reducing balance method has the advantage that it allows a greater rate of depreciation in the earlier years of the life of an asset, reducing as its life proceeds. This means that the total annual costs of an asset are likely to be evened out to a uniform sum, since maintenance and repair costs are almost always low in the early years of its life and then tend to increase as time goes by.

Valuations of assets calculated in the above fashion can eventually result in book values which have little or no relationship to real market values. A good example is that of property where, in times of inflation, the market value will tend to increase rather than depreciate. Hence, periodically, a company's assets will be revalued so that their values in the company's books of account will reflect their true market worth. This is largely due, among other things, to the fact that the estimation of an asset's life is implicit in setting the rate at which it will be depreciated. The life of assets, particularly for example buildings, is often grossly understated when the rate is set initially. This means that in addition to generally inflationary values, the book values of assets can be widely at variance with the true value.

The questions of replacing and retiring assets place a different emphasis on the appraisal of an investment and take into account the time-value of money. Firstly, because the asset to be replaced already exists (and presumably its original purchase and continued existence has therefore been justified by the appropriate methods), there is rarely the inclination to examine with such exactitude the financial and other characteristics of all possible replacement assets. Secondly, the replacement must compete against an asset which already exists. The analysis of such a situation can be more complex than that of one in which an investment is made to expand a business or process.

The general replacement situation is one in which the increasing maintenance and repair costs of an asset, coupled with reducing production efficiency, make it questionable as to whether the asset can be continued in service in view of the other modern equipment which is available for purchase. The principal objective of a replacement study is to locate the best possible point in time when the replacement, or possibly the retirement, should take place.

In locating the point in the life of an asset at which it should be replaced with one that provides a similar service (on the assumption that the opportunity is not taken to provide other facilities, as well as simply making a true replacement with like for like) it is necessary to have knowledge of the resale market value and the operating or running costs for the asset.

Example 8.9

A machine tool plays an essential part in the production process of a standard part for a motor car. It is unlikely that the function of the machine tool will be superseded in the foreseeable future. The tool costs a sum of £10 000 and studies of its maintenance costs (including an allowance for reduction in efficiency with age) show that these are £300, £400, £500, £1000, £1500, £3000, £4000 and £4000 for Years 1 to 8 of life, respectively. The market value of the machine tool is estimated to be £8000, £6000, £4000, £3000, £2500, £2000, £1500 and £1000 at the end of Years 1 to 8, respectively. If capital cost the company 10 per cent, when should the machine be replaced?

The problem is easier to deal with using a tabulation of the form used in Table 8.14. In Column 1, the calendar of the asset's life as it is known is listed, and in Column 2 are listed the cash expenditures, commencing with the initial capital outlay and followed by the estimated maintenance and repair costs. It will be noted, as is usual in such cases, that the maintenance and repair costs tend to increase year by year with a larger than usual increase coinciding with a major overhaul of the machine at a later stage of its life. The third column contains further information which is given as far as this particular problem is concerned – the estimated resale values of the machine at the end of each year. For example, it is believed that such a machine will have a resale value of £2500 at the end of five years of its normal life.

The remainder of Table 8.14 consists of a series of valuations, as have already been demonstrated, on a similar basis to those carried out in Example 8.6, assuming that the machine might be replaced at the end of each year in sequence. The first stage is to calculate the present worth of one cycle of cash flows, that is capital investment plus maintenance and repair costs less resale value, where the cycle is from 0 to 8 years (in this case) in increments of one year. In Column 4 are listed the present value factors for the company's cost of capital of 10 per cent and in Column 5 are the present values of the capital and maintenance costs listed in Column 2 at Year 0. For

Table 8.14

Year	Cash Expenditure (£)	Resale Values (£)	Present Value Factors 10%	Present Value of Operating Costs (Cols 2 × 4) (£)	Progressive Cumulative Total of Col. 5 (£)	Present Value of Resale Values (Cols 3 × 4) (£)	Present Value of Life Payments (Cols 6 + 7)	$1 - (1 + i)^{-t}$	$\dfrac{A}{1 - (1 + i)^{-t}}$
1	2	3	4	5	6	7	8	9	10
0	−10 000	10 000	1·000	−10 000	−10 000	10 000	—	—	—
1	−300	8 000	0·909	−273	−10 273	7 272	−3 001	0·0909	−33 014
2	−400	6 000	0·826	−330	−10 603	4 956	−5 647	0·1736	−32 529
3	−500	4 000	0·751	−376	−10 979	3 004	−7 975	0·2487	−32 067
4	−1 000	3 000	0·683	−683	−11 662	2 049	−9 613	0·3170	−30 325
5	−1 500	2 500	0·621	−932	−12 594	1 553	−11 041	0·3791	−29 124
6	−3 000	2 000	0·565	−1 695	−14 269	1 130	−13 139	0·4355	−30 170
7	−4 000	1 500	0·513	−2 052	−16 341	770	−15 571	0·4868	−31 986
8	−4 000	1 000	0·467	−1 868	−18 209	467	−17 742	0·5335	−33 255

example, the present value of the maintenance cost in Year 5, £1500, is £932.

Column 6 shows a progressive total of the initial capital cost plus the present worth of the maintenance costs which would have been incurred up to that time if the machine were then replaced. As an example, if the machine were replaced at the end of Year 4, the present worth of £10 000 at Year 0, £273 at the end of Year 1, £330 at the end of Year 2, £376 at the end of Year 3 and £683 at the end of Year 4, amounts to £11 662.

Column 7 lists the present values of the resale values, obtained by multiplying Columns 3 and 4 together. The equivalent lump sum which will occur at the regular interval of each life chosen for the machine can now be established. This is obtained, as in Example 8.6, by adding Columns 6 and 7 together to give Column 8.

By using Equation (8.3), the sum listed in Column 8 can now be converted to a capitalized cost varying with the life cycle of the asset. The capitalized cost gives the present value of having the asset in service in perpetuity. The denominator of the right-hand side of Equation (8.3) for varying values of *t* is listed in Column 9 and the whole of the right-hand side of this equation is evaluated in Column 10. It will be seen that the capitalized costs fall up to a life-cycle of five years and then begin to increase. Clearly it is desirable to possess assets which can be financed at a minimum and the recommendation for replacement in this example would therefore be after five years.

Very often it is unnecessary to make the calculations for a whole series of life cycles, since by looking at the cash expenditure column the likely replacement life can be identified. This often occurs where a major replacement or overhaul of a machine takes place, causing a sudden jump in the maintenance and repair expenditure for that time. For example, in Column 2, of Table 8.14, it is highly likely from inspection that the life cycle would be four, five or six years and the calculations would be carried through in the first instance for those figures only. By displaying the figures graphically, the point of replacement is illustrated much more clearly. Figure 8.17 shows the variation in capitalized cost of the asset analysis in Example 8.9 with a varying life cycle. The low point of the curve is clearly the period of time after which one should replace the asset.

8.15 Replacement – lease or buy?

No two owners are likely to agree in every detail about the costs of owning and operating mechanical equipment. All owners will have a different method of calculating plant charges, in which depreciation plays a large part, and of collecting the costs which are relevant to the charges. In addition to the well-defined costs, such as capital cost, it is frequently very difficult to establish the true cost of down-time delays and their true effect on other aspects of the work, such as delays in a construction programme. Obsoles-

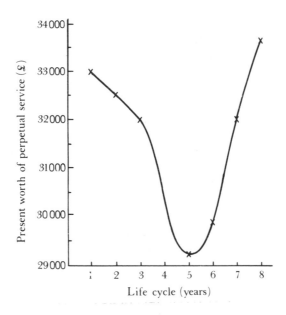

Fig. 8.17 Capitalized cost curve for replacement decision.

cence is a difficult aspect of cost to establish, as is the general lack of availability of some of the older items of equipment, which in turn often hinges on the type of work which the equipment is expected to carry out. It is essential, therefore, that the users of mechanical equipment of this nature should make every effort to collect and record as much cost and time data as possible.

In examining the appropriate time at which an item of equipment should be replaced, the basic premise that the capitalized cost of the equipment should be minimized was accepted. An alternative objective in setting a replacement target can just as readily, and possibly more appropriately, be the optimization of profits which arise from the use of the equipment. To be realistic, profits must be calculated after the payment of tax. James Douglas has carried out some interesting and useful research into the optimum life of equipment of this nature, based on the objective of maximizing profits. He has developed a mathematical model which clearly demonstrates that the economic life for maximum profit is often much shorter than the economic life based on minimizing total investment and running costs. Of course, it is not always possible to look at each individual machine and isolate costs and profits even with the most sophisticated of control methods. Nor is it always politic, because of liquidity problems for example, to purchase new equipment at a specific time. Douglas includes in his model facilities for taking account of the time-value of money, taxation, inflation, increasing costs of borrowing, replacements to provide perpetual availability and obsolescence.

Having made an analysis of the costs associated with a particular item of equipment Douglas is able to produce a curve of the general form shown in Fig. 8.18. The curve is drawn so that the present value of profits is represented by the ordinates and the age of replacement on the abscissae. The shape of this curve clearly illustrates how many items of equipment may be retained beyond the period at which it maximizes profits. In fact Douglas draws as one of his conclusions, the fact that an item of mechanical equipment has an economic life which is much shorter than that over which it can return profits as a result of its operation. Such a model as this enables a thorough investigation to be made of the reactions of the variables involved on the outcome of each situation.

With the replacement of equipment comes an examination of whether the replacement should be leased or bought outright. The popularity of leasing buildings, plant, equipment and machinery has increased in recent years. The principal reasons for this are twofold. Firstly, leasing assets does conserve liquid capital which might otherwise be absorbed by outright purchase. Secondly, rental payments made under leasing conditions are usually chargeable immediately against taxation, whereas writing-down allowances against profits for purchased assets do not, if ever, completely cover the whole cost of those assets.

There are, however, disadvantages in leasing equipment and other assets. In some countries, investment incentives can ultimately lead to more than the initial cost of the asset being charged against taxation where it is purchased

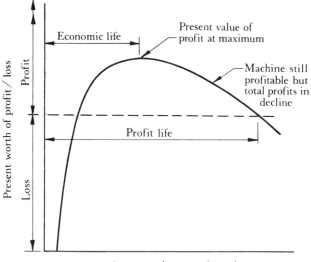

Fig. 8.18 Profit against replacement age.

outright. Probably more important, however, is the effect that leasing can have on a company's balance sheet. No indication is given in the balance sheet of a quoted company as to the nature or magnitude of assets which are leased. Because ownership of them is not vested in the company itself, they do not appear as fixed assets. Neither is the rental identified as such in the profit and loss account of a company though it does effectively reduce that profit, which should be available at some later date for distribution by way of dividends and interest. The fact that leased assets do not appear as fixed assets of a company in its balance sheet or in a company's books, may make it difficult for the company to borrow or raise money at some later date. Effectively they reduce the substance of the company. These factors are some of those that must be taken into account when making a decision as to whether to lease or buy plant, equipment, buildings, etc., for use by a company.

Care must be taken, when assessing a lease or buy situation, to ascertain whether for example, a comprehensive maintenance service is provided free or alternatively at low cost to the user. It may be expensive, or even impossible for the company requiring to lease the equipment to provide suitable facilities for such a service. There may be long-term advantages by way of cheaper replacements for leased equipment on the first and sub-sequent renewals which result from the agreement. However, consideration of the advisability of leasing as opposed to buying, should only come after a realistic analysis, in the first place, of whether the asset is required at all. One popular method of raising finance is the sale and lease-back type of arrangement. The user owner of a building, of land, of equipment, sells his asset to a financial institution and immediately executes an agreement to lease the asset back for a specific number of years. The owner then has the use of the capital, and the financial institution has an investment which will return its money plus a calculated rate of return.

Tax consideration must, of course, be taken into account in arriving at decisions as to whether equipment should be leased or bought. This is implemented in the following example, which is designed to illustrate and explain the principles of the method.

Example 8.10

A company wishes to invest in an industrial factory building having an initial cost of £300 000. A property development company offers to provide a similar and equivalent building on a 21-year lease for a rental of £90 000 per year, payable in advance. In addition to the lease rental the company will have to pay taxes of £300 per year, and the cost of repairs and maintenance is estimated to be £5000 per year. However, these items of taxes and repairs and maintenance are common to both the purchase scheme and the rental scheme.

If the factory is purchased, the owning company will be able to write-down the cost against profits at the rate of 4 per cent per year, straight line. Corporation tax is currently payable on profits at a rate of 40 per cent and it may be assumed that this rate will not change in the future. Corporation tax is payable one year in arrears. Should the company lease the building, the rent qualifies as a deductible expense for tax purposes. The estimated resale value of the factory after 21 years is considered not to be more than £150 000 because of the rather specialized structure that is involved. The cost of capital to finance the purchase of the factory amounts to 10 per cent.

Since the taxes, repairs and maintenance are common expenditure, these items can be ignored in the analysis. The analysis really becomes a choice between two alternatives and may be classified as a cost-reduction exercise. If the factory is purchased outright, then the saving will be the reduction in rental over the period of the lease. By using the present value method the alternatives can be evaluated as follows:

Present value of purchase scheme

$$= 300\,000 - \left(\frac{4}{100} \times 300\,000 \times \frac{40}{100}\right) \times (P/A, 10\%, 22) - (P/A, 10\%, 1)$$

$$- 150\,000 \times (P/F, 10\%, 21) + {}^{*}(150\,000 - 60\,000)\,\frac{40}{100}\,(P/F, 10\%, 21)$$

$$= 300\,000 - 4800 \times (8.7715 - 0.9090) - 150\,000 \times 0.13513$$

$$+ 36\,000 \times 0.13513 = \underline{£246\,855}$$

Present value of rental scheme

$$= 90\,000 + 90\,000(P/A, 10\%, 20) - \left(\frac{40}{100} \times 90\,000\right)\,(P/A, 10\%, 21)$$

$$= 90\,000 + 90\,000 \times 8.5135 - 36\,000 \times 8.6486$$

$$= 90\,000 + 766\,215 - 311\,350$$

$$= \underline{£544\,865}$$

* Balancing charge

The purchase scheme is clearly the better of the two proposals. If there were little to choose between the two schemes on the basis of present value, as the cost of each is much the same, other considerations particularly with regard to the risks involved must be taken into account. One might reasonably expect to obtain two similar present values if the property development company and the industrial company view risk of investment in similar terms. Other factors that might now be expected to play a part in the decision are, for example, whether the firm would like the opportunity of rethinking its asset distribution in 21 years' time; whether the cost of capital is going to increase a great deal over the next 21 years; whether inflation is going to have a significant effect on either alternative, and so on.

A number of factors to be considered in making the decision of whether to lease or buy are set out above and it will be noted that some are quantitative and others are qualitative. In summary the key considerations are:

(a) the immediate availability of capital for investment;
(b) the anticipated period of use – the shorter the period the more likely is leasing to be the appropriate solution;
(c) the effect of interest rates both current and future and the required rate of return;
(d) the effects of present and possible future Government taxation policies;
(e) the state of technology in machine, etc., design and practice – equipment subject to rapid developments in technology, e.g. computers, may be better rented or leased;
(f) potential further use of the equipment beyond the present needs – highly specialized equipment often has little reuse value and, in any event, is frequently unavailable for rental;
(g) whether an option to purchase after a minimum period of rental or lease can be negotiated at the outset in the case of new models or technologically advanced equipment;
(h) whether the equipment needs specialized skills for maintenance and repair that may be more readily available under a leasing agreement;
(i) whether the equipment, particularly small items, will be needed on sites which are long distances apart and will hence involve heavy transport costs if moved between them in relation to their value.

The quantitative evaluation of hiring versus purchase is particularly important where interest rates are high with a possibility of fluctuations. In these cases the timing of the various cash flows is of critical importance. The qualitative judgements are no less important in the overall context but a thorough quantitative assessment may well eliminate any need for them.

Problems

8.1 The XYZ Company Ltd borrows £5 000 000 on 1 January 1990 at an interest rate of 8 per cent per year payable half-yearly, in order to construct a large factory. In order not to have the money idle it is all deposited in a bank account which brings a return of 8 per cent. Funds are withdrawn from this deposit account as they are required to pay the contractors who are building the factory and supplying the production equipment.

The factory takes three years to complete to the stage where it can be used for production purposes. The payments to be made to the contractors from the account are estimated to be as follows:

	£
30 June 1990	500 000
30 June 1991	1 000 000
30 June 1992	2 500 000
30 June 1993	1 000 000

During the year ended 31 December 1993, the firm expects to make a profit of £200 000 on production from the new factory. In each subsequent year it is expected that the net profit will be £700 000 per year.

How soon will the firm be able to pay back the loan and the interest thereon assuming that it devotes the whole of the net profits as they arise each year, to doing so? Illustrate your answer graphically.

8.2 A company is in need of approximately £750 000 of capital. The Board of Directors considers a number of ways of raising this amount of money, the first of which is to issue more equity stock. It is estimated that capital raised in this way will cost the company $8\frac{1}{4}$ per cent.

The only other viable alternative at this time is considered to be the sale of the company's Head Office building on the basis that it can be leased back from the purchaser for the remaining 30 years of its estimated useful life. This situation is investigated further and an offer of £750 000 is received for the building. The arrangement for lease-back rental is offered in two alternative forms:

(a) Rent for each of the first ten years is to be £100 000. The rent is then to be increased to take into account inflation in the previous ten years at the rate of 3 per cent per year; the adjusted rent to remain in force until the end of the 20th year when a further increase will be applied, again taking account of inflation at 3 per cent per year for each of the previous ten years. This third rate of rental is then to remain in force for the rest of the building's 30-year life.

or

(b) A flat rate rental of £120 000 per year for the whole of the 30 years.

The company, if it sells and leases back, is then allowed the rental of the building against profits for taxation purposes. It is anticipated that there will always be sufficient profit to take full advantage of this. A corporation tax at the rate of 45 per cent of profits is payable one year in arrears. It is estimated that it will remain at this rate for the future duration of the lease.

If the rent is payable on an annual basis in advance, which is the cheapest source of capital for the company?

8.3 The following table sets out the data provided by the administrative section of a project manager's office concerning the financial position of the project at the end of each of three consecutive months during the middle of the programme. Produce a variance analysis for the project at the end of each of these three months and comment on the analysis.

End of month	Budgeted value of work to be completed (£)	Budget value of work actually completed (£)	Budget value of work to completion (£)
14	62 720	60 320	65 000
15	75 000	70 000	55 000
16	90 000	89 000	42 000

Total budget of work as planned (£)	Actual cost of work completed (£)	Revised cost estimate of work to be completed (£)	Revised total estimated cost of project (£)
127 720	62 000	68 000	130 000
130 000	71 000	61 000	132 000
132 000	89 000	53 000	142 000

8.4 A road haulage company owns a fleet of 15 vehicles each of which has a current market second-hand value of £2500. The time has come for these vehicles to be replaced and quotations for new vehicles are received at £17 500 each. The company has a policy to keep each vehicle for three years and at that time it expects that if it buys new ones, they will in turn be worth £5000 each on the second-hand market. The running costs of each new vehicle are expected to be £2000 per year less than the existing ones.

On enquiry, it is found that each new truck can alternatively be leased at a rate of £7300 per year. If the rate of return for the company is 9 per cent, which method should it adopt?

8.5 Draw curves to show the variation in book value of an asset over its life when depreciated by the straight line and diminishing balance methods. Assume that the asset has an initial cost of £10 000, a salvage value of

£2000 and a life of eight years. Draw curves for the same asset and depreciation method showing annual depreciation against the age of the asset.

8.6 On examining the records of one particular type of compressor (costing £15 500 when new) it is established that annual maintenance costs increase approximately in accordance with the expression

$$c_y = 300 + 100(1.50^y)$$

where c_y is the annual maintenance cost (£) for any particular year, y, of the excavator's life. Establish the value of such a machine when it is three years old, on the basis that its equivalent replacement at that time will cost £18 000 and will have estimated uniform maintenance costs of £650 per year. The working life of the existing compressor is estimated to be seven years and that for the possible replacement to be six years. Each machine can be sold for £1000 at the end of its working life. Cost of capital to the company is currently assessed at 12 per cent.

Assuming that the resale value of the existing machine reduces by £3000 in each of the first three years and then £1000 for each remaining year of its life, establish the optimum replacement time of the machine.

8.7 A firm of consulting engineers has been using the same equipment for the reproduction of drawings over a period of ten years and at the end of this period takes a decision to buy new equipment because the maintenance costs of the existing machine appear to be getting excessive.

Table 8.15 lists the annual maintenance costs over the period of ten years and the estimated resale value for the equipment at the end of each year. If the equipment had an initial cost of £50 000 when installed what would have been the optimum time to replace it if the cost of capital to the firm is 10 per cent?

Table 8.15

Year	Maintenance costs	Resale value
1	3 000	45 000
2	3 500	40 000
3	4 000	34 000
4	4 700	26 000
5	6 500	22 000
6	8 200	20 000
7	10 000	17 500
8	12 000	15 000
9	14 000	12 000
10	16 500	9 000

8.8 A limited liability company operating in plant hire wishes to carry out a valuation of some of the equipment it owns which has been in operation for some years.

One particular item of equipment has been in use for two years of its estimated economic life of nine years. At the end of its economic life it is estimated that it will have a resale value of £10 000. A new replacement for this machine at today's price will cost £55 000 and it will have a similar economic life and resale price to that of the existing machine.

The annual operating and maintenance costs of the existing machine are estimated to be £15 500 for the third year of service and thereafter to increase by £1000 per year until the time comes to dispose of the machine. A new replacement machine to provide the same service has the advantage that annual operating and maintenance costs are estimated to be lower than the existing machine and to start at £12 000 for the first year increasing by £1200 per year throughout the machine's life. If the cost of capital to the company is 10 per cent, what is the value of the machine to the company at the present time?

8.9 A contractor wishes to obtain a road-roller for use on a contract which has an expected duration of three years. He can purchase a new roller outright for £80 000 and in this case his estimated annual operating costs will amount to £32 000. The expected second-hand value of this machine is £60 000 at the end of the contract. His alternative is to buy a second-hand machine which cost £100 000 when it was new ten years ago. The purchase price now of this machine is £25 000 with an expected sale price of £5000 at the end of the contract. If the company's required rate of return is 12 per cent and the operating cost of the second-hand machine is £54 000 per year, what is the value of the second-hand machine to the contractor?

Chapter 9
The Contractor

9.1 Working capital

A considerable amount of attention needs to be paid, in all companies, to the management of the working capital and liquid resources of those companies. The working capital of a company is defined as a company's total current assets. Current assets are made up of cash, stocks or raw materials and those incorporated in work in progress, the debtors of the company or current accounts receivable and short-term securities. Short-term securities may, for example, include tax reserve certificates bought against the future payment of company tax. Working capital is required by a company for many of its regular short-term commitments; to pay salaries and wages, to buy raw materials for the construction of the works, to pay for hiring and operating plant and to provide the money required for the payment of interest, dividends and taxes as they fall due. In the case of some companies, the working capital in use or available will exceed the investments of that company in fixed assets. The need for working capital arises because of a *delay* between the expenditure on resources and the payment for goods subsequently provided.

Each of a firm's major undertakings, whether a capital investment project for the production of a consumer good or a contract undertaken by a building contractor, should be appraised and sanctioned by the company's senior management. In this respect it is vital that an overall view of the firm's total activities is taken and that each major commitment should fit into a firm's long-term and short-term plans. The restricted availability of working capital may often limit the progress which the firm can make in expanding its business by taking on new work. Frequently it will be necessary for a firm's senior management to make a decision between the possibility of earning profits immediately or looking more to the long-term improvement in the firm's financial soundness. Care must be taken to avoid overtrading which can happen so easily at a time when rapid growth and expansion of a company are in progress. Sometimes this is as a result of too much capital expenditure on fixed assets which in turn, when they are put into production, necessitate a sudden increase in the cash required to finance greater stocks of materials, wages, etc. Without care, this situation can lead to uneconomic practices such as the forced realization of fixed assets at the wrong time, or seeking excessive credit to tide the company over the difficult period.

It is relatively easy to correlate the need for working capital in, for example, a contracting business to the growth of a firm. A construction contractor is paid in arrears for the work that he carries out. In normal practice, a contractor applies at the end of each month to be paid for the work he carries out in that month. He then gets paid at some time during the ensuing month or later depending on the contractual terms. The delay in payment will vary and will depend on many other factors such as the outstanding claims for variations to the work. A contractor, therefore, needs to have working capital to pay wages for work that has been done but not yet paid for by an owner. He needs to pay for materials promptly in order to take advantage of discounts as they are due, and he has other administrative and company expenses to pay as time goes by. If his tempo of work is reasonably uniform, in other words his monthly requests for payment and their receipt over all the company are reasonably uniform, then he will always require an approximately similar amount of liquid assets for use as working capital. If his annual turnover suddenly doubles, then he is likely to require perhaps twice the amount of working capital.

In order to have a ready and continuous supply of working capital it is desirable that a firm is profitable. However, it does not necessarily follow that a firm operating on a profitable basis is safe from overtrading. A deterioration in the promptness of payments for work done, a sudden increase in the rate of work being carried out, rapid increases in the cost of resources that are not recoverable, all can have a drastic effect on the firm.

Accurate and detailed consideration of working capital requirements can be of particular concern to a small firm in the construction industry. Very often such firms have relatively little by way of capital investment in fixed assets, since office accommodation, plant and equipment can all be hired or leased. However, wages and other suppliers have to be paid as work progresses, meaning that firms having relatively poor access to the long-term capital markets have to turn to trade credit and bank overdrafts for finance purposes. Their current liabilities, therefore, tend to rise steeply under even modestly increasing workloads.

If a company is to control its capital expenditure (including working capital) then it is necessary to have a plan which provides the framework for future operations. The more keen the competition, the more necessary becomes the use of an accurate and soundly based forecast for the future. The conviction with which forecasts can be made depends, to a large extent, on the future period which it is intended to cover and the depth and extent of relevant information which is available about the economic, commercial and governmental activities influencing the industry. In the long term, it is probably not feasible to go beyond five to ten years in the future with a forecast that is at all rationally based, though predictions of this nature do sometimes cover longer periods. The contribution of such plans is probably more the stimulation of an awareness of the future than in knowing, with absolute accuracy, what will happen in the future. Five years is likely to be a realistic period for most situations in construction where the long term is being considered, although there are instances in some industries where a particular development involves the commitment of funds for much longer periods. These situations have to be taken into account as they arise and an assessment must be made on the merits of each case.

The extent to which cash is needed as working capital in a business results from the implementation of a series of decisions and policies within that company. A cash forecast can only be prepared when it has been decided, for example, what level of turnover or sales should be the objective for a given period. Furthermore, it is as the result of predicted production figures, and possible decisions to invest more capital in fixed assets of all kinds, rather than being based on established facts. In turn, on the preparation of a cash forecast, the ability of the firm to provide or to obtain the necessary funds for all of the forecast programmes will be confirmed or otherwise. In some cases it may obviously be necessary to adjust what was previously thought to be feasible in the light of the probable availability of working capital. The cash forecast may lead to a reduction in anticipated turnover if it appears that liquidity will be a problem, though on the other hand, if there is a surplus of cash available, it may well lead to the consideration of further projects within the firm or profitable investment external to the company.

Much more detail needs to be prepared in a short-term forecast of cash requirements than for the long-term predictions. Such a forecast will normally cover a period of 12 months, although updating and extension may well take place at intervals of one to three months depending on how critical is the liquidity position of the firm or the fluctuations in the level of business over the current period of time.

The need to appreciate the nature of the constantly changing requirement for working capital with fluctuations in level of trading can best be demonstrated by using figures in a simple example. An example of a small construction company with an expanding business will be used to review the estimated requirement for working capital over two future years. Assume that the turnover for the current year is £1m and that is forecast to be £1.4m next year and £2m in the year after. In looking at this example, three instantaneous and static photographs of the situation at three specific points in time are being taken. Nothing is revealed about the fluctuations in between each view and the detailed changes that may have taken place in each item during that time. The principle involved in establishing the requirement for working capital is to calculate or estimate the cash receipts together with the payments which are outstanding.

Assume that clients of the company are allowed an average of eight weeks' credit. This means that at any one time, the business with a turnover or sales of £1m per year, will have an average of 8/52 × 1 000 000 or (in round figures) £154 000 owed to it by the company's clients. A *debtor* is created when work or goods are delivered to a client. The debt is discharged when the customer pays the firm for those goods. (In a small business dealing only in cash payments for goods sold, there should not be any debtors to finance.) It is worth noting that if the credit period allowed to customers is increased to, for example, ten weeks, then the average amount owed by trade debtors, or what are known as the *receivables*, increases to £192 500 and hence the working capital required must also increase by at least a similar amount. In addition to debtors there may also be *prepayments* which arise in cases where certain payments are made in advance for expenses which may, in whole or in part, be strictly applicable to the following accounting period. Examples of prepayments are rates, insurances and rents when paid in advance. In practice, these prepayments can be estimated with reasonable accuracy from previous experience. For the purpose of this example it will be assumed that, for the first case, the prepayments amount to £6000.

The most important and a very significant item in which working capital can be tied up is that of *stock*. Stock, for the purposes of accounting, is recognized as being one of three kinds – raw material, work-in-progress and finished product. The first two are of principal importance to a construction company and the third features largely in the accounts of a company manufacturing finished goods for stock or with a similar operation. Raw material stock is self-defined and thus its cost represents the investment required to ensure that sufficient raw materials are available, at the place of work, so as not to interfere with the progress of the work. Again, historic records are the guide in the calculation. For the sake of the example it will be assumed that 35 per cent of the turnover is in raw materials and that eight weeks' consumption needs to be on hand at any one time. The raw material stocks will therefore absorb

$$\frac{35}{100} \times \frac{8}{52} \times 1\,000\,000 = £53\,850, \text{ say } \underline{£54\,000}$$

Stock in the form of work-in-progress is a very difficult figure to assess in construction. The required figure for work-in-progress stock working capital is broadly the cost of all that work carried out but for which the owner or customer has as yet not reimbursed the contractor. The cost includes that of the raw materials in the work finished but not yet valued, the cost of direct labour on a similar basis and the cost of the overheads which can be attributed to the production of these works. For the example, if the work-in-progress is represented by six weeks' turnover,

$$\text{Work-in-progress} = \frac{6}{52} \times 1\,000\,000 = £115\,380, \text{ say } \underline{£115\,000}$$

The total firm's capital tied up in the form of working capital on the debtor side can now be summarized as follows:

	£	£
Debtors		£154 000
Prepayments		6 000
Stocks: Raw material	54 000	
Work-in-progress	115 000	
		169 000
Total		£329 000

The above tabulation only provides one side of the story. Just as in manufacturing, a construction company has debtors and it also has creditors, sources to whom it owes money and whom it must pay within a comparatively short time. Creditors are one class of a firm's liabilities. Therefore, in the calculation of working capital it is necessary to offset against the total debts, etc. (in our example £329 000) the credit which is likely to be available at the point in time for which the calculation is being made. The firm might aim to have a slightly longer period from its creditors than it allows to its debtors and an average of ten weeks is used for our example. Creditors will therefore be owed, at this particular point in time an average of:

$$\frac{10}{52} \times 1\,000\,000 = £192\,300, \text{ say } \underline{£192\,000}$$

Prepayments have their opposite numbers in *accruals*. These are liabilities for expenses such as electricity, gas, water, telephone, etc. which have already been incurred for the accounting period under consideration but for which the sums of money have not yet appeared in the company's accounts. Again, accruals can be estimated on the basis of experience; a sum of £10 000 will be used in the example and the full calculation for the working capital requirement now appears as follows:

	£	£
Debtors		154 000
Prepayments		6 000
Stocks: Raw materials	54 000	
Work-in-progress	115 000	
		169 000
		329 000
Deduct	£	
Creditors	192 000	
Accruals	10 000	202 000
Working capital use		£127 000

On a similar basis, calculations can be made for a turnover of £1 400 000 per year and then £2 000 000 per year, from which the changing requirements for working capital can be gauged. The three sets of calculations are summarized in Table 9.1:

Table 9.1

Turnover	£1 000 000	£1 400 000	£2 000 000
Debtors	154 000	215 000	308 000
Prepayments	6 000	8 000	10 000
Stocks	169 000	236 600	338 000
	329 000	459 600	656 000
Deduct			
Creditors	192 000	269 000	384 000
Accruals	10 000	11 000	13 000
	202 000	280 000	397 000
Working capital in use	127 000	179 600	259 000

The above analysis points to a number of important factors to be taken into account in planning investment projects and the related construction work.

Firstly, fluctuating trading conditions can and do affect the working capital required. Whether these fluctuations are brought about deliberately or whether they are the result of chance, there is a need to be aware of their likely effects.

Secondly, the length of time over which credit is allowed and received can cause wide fluctuations in the requirement for working capital. In construction, for example, retention moneys may well have a considerable influence on working capital requirements. From Table 9.1, under the turnover of £1 000 000, it will be seen that debtors' credit time does have to double before the working capital required doubles, but that this relationship depends on a number of other relationships.

Thirdly, the amount of stock or inventory carried can also have considerable influence on working capital requirements. Every attempt must be made to rationalize stock-holding compatible with smooth and satisfactory production. Often the pre-purchase of stocks in anticipation of inflationary increases in price does not withstand critical examination in times of high interest rates. Related to this is the fact that changes of price level of materials in stocks can also have a great effect on the working capital or cash forecast.

In the event that sufficient cash is not available as and when required, then from the above, three broad measures can be taken:

(a) finance can be obtained from external sources, either short- or long-term in the light of the situation as it is forecast;
(b) turnover can be cut back or the level of production can be temporarily reduced. Neither of these alternatives is particularly easy in the short-term for construction projects where a commitment is sometimes made for many years ahead and the incentive is almost always present to produce at as high a rate as possible;
(c) reduce stock levels by not renewing stocks until nearer the time of use. Reduce work-in-progress by speeding-up the claims for payment as work is completed.

In summary, it will be seen, therefore, that the need for working capital is largely controlled by two types of delay; the delay caused by the production/conversion process itself (though this can be minimized by scheduling and programming efficiently) and the delay caused by decisions by management, such as when to pay suppliers or when to press for payment of outstanding accounts.

9.2 Working capital at the project level

If a close and tight control of a contractor's working capital is to be achieved, it is necessary to look at the level of each individual project.

The method of forecasting the cash flows of a project that makes use of an S-curve has already been described in Section 8.6.

Figure 9.1 illustrates the principles behind the construction of an S-curve for the direct cost of carrying out a contract. It is an example of a curve that can be prepared by a contractor in order to obtain an overview of the pattern of his cumulative direct cost expenditure and as a basis for other controls. The curve itself is a forecast, or budget, by which progress can be measured and hence controlled. Once a time-based programme has been drawn up for each operation or activity, an allocation of the cost for each unit of time can be made. In Fig. 9.1 the time-base has been drawn up in months and, initially, the programme is set out in the form of the familiar bar-chart. The programme is in its preliminary form and consists of six main activities. It is for example purposes only and the precise detail that is required in a practical situation is largely a product of the experience and the judgement of the compiler. Against each bar is placed the estimated value of the work that is planned to be carried out in the month in question (in this case in thousands of pounds). The total value of work to be carried out in each month can then be established by summing the values in each column. The total overall value of the work amounts to £658 000. A curve of cumulative total value can now be drawn on the time-base for the project, nine months in this instance. The pattern of expenditure for each time unit and on a cumulative basis can be

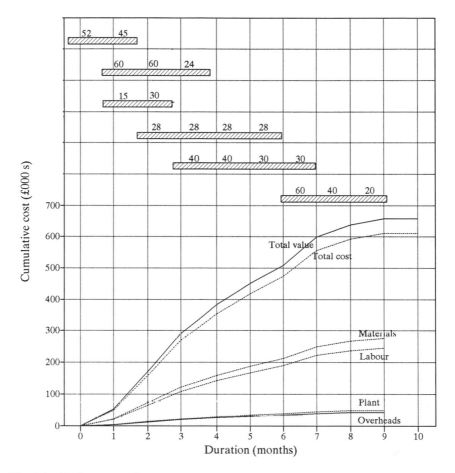

Fig. 9.1 Project expenditure curve.

examined and the *S*-curve gives a clear picture of this pattern and the various rates of working.

This device can be used as a means of preparing cash forecasts and reviewing the project requirements for working capital from a contractor's point of view. Usually, a contractor alone has the detailed information to prepare such charts in significant detail and, in any event, it is the contractor who is primarily concerned with reviewing working capital at the project level. The first step in the process is to produce a total expenditure curve reflecting the quantity and timing of actual cash outflow throughout the contract.

It can sometimes be of assistance to break down the total cost curve into its main constituents and plot each to provide a composite picture, as in Fig. 9.1. Direct costs may be broken down into, for example, labour, plant, materials, sub-contractors and overheads or by trade or characteristics of the work into, say, concrete, excavation, formwork, etc. Due account must be taken of the phasing of the costs. For example, labour is paid on a weekly basis but plant hire charges may involve credit of up to two months or more. Experience may be the best guide in the preparation of a budget curve of this nature. The sum effect of the budget curve, however, is to have a clear picture of the size and timing of the anticipated outgoing costs and overall progress of the project.

Working capital is the requirement for cash to finance work from day-to-day, having paid out the costs of production but not having received payment for work done. Essentially, it is the difference between expenditure and receipts where expenditure exceeds receipts at any one particular point of time in a project's life. If now a curve representing the payments received is superimposed on the cumulative cost expenditure, the extent to which working capital is required, and its cost, can readily be calculated. Figure 9.2 shows a curve of cumulative expenditure and the income curve related to it.

The exact form of the repayment curve will clearly depend on the contractual conditions related to the terms of payment for work completed. The cost curve of Fig. 9.2 is based on the example used to compile Fig. 9.1. It is assumed, for the sake of the example, that the margin of value over cost is 7 per cent of the value. In addition, it is assumed that 5 per cent of each interim claim for payment for work completed is retained and that each payment will be made 28 days after the claim is received, on the first of each month. Half of the total retention will be paid back at the end of the nine-month programme and the remainder, one month later. The complete data upon which Fig. 9.2 is based are set out in Table 9.2.

The revenue curve will take the form of a stepped curve as shown in Fig. 9.2, wherein each step represents the receipt of the certified value of the latest application for payment. It must be remembered that the use of the curves in this instance is for the purposes of prediction or forecasting the likely cash requirements. It will be assumed, therefore, that the project will be profitable and thus the total payments will eventually exceed the total costs by the margin which has been added to the estimated costs before submitting the tender. However, with the given conditions of retention and the delay in payment after completion of work, the revenue will not catch up with the expenditure until the completion of the project. Until this happens, working capital of varying amounts will always be required on the basis that the cumulative cost curve is actual cash expended.

The cut-off on the ordinate at any point between the cumulative cost curve and the revenue curve, where the revenue curve is below the cumulative cost curve, represents the actual amount of working capital required at any one time. In the example illustrated in Fig. 9.2 the maximum working capital required is £191 900. This occurs at the end of the fourth month after the commencement of the contract. As a payment of nearly £112 100 is due at

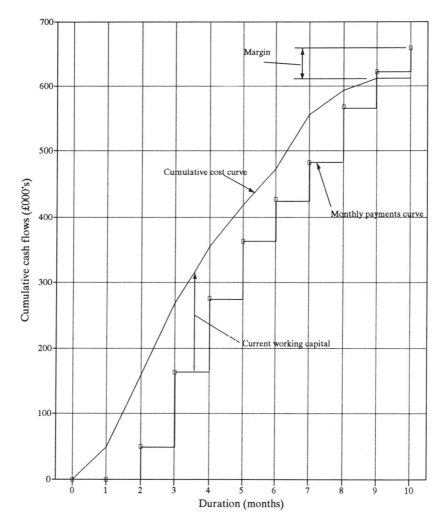

Fig. 9.2 Financial curves for construction project.

that time, this peak occurs only for a very short time. The working capital requirements can be extracted from the information shown in Fig. 9.2 and plotted separately as shown in Fig. 9.3. The curve of working capital in use is composed of vertical and sloping lines. The sloping lines represent the gradual increase in the project spending over the current month which is assumed to be linear. The vertical lines represent the monetary decrease in outstanding working capital which occurs at the time that a payment is received.

While this analysis has been carried out for a single project, the information which is derived from it can be very useful in establishing the working

Table 9.2

£000

Month end	0	1	2	3	4	5	6	7	8	9	10
Cumulative cost	0.0	48.4	160.0	269.7	355.3	418.5	472.4	556.1	593.3	611.9	611.9
Cumulative value	0.0	52.0	172.0	290.0	382.0	450.0	508.0	598.0	638.0	658.0	658.0
Cumulative paid	0.0	0.0	49.4	163.4	275.5	362.9	427.5	482.6	568.1	622.0	658.0
Working capital after payment	0.0	48.4	110.6	106.3	79.8	55.6	44.9	73.5	25.2	-10.1	-46.1
Working capital per month at 1%	0	242	1042	1654	1491	1114	826	868	921	345	-191
Cumulative retention	0.0	0.0	2.6	8.6	14.5	19.1	22.5	25.4	29.9	16.0	0.0

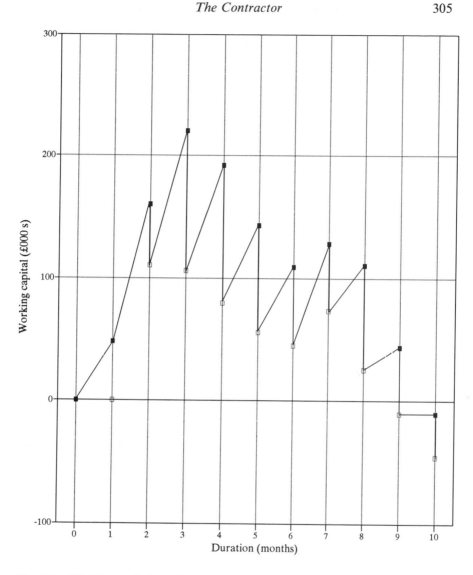

Fig. 9.3 Working capital requirements.

capital requirements over a company's operations as a whole. Figure 9.4 illustrates a chart summing up the current situation in a company and adding the effect of a possible addition to the list of projects currently being undertaken or awarded but not yet started. The present time is taken to be the zero/one point and the review is for a period of 18 months ahead. Projects 1 and 2 are currently under construction and Project 3 is expected to commence in one month's time. Project 4 is programmed to commence in seven months' time and an enquiry has been received for a project which, if

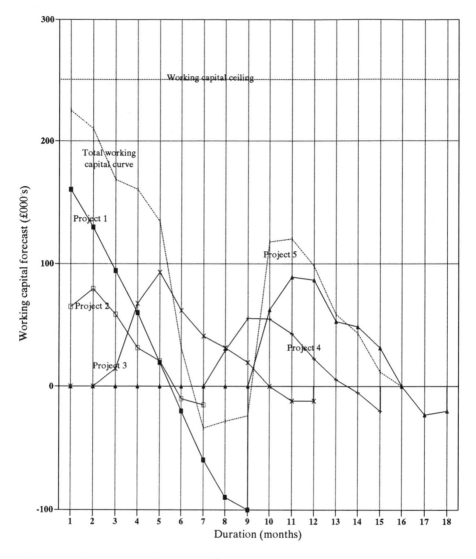

Fig. 9.4 Company working capital forecast.

awarded, will commence in nine months' time. The maximum available working capital during this period is estimated to be £250 000. By summing the various project curves and taking into account the fact that each project should be contributing a surplus to the general pool eventually, a check can be made quickly as to the state of the overall system.

Whether cash resources are available within a company or some special provision has to be made to obtain the cash from external sources, such as a bank overdraft, the finance involved will incur a cost. On the one hand the finance will be denied to the company for investment in other things; on the other, a charge will be made in the same way as other borrowed money incurs the cost of interest. This cost of interest then becomes an additional charge to the company and in the case of a contractor constructing projects for various owners, the cost will need to be passed on to them by including a sum for it in tenders.

Returning to Fig. 9.2 showing the curves for the cost and income of the example project, it can be seen that the working capital in use at any stage in the project can be established. Since the horizontal axis is that of time and the vertical axis represents expenditure and/or income in money terms, then the area between the cumulative cost curve and the revenue curve (before the revenue curve crosses above the cumulative cost curve) must represent the interest (time multiplied by outstanding cost) that accrues on working capital.

Given Fig. 9.2 and Table 9.2, it is possible to calculate the cost of the working capital required up until any point in the programme. For example, the working capital required in the first month is 0.5(0 + 48 400) 0.01 = £242 if it is assumed that the interest rate is one per cent per month. For the second month the calculation is 0.5(160 000 + 48 400) 0.01 = £1042 and for the third month it is [0.5(269 700 + 160 000) − 49 400]0.01 = £1654. If this process is pursued until the *monthly payments curve* breaks through the *cumulative cost curve* then it will be found that the total interest that is involved is £8503 or 1.29 per cent of the contract value. On many occasions the *month payments curve* will cross and re-cross the *cumulative cost curve* during the course of a project. However, the principle is the same; it is the total area under the *cumulative cost curve* that is in proportion to the working capital.

Clearly, one of the significant factors that affects the cost of the working capital is the timing of the payment from the owner. As the payment from the owner is delayed, so the revenue curve in Fig. 9.2 moves towards the right, the break-even point in the contract comes later (if at all) and the area between the two curves gets larger. If, for example, the owner did not pay until two months after the end of the month at which the valuation is made, the working capital would increase by £611 940 × 0.01 = £6114. This would make a revised total working capital requirement of £14 622 or 2.22 per cent of the value of the contract. There is, therefore, scope by adjusting the programme, by adjusting credit terms on both sides and by negotiating payment terms, to optimize generally the working capital that will be required at any one stage of the company's activities.

These interest costs will, of course, be affected by the payment terms as set out in the contract documents. In many cases delays in payment of certified sums beyond specified limits will oblige an owner to pay interest on the outstanding amount. Claims for variation made by a contractor but not yet certified are another complication to the general calculation.

The examination of *S*-curves drawn to aid the calculation of predicted working capital required by a contractor throughout the construction programme emphasizes the need to obtain revenue at the earliest possible moment in the project's duration. Sometimes, and this is more frequent in the construction of overseas projects, it is possible to arrange for conditions of contract that require an owner to make an advanced payment (usually about 10 per cent of the contract cost) at the commencement of the work. This payment covers the cost of the contractor's mobilization of his resources in what is often a very difficult environment. The effectiveness of the advanced payment from the point of view of reducing working capital is shown by the examination of Fig. 9.3 with Fig. 9.5. Both are graphs of working capital against project duration – the former has already been discussed and the latter is the same example but with the client providing a mobilization payment of £100 000 at the outset of the project. This is subsequently set off against the initial certified payments, thereafter following the previous practice with regard to payments. It will be noted (and expected) that the contractor does not start incurring costs for working capital until almost the middle of the second month. The effect of this initial payment is to reduce the contractor's working capital to £6872 or 1.04 per cent of the contract value. The client of course has to bear the additional cost of providing the £100 000 in the first place as well as taking the risk that the contractor will not default before £100 000 worth of work has been completed.

Having established the *S*-curves which represent the estimated performance by a contractor on his project, they can then be used for control. However, there is almost always at least one significant problem in using such curves or other methods of control, which in any event tend to be an approximation of what is likely to happen and which, as with all estimates, are likely to be inaccurate. This is the problem of the variations to the scope of work originally defined in the contract. That a contractor is requested to undertake a variation will almost certainly affect all of the normal curves used within an *S*-curve diagram. Whilst a contractor is required to undertake the work involved in the variation at the time that he is instructed to do so there can be a dispute as to where the cost of the work involved lies, whether it be with an owner, with the contractor or whether it is some division between the two. Until the matter is settled a contractor needs to meet the cost of the resources required to undertake the variation of work and to meet the changed requirement in working capital as it arises. In the case of some variations these additional demands, especially when there is a considerable delay in agreeing the apportionment of costs for the variation, can make a considerable difference to the overall picture of the need for financial resources.

An *S*-curve can be useful to a contractor as a means through which control of the progress in both time and cost of work can be effected. The general arrangement of a curve used for this purpose is set out in Fig. 9.6. The *y-axis*

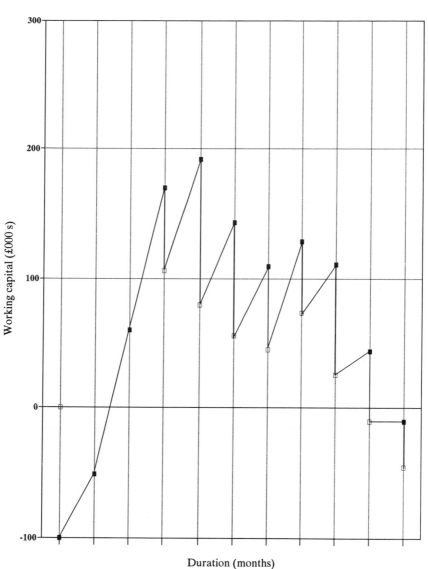

Fig. 9.5 Working capital requirements with advance payment.

is graduated in terms of cost and the *x-axis* shows the project duration in months. The curve illustrating the original cost programme is shown as a full line throughout the project duration. This is the curve showing the budgeted expenditure for the contractor in the example used earlier in this chapter. The review date for the progress of the work is the end of the fourth month of the

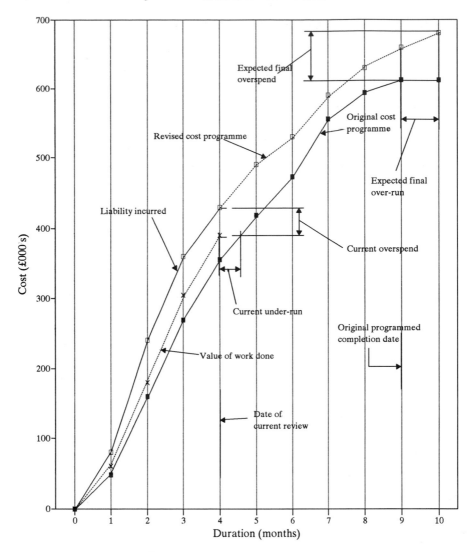

Fig. 9.6   *S*-curve for control.

project duration. The work was originally programmed to be completed by
the end of the ninth month.

The curve labelled *liability incurred* represents the *actual expenditure* so far
incurred by the contractor. This is situated at a higher cost level above the
cost programme curve and, therefore, indicates that more cost has been
incurred than was budgeted for the work actually carried out. The key curve
enabling the current position to be analysed more fully is the third one, *value
of work done*. As indicated in Fig. 9.6, this falls between the *liability incurred*

and the *original cost programme* curve for this particular example – it will not necessarily do so in other cases. This curve sets out the valuation of the completed work in relation to the cost that has been expended on it and the duration that has been expended in relation to that programmed. It will be noted that the contractor has expended greater cost on the completed work than value he will receive in return and that the programme has underrun the original anticipated programme for the amount of work completed in the valuation. It may well be, for example, that the contractor has incurred more cost on longer working hours than was originally budgeted. The important aspect of the *S*-curve, at this stage, is that a revised cost programme to the end of the project should be constructed so that a forecast of the ultimate situation can be constructed. This is shown in Fig. 9.6.

9.3 Profit and profitability

The early chapters of this book were concerned with the methods which can be used to establish the yield which apparently accrues when an investment in a project and the resulting net income are related. In the contracting field of construction business, the major source of income arises from a series of contracts of a one-off nature which are usually physically and contractually quite separate, one from the other. The organization of each construction project is virtually self-contained at the site of the work, with exception that each is linked to the other by a common organizational structure on either a national or a regional basis that usually operates from a central office. In a general sense, each project may be looked at as though it were a separate business and, in itself, it is required to be profitable on the basis that, if all contracts are profitable (after supporting all central overhead costs, etc.), then the whole company will be profitable. This philosophy, however, rarely maximizes overall profit because of the relatively poor use which is made of central services of a specialist nature.

It is necessary, before developing the discussion of what constitutes contract profitability, to define the terms *profit* and *profitability* in this context. Profit is simply what is left from what is received as a result of carrying out the work after paying all the charges that arise out of a project. Profit/loss is expressed in absolute terms as money and it gives no measure of the productivity or efficiency of operations on a site, in this context, except to indicate that either more has been received than was paid out or less has been received than was paid out. In the latter case, the project is said to have carried out at a loss. Profitability is a *rate* of making profit in relation to a company's *investment* (though it may be a temporary profit) that is required to carry out the work. Since it is a rate of making profit, it is implied that profitability is related to a specific contract duration. Therefore, the three factors which determine profitability must be the amount of profit, the duration of the contract and the company's investment that is required.

Since there is now a means of establishing profitability, the means is available, therefore, for comparing the desirability to a contractor of different projects, that is if profitability is his prime means of making the comparison. Assume that a contractor makes £5000 profit on one contract and £6000 on a second. Contract 1 requires an investment of £10 000 and Contract 2 requires an investment of £8000. The contracts have durations of 1 year and 18 months, respectively.

In the case of Contract 1, the return on the capital employed amounts to £5000/10 000 for the contract duration, or 50 per cent on an annual basis. In the case of Contract 2, the return on the capital employed is 75 per cent for the contract duration of 18 months or also 50 per cent on an annual basis. It can be seen therefore that with quite different contract conditions, the two projects are equally profitable. If Contract 1 had a duration of 18 months, or was extended to that time because of slow working, the profitability would then fall to $33^{1}/_{3}$ per cent, so long as the other two factors, investment and profit, remained the same. Profitability, therefore, is dependent on these three factors and any variation in one or more of these (assuming that it is not self-compensating) will lead to a change in the profitability. However, while the profit and the duration can ultimately be established for a contract, there is a need to look a little further into what is meant by the investment that is required in order to undertake a contract.

The means of establishing the investment in carrying out a contract has already been discussed at some length in another context. It is the use of a cumulative *S*-curve of actual cost against which is plotted the income curve, in a stepped form for most construction projects. The method of establishing from this curve the cost of finance for a project has already been demonstrated. It will be remembered that the area between the two curves was established and then multiplied by the cost of finance, in order to establish the cost of working capital. In this case the total investment required to finance the contract over its duration is represented by the area between the two curves. For a shorter period of time during the contract duration, the relevant area is that cut off by the two ordinates representing the beginning and the end of the period of time to be considered. It is doubtful, however, whether the calculation of profitability on this basis is very meaningful for a period of less than the project duration. This is due to the delays which occur in establishing total actual expenditure and to the great variations in the rate at which, overall, the work is carried out. As before, the emphasis must therefore be on minimizing this area if the profitability is to be as high as possible. It should be noted that the actual cost curve in such a diagram represents the actual cash outlays at the site that occur directly as a result of carrying out the work. Head office overheads and certain staff salaries, for example, would not form part of the curve establishment of this because, by the nature of the calculation, these are costs which will necessarily occur and will have to be borne by the company, at least in the short-term, whether the work is carried out or not.

The actual cost curve will consist of the expenditure on materials, wages, plant, sub-contractors and that part of site oncosts that would not have been incurred if the work had not proceeded. The overheads and oncosts that would have been incurred in the event of no contract being undertaken have to be dealt with in a different way. With regard to plant costs, these will normally take the form of accumulated hire rates or costs, whether the plant is hired externally or owned. If plant purchases are included in the actual cost calculation, then care must be taken to credit any residual or second-hand value which accrued when the contract ends.

The company overheads, largely made up of the costs of the central organization which exists to provide a service to projects, obtain work and generally administer the company, together with the accommodation of a permanent nature which is required to house them, have to be related to the overall turnover of the company as a whole. Each project must make a contribution towards these costs and it is worth looking at the relationships which exist between the various elements of cost involved.

The nature of costs involved in project work has already been discussed at some length and it has been seen that they may be classified into *variable*, *semi-variable* and *fixed*. This is one method of analysis that contrasts with a number of others. Overhead costs (to include site oncosts and head office overheads) may, in their turn, be similarly classified. In other words, some components making up the overheads will vary with the rate of working, whilst others will remain fixed, whatever the rate of total production. One technique by which the relationship between production, overheads and profit can be examined is that of *marginal costing*. The marginal costing technique will be discussed here, only so far as it can be used to explain this relationship.

It is possible to determine from a break-even analysis the effect of variable and fixed costs on the unit cost of production or turnover. The *marginal cost* of a unit of production may be defined as the total cost which is incurred in increasing or decreasing production by one unit from a given level of production. It will be appreciated, therefore, that the marginal cost of one unit of production will usually consist of the direct costs for its production plus the variable overheads associated with it.

The relationship of turnover (sales), direct costs, variable costs, fixed costs and profit may be expressed in a formula as follows:

Turnover = direct and variable costs + fixed costs + profit

By transposing the equation we get:

Turnover − direct and variable costs = fixed costs + profit

In this formula, the left-hand side of the equation is known as the *contribution.* Contribution may be looked upon as the difference between the marginal cost of a unit of production and the value or selling price of that unit. If the above formula is re-examined, it will be appreciated that the left-hand side of the equation is a contribution towards the fixed costs and profits of a contract, a project or a business. It is very important to keep a close watch on the ratio, contribution/turnover (more conveniently expressed as a percentage) since this highlights changes in the rate of contribution with changes in the volume of turnover, thus indicating whether fixed costs and profit are being recovered at least at the necessary rate. This ratio is known as the *profit/volume ratio* and can be expressed very clearly in the form of a profit/volume chart.

A profit/volume chart is a simplified version of the break-even chart. Figure 9.7 illustrates the profit/volume chart for a business which, at 100 per cent output, has a turnover of £100 000. The fixed costs at this full turnover amount to £10 000 and the variable costs amount to £75 000. It will be seen from this chart that the turnover has to be $33\frac{1}{3}$ per cent of the full turnover if a break-even situation is to be reached. From this chart, the likely profits at other levels of turnover can be established, providing other factors remain the same.

The concept of the profit/volume chart can be extended to cover a number of projects being constructed by a single contractor so as to give an idea of

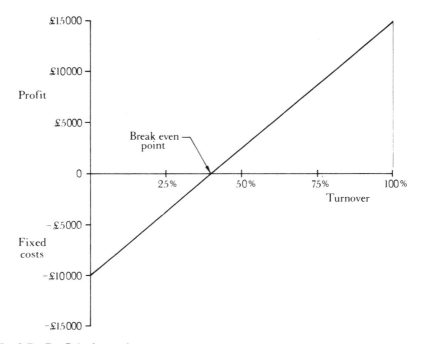

Fig. 9.7   Profit/volume chart.

Table 9.3

Contract	Value completed (£)	Contribution to date (£)
1	240 000	28 000
2	120 000	10 000
3	100 000	3 000
4	340 000	25 000
5	90 000	(6 000)
	£890 000	£60 000

the *relative* rate at which contribution is being made by each. For example, a contractor has five contracts in progress as in Table 9.3.

Figure 9.8 illustrates the graphical analysis if the fixed overheads of the company for the period concerned amount to £45 000. By plotting the overall company profit/volume line, the rate at which individual contracts contribute can clearly be seen. Contracts 1, 2 and 4 are contributing at a

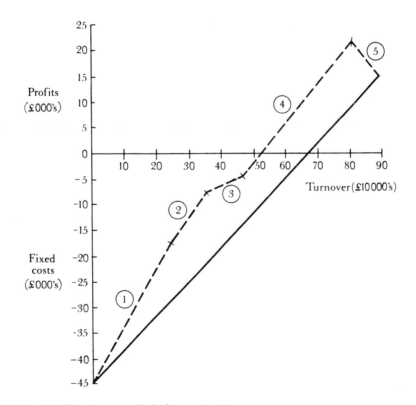

Fig. 9.8 Profit/volume analysis for contractor.

higher rate than average. Contract 3 is also contributing but at a lower rate than average, and Contract 5 is not contributing. This tool, as with break-even analysis, tends to be static in its concept rather than dynamic, as are some of the other methods of control.

An example will illustrate the application of the foregoing relationships to the establishment of the mark-up or addition for profit, risk and fixed overheads that is required for an estimate of cost for a project, if a specific rate of return on capital employed is to be achieved.

Example 9.1

The Board of Directors of a construction contracting company requires an annual return of 17 per cent on the £20 000 000 capital employed in the company. Of this capital, £13 000 000 is invested largely in fixed assets and the remainder is available as working capital to finance all contracts in progress. In the coming year, the turnover on contracts is estimated to be £40 000 000 based on actual cost of work done. What percentage must be added to contract costs for the year by way of mark-up if the company's overheads, etc. are expected to amount to £1 600 000?

The work on contracts must contribute £1 600 000 for the company's overheads, etc. and a further 17% × £20 000 000 for the return required in the capital employed in the company. This makes a total of £5 000 000.

This contribution as a percentage of the capital employed in operating the contracts = (5 000 000/7 000 000) × 100 = 71.4 per cent.

It is necessary, therefore, to aim at an average annual return on investment of nearly 72 per cent if the contracts being undertaken by the company are to make the required contribution to the company's overheads and return on capital employed.

As a percentage of the turnover, the contribution amounts to (5 000 000/ 40 000 000) × 100 = $12\frac{1}{2}$ per cent which is the average mark-up to be added to contracts in order to generate the required contribution.

From the foregoing it will be seen that profitability of individual contracts is a function of the profit, the duration and the amount of investment that is required to keep them going. Management attention must therefore be paid to each of these aspects rather than being concerned with absolute profit alone. The area between the *S*-curve of actual cost and the curve of payments received must be minimized as far as is reasonable by delaying outgoing payments and enhancing payments received. Attention must be paid to the programme of work so that return on investment does not drop too severely as a result of a slow rate of progress and hence lack of turnover. As far as the company view is concerned, every contract must make an average contribution to the company's overheads and the return that it requires on the capital employed.

This contribution should be watched on a monthly basis by looking at the contribution required from the contract as a percentage of the investment during that month. If, for example, the company's annual turnover fails to reach the level of £40 000 000 by, say, £5 000 000, then the contribution will be short by $12\frac{1}{2}\% \times £5\,000\,000 = £625\,000$. This can only be made up by additional contributions from other contracts, otherwise the annual return on capital employed by the company

$$= \frac{(4\,375\,000 - 1\,600\,000)\,100}{20\,000\,000} = 13.9 \text{ per cent.}$$

The calculation of this return is clearly sensitive to the proportion of capital employed in the company and the size of the company's overheads.

9.4 Return on a construction project

The profitability of a construction project is frequently calculated in absolute terms by subtracting the total expenses incurred in carrying out the work from the total revenue accruing as income. These measurements are made in accounting terms. Sometimes the profit so calculated is expressed as a percentage of the contract value. With large projects having durations of several years such calculations can be very misleading and, more importantly, do not permit a ready comparison between the profitability of one project and that of another. A further important influence on such profitability is that of inflation and a means of calculating a project return needs to have a facility for examining whether or not inflation is an important factor in the assessment.

In undertaking a project a contractor needs to make a conscious decision as to his real objectives. He may be seeking to maximize his profit at any cost, or to do so by enhancing his reputation for a high quality of performance, or to utilize existing idle resources with profitability as a secondary consideration, or to gain experience of work of a different nature, or perhaps to gain a project with high publicity value. There will be other objectives, but all will need to be evaluated in financial terms with due allowance being made for each and, in either the long or the short term, the profit to be made must be reflected in the organization's total financial resources in real terms. The latter point is paramount. In measuring the net assets of an organization in money terms, the effect of the inflationary process is to erode the real value of those assets. In order to maintain their real value over the duration of a project, that project's profitability must include an element of money for this purpose related to the total equity which is being made available to finance the project.

In addition to the inflation element, equity resources are only available if the shareholders providing it are compensated adequately in terms of the risk that they take and in line with the return they would expect to receive from investments elsewhere which offer a similar risk.

In addition to equity funds, a contractor may expect to use loan finance for the purpose of working capital, the cost of which will fluctuate with varying conditions. In many instances a mixture of both equity and loan funds will be utilized – the mix ratio must be known since it is likely that the cost of each of these two elements will be different.

The profit assessment of a project will need to provide for a number of other aspects of a contractor's operations, such as the extent that he will be required to bear contract risks, but it is proposed only to examine financial aspects at this point.

For the purposes of demonstration a hypothetical project will be assumed giving rise to the cash flows set out in Table 9.4. The simplest contract has been assumed so as to concentrate attention on the question of return. The project has an estimated cost of £720 000 to which has been added a mark-up of 10%. The gross operating profit for the project is equivalent to 9.09% of the tendered price. In Column 4 of Table 9.4 is shown the project net cash flows which total to the mark-up of 10% of the cost. Column 5 lists the cumulative net cash flows. The maximum negative cash outlay of the contractor is £44 000 at the end of Month 6. The project does not achieve a positive cumulative net cash flow until the end of Month 10.

The internal rate of return (IRR) can be calculated for the net cash flows of the project (Column 4). These are the cash flows which take account of the revenue payments from an owner and offset the expenditure by the contractor. The IRR is the rate that will discount the cash flows to zero taking account of their timing and whether they are positive or negative. However, the calculation of the IRR in this way assumes that the working capital for the project is provided from the contractor's equity or share-

Table 9.4

1	2	3	4	5
End of month	Expenditure (£)	Revenue (£)	Net cash flow (£) (3 − 2)	Cumulative net cash flow (£)
1	40 000	0	−40 000	−40 000
2	45 000	75 000	30 000	−10 000
3	78 000	72 000	− 6 000	−16 000
4	95 000	81 000	−14 000	−30 000
5	120 000	108 000	−12 000	−42 000
6	162 000	160 000	− 2 000	−44 000
7	50 000	62 000	12 000	−32 000
8	80 000	91 000	11 000	−21 000
9	40 000	20 000	−20 000	−41 000
10	10 000	58 000	48 000	7 000
11	0	65 000	65 000	72 000
Totals	£720 000	£792 000	£72 000	

holders' funds since it makes no allowance for the charges of a loan element of funding. As calculated in this way, the IRR amounts to 13.73%.

If the plans for providing working capital were changed to include finance by borrowing and/or taking inflation into account, the IRR will change. It is not therefore a very useful measure for the profit assessment of a project. Another possible defect with the calculation is that referred to in Section 4.11, where multiple yields may well result as the solution to the calculation of IRR given reversals of sign in the cash flows under consideration. There are five such reversals in this example.

If the assumption is now made that a contractor finances the whole of his working capital by borrowing, another set of net cash flows can be derived and two examples of this are set out in Table 9.5.

In one case the interest rate on borrowed money is assumed to be 1 per cent per month; in the other a rate of $1\frac{1}{2}$ per cent is used. The simplest assumption is made that a contractor borrows at the end of each month sufficient to cover the negative cash flow of the previous month and then pays interest on that amount until the principal is reviewed at the end of the next month. The IRR for each of these projects now becomes 13.15% and 12.85% for 1% and $1\frac{1}{2}$% interest rates, respectively and the corresponding returns measured as gross operating profit to tendered price become 8.74% and 8.57%.

If the net cash flows for the project, either with or without interest, are examined, it will be seen that they are particularly favourable to the contractor. Few contracts will show a relatively large positive net cash return in the project duration but the resultant enhanced IRR calculations demonstrate the advantages of doing so. This can sometimes be achieved by increasing the tendered unit rates for work which occur early in the project

Table 9.5

1	2	3	4	5	6	7
End of month	Net cash flow without interest	Cumulative net cash flow	Interest at 1%/month	Interest at 1½%/month	Net cash flow incl. interest at 1%	Net cash flow incl. interest at 1½%
1	−40 000	−40 000	0	0	−40 000	−40 000
2	30 000	−10 000	−400	−600	29 600	29 400
3	− 6 000	−16 000	−100	−150	− 6 100	− 6 150
4	−14 000	−30 000	−160	−240	−14 160	−14 240
5	−12 000	−42 000	−300	−450	−12 300	−12 450
6	− 2 000	−44 000	−420	−630	− 2 420	− 2 630
7	12 000	−32 000	−440	−660	11 560	11 340
8	11 000	−21 000	−320	−480	10 680	10 520
9	−20 000	−41 000	−210	−315	−20 210	−20 315
10	48 000	7 000	−410	−615	47 590	47 385
11	65 000	72 000	−	−	65 000	65 000
Totals	£72 000		−£2760	−£4140	£69 240	£67 860

and reducing others that will be effective at a later stage in order to compensate. There are obvious dangers in this, not the least that the measured quantities may change to a contractor's disadvantage. It has the added disadvantage to the owner that he may unwillingly use unbalanced rates for forecasting the cost of future work.

9.5 Influence of inflation on project profitability

The purpose of this section is to indicate the possible effect of inflation on project profitability. There are, of course, many projects in which the compensation for increasing or decreasing resource prices is clearly specified, usually by the use of an indexing method. Certainly a contractor will know the conditions of tendering beforehand and will be able to make due allowances accordingly. By examining the effect of not making these allowances when required it will be demonstrated how important the matter becomes in times of high inflation.

Referring to the hypothetical actual project cash flows shown in Column 2 of Table 9.6, these may be converted to time zero values by dividing each value given in the table by $(1 + d)^n$ where

d = the constant inflation rate per month
n = the time period (month)

Table 9.6 shows the adjusted cash flows (ignoring interest payments) in terms of time zero units of money, that is less inflation, for values of inflations of $^3/_4$ and $1^1/_2$ per cent per month.

It will be seen from the table that, without compensation for inflation (and on the assumption that nothing was included in the estimate), the return as a

Table 9.6

End of month	Net cash flow	Time zero net cash flows ¾% inflation p.m.	Time zero net cash flows 1½% inflation p.m.
1	−40 000	−39 702	−39 409
2	30 000	29 555	29 443
3	− 6 000	− 5 867	− 5 738
4	−14 000	−13 588	−13 191
5	−12 000	−11 560	−11 139
6	− 2 000	− 1 912	− 1 829
7	12 000	11 388	10 812
8	11 000	10 362	9 765
9	−20 000	−18 699	−17 492
10	48 000	44 544	41 360
11	65 000	59 871	55 181
Totals	£72 000	£64 392	£57 763

percentage of the tendered price has declined from 9.09% to 8.13% and 7.29% for monthly inflation rates of $^3/_4$ and $1^1/_2$%, respectively. This example has taken a simplistic view of inflation in construction and the way it may be represented. In practice, there will be a number of different cost components making up the construction process cost, such as labour, plant and materials. Each, and others, will probably have different rates of price change and many of them will have a significant influence on the overall inflationary effects. In many cases it may be necessary to treat the cost components for each resource individually in order to make an accurate statement of its effects.

9.6 Company budgeting for overheads and profit

It is essential for a construction contracting company to establish an annual budget which sets its objectives as far as its construction programme is concerned together with the contribution to the head office overheads and the profit of the company. Each contract which is being operated by the company will receive certain services that are centralized in the head office of the company and which have to be supported by all of the activities that the company undertakes. Such services may well include centralized purchasing, company directors and senior managers, some design of temporary works, estimating and tendering services, possibly a centralized computer undertaking the production of various kinds of reports and administrative processes such as the preparation of wages sheets and the like. The cost of the head office overheads is normally expressed as a percentage which is added to the net cost of an estimate for a project as part of the mark-up. A simple company statement may follow approximately the pattern set out in Fig. 9.9 which is a company budget for one financial year and shows the anticipated receipts that will occur during that year. The figures are summarized in thousands of pounds. In the preparation of such a budget it is necessary to make forecasts of the likely performance of projects which are already underway, to look at those that may be finishing during the financial year and to attempt to anticipate what work, or volume of work, will be gained during the course of the year.

With reference to Fig. 9.9 it will be seen that there are three contracts, the first three on the list, which are in progress or will be commencing on 1 January 1994, and it is further anticipated that another contract will be awarded which will start during February with a second new contract starting during March. The figures included on the line of each contract are the total estimated value of work done on the contract for each respective month. The totals are then arrived at horizontally and vertically to give an indication of the value of work which will be carried out on each contract and the total value of work which will be carried out on these contracts by the

Company budget for one financial year showing anticipated receipts (£000s)

Contracts	1994												Totals
	Jan	Feb	Mar	Apr	May	June	July	Aug	Sept	Oct	Nov	Dec	
CB/121	100	250	300	300	400	300	200	100	50	–	–	–	2000
CB/124	600	600	400	200	250	300	300	200	200	50	–	–	3100
CB/130	100	100	100	150	300	400	400	600	600	400	200	100	3450
CB/142	–	50	200	200	250	300	400	450	500	500	500	500	3950
CB/150	–	–	100	100	200	300	400	600	600	400	200	100	3000
Totals	800	1000	1100	950	1400	1600	1700	1950	1950	1350	900	800	15 500

Of the above totals: (i) Direct costs and site oncosts = £12 710 000 (82%)
(ii) H.O overheads = £1 240 000 (8%)
(iii) Profit = £1 550 000 (10%)

Fig. 9.9 Company financial budget.

company during each month of the financial year, respectively. Review of the bottom line totals for the monthly amounts of work will give some indication of any wild fluctuations in the value of work undertaken which may place some particular stress on various aspects of the contracting organization such as, for example, the technical personnel that the company has at its disposal. This will allow some forward planning in this respect.

The total value of work to be carried out for the financial year amounts to £15 500 000. This represents value of work and therefore not only includes the direct cost and site oncosts but also includes the various contracts' contribution to the head office overheads and the profit. The breakdown at the bottom of Fig. 9.9 shows how these figures are built up. The direct costs and the site oncosts amount to 82 per cent of the total value, the head office overheads amount to 8 per cent of the total value and the profit to 10 per cent. The progress throughout the year can then be measured against these absolute sums and the percentages to ensure that the contracts overall are making the requisite or anticipated amount of profit and are also making their appropriate contribution to the support of the centralized head office of the company by way of its overheads.

In order to explain the functioning of this company budget and the way in which progress can be measured against it, it will be assumed that a cumulative report is received by the Board of Directors showing the progress until the end of April. This cumulative report is set out in Fig. 9.10. The report summarizes the position with regard to direct costs, site oncosts, head office overheads and profit. Firstly the cumulative report indicates that Contract CB/142, though it was anticipated it would commence in February, was not awarded. Contract CB/150 was, however, awarded.

The third column of the cumulative report is a statement of the budgeted values that appear in the company budget set out in Fig. 9.10. For example, the company budget anticipated that £950 000 worth of work would have been completed in the first four months of the financial year on Contract CB/121. In the cumulative report this £950 000 is broken down into the 82 per cent for direct costs and site oncosts, the 8 per cent for head office overheads and the 10 per cent for profit. The analysis is undertaken for each of the four contracts which are now current. In the fourth column of the cumulative report is the statement of the value of work which has actually been undertaken. It will be noted that two of the projects, CB/121 and CB/150 are both behind programme when measured on the value of work done; the other two are more or less on programme or a little in advance of it. The danger of falling behind the budgeted volume of work is that insufficient work is carried out so as to make the contribution of head office overheads and profit that was anticipated in the company budget. One of the lagging projects, CB/150, has clearly started later than was anticipated and work to the value of only £50 000 has been undertaken in place of the £200 000 value which was budgeted. This means that, as far as the budget is concerned, only

Contract	Item	Budgeted value (£)	Value of work done (£)	Actual cost of work done (£)	Variance (£)
CB/121	Direct costs and site oncosts	779 000	738 000	758 000	(20 000)
	H.O. overheads	76 000	72 000	76 000	(4 000)
	Profit	95 000	90 000	95 000	(5 000)
		950 000	900 000	929 000	(29 000)
CB/124	D.C. and S.O.	1 476 000	1 533 400	1 540 000	(6 600)
	H.O.O	144 000	149 600	144 000	(5 600)
	Profit	180 000	187 000	180 000	(7 000)
		1 800 000	1 870 000	1 864 000	(6 000)
CB/130	D.C. and S.O.	369 000	379 660	354 000	(25 660)
	H.O.O.	36 000	37 040	36 000	(1 040)
	Profit	45 000	46 300	45 000	(1 300)
		450 000	463 000	435 000	(28 000)
CB/142		Contract not awarded			
CB/150	D.C. and S.O.	164 000	41 000	40 000	(1 000)
	H.O.O.	16 000	4 000	16 000	(12 000)
	Profit	20 000	5 000	20 000	(15 000)
		200 000	50 000	76 000	(26 000)

Fig. 9.10 Cumulative report for all contracts as at end of April, 1994.

£400 (8 per cent of £50 000) is to be contributed towards the head office overheads and £5000 (10 per cent of £50 000) is to be contributed towards the company profit. A shortfall of £12 000 and £15 000 respectively results.

Using a similar methodology the remainder of the cumulative report can be analysed and the effects of a shortfall or an excess of value of work being undertaken can be examined and the cumulative effect calculated. In this particular example there is obviously going to be a serious shortfall on the contribution made by the contracts to head office overheads and to the company profit in spite of the better performance than budgeted on two of the contracts. In addition, there will also be the shortfall on the contribution from Contract CB/142 which did not materialize. Since this amounts to 18 per cent of £3 950 000 (or £711 000 in absolute terms) there is an urgent need to obtain more work and also to increase the tempo of the work already in hand.

The following is a formal analysis of the company budget statement:

Analysis of company budget statement:	£	£
At end of April total budget sales		3 850 000
Budget sales from CB/121, 124, 130 and 150	950 000	
	1 800 000	
	450 000	
	200 000	3 400 000
Shortfall		450 000

$\therefore$ Sales variance $= -£450\,000 \times 18\% = -£81\,000$

Calculation of profit

	£	£
Total receipts	900 000	
	1 870 000	
	463 000	
	50 000	3 283 000
Less total actual direct costs and	750 000	
site oncosts to date	1 540 000	
	354 000	
	40 000	2 692 000
Less H.O. overheads for four months		591 000
$\left(\dfrac{3850}{15\,500} \times 1\,240\,000 \right)$		308 000
Profit to date		£283 000

There are two different methods of reconciling the profit derived in the above analysis.

Reconciliation (A)		£
Budgeted profit $= \dfrac{3850}{15\,500} \times 1\,550\,000 =$		385 000
Total contract variances	−29 000	
	+6 000	
	+28 000	
	−26 000	−21 000
		364 000
Sales variance		−81 000
Profit to date		£283 000

Reconciliation (B)

Contract	Operating variance (£)	Volume variance (£)	Total variance (£)
CB/121	−20 000	−9 000	−29 000
CB/124	−6 600	+12 600	+6 000
CB/130	+25 660	+ 2 340	+28 000
CB/150	+ 1 000	−27 000	−26 000
Totals	+60	−21 060	−21 000

$$\text{Total variances} + \text{ sales variance} = -£21\,000 + (-£81\,000)$$
$$= -£102\,000$$
$$\text{Profit to date} = £385\,000 - 102\,000 = \underline{\underline{£283\,000}}$$

9.7 Price changes and the use of cost indices

The higher the rate of inflation, the greater the need to employ reliable methods of forecasting future cost increases. It is important for both an owner and a contractor to be able to predict with reasonable accuracy the future effect of inflation on prices in general; an owner will often need to budget for a period of at least five to ten years ahead; a contractor will certainly need to be able to forecast for the future duration of his longest project and possibly beyond. One such tool for use in these circumstances is the cost or price index.

Construction costs and prices change over time for a number of reasons other than because of the effects of inflation. There are important differences between the ways in which these influences affect future costs. Inflationary economic tendencies and potential markets tend to undergo comparatively rapid fluctuations which cause short term, quite rapid and perceptible effects on costs and prices. At times, the effects may be considerable. On the other hand changes in productivity due to the results of the evolution of new construction methods, training schemes, alterations in labour/equipment ratios and the development of more sophisticated equipment, will take place relatively slowly. Often, because the rates of changes arising from these actions are almost imperceptible, no special attention needs to be paid to them and, in collecting data on a continuous basis, the effects can be absorbed with other trends.

The adjustment to the value of money and to cost/price data generally is best carried out by the use of an index. If, for example, a cost index was established in 1987 (its *base date*) at the value of 100, by 1990 it had reached 120 and by 1994 it had risen to 134, then the cost index change between 1987 and 1990 would be $120 - 100 = 20\%$ and between 1987 and 1994 it would be $134 - 100 = 34\%$. The calculation is not difficult; if there is difficulty it usually arises in the compilation of a representative model to form the index. The use of an index can only be reliable if it is based on a large amount of accurate data which have been properly classified and defined and if it has been carefully constructed in order to reflect truly the purpose for which it is required.

There are many types of index created and maintained for specific purposes. Some are established and updated by official organizations such as government agencies, others are created by private organizations, such as trade federations and the like. The resources required to maintain an index, particularly of the kind that is created by a government organization, are extensive. Models of such indexes, such as the price adjustment formulae used in the UK for building and civil engineering contracts, have many constituent items of labour, plant and materials in the overall model and it requires a vast input of time and financial resources to review and maintain them on a regular basis. Frequently constituent elements of such models will be maintained on confidential data that are only available to restricted users.

A brief description of the principles involved in developing indices will be helpful in understanding their application. (For a deeper insight into the construction and use of indices, reference may be made to one or more of the many standard textbooks on statistics or to the construction orientated texts such as those of Beeston and Tysoe.)

There are two general approaches to the development of a cost/price index. The first is to use historical prices, preferably derived at tender time, for a unit quantity of a well-defined product. For example, the price of a two-storey domestic house is often quoted, initially, as a price per square metre of floor area; an average price may be established from historic records for a cubic metre of excavation.

This approach may be extended to become more complex by using a model that consists of several unit items, each of which is given a relative weight in the compound index. An example is that of combining the average price of each of the following components with their respective weightings in order to provide an index for excavation in cofferdam:

(a)	1 m^3 of excavation	31%
(b)	1 tonne of driven steel sheet piling	50%
(c)	1 m^3 of baulk timber	15%
(d)	1 m^3 1:3:6 concrete	4%
		100%

Such models have the advantages that they:

(a) are relatively easy to understand;
(b) require data which can be retrieved relatively easily, for example from priced bills of quantities.

They have the disadvantage that:

(a) they are based on unit rates which in themselves may be made up of a wide range of cost components;
(b) it is usually difficult to identify whether the unit rates are strictly comparable from one project to another;
(c) it is frequently difficult to identify the true reasons for changes in price.

The second approach is rather more complicated and usually requires a supply of data which is not always readily available. A model is constructed on the basis that the index reflects the relative quantities of resource that are required to produce a unit of construction, such as one cubic metre of reinforced concrete. The resources may include all or some of the relative amounts of plant, labour and materials, possibly plus other factors that may have an influence on the unit cost. This approach overcomes many of the disadvantages of the first method, but its disadvantages are the reverse of the first method's advantages. It has the distinct advantage that it is easier to detect the reasons for changes in the index. It is necessary, however, to check from time to time that the relative weights/values of the components for a specific index do not change significantly.

A contractor has a need to forecast the possible future behaviour of prices when preparing an estimate prior to the submission of a tender for a project. At the time the contractor has no alternative but to use the current price levels for the resources that will be used adding on an allowance for future changing levels of prices. This process may well be further complicated by conditions of contract which specify what methods of price adjustment will be used and perhaps how inflationary effects will be measured. In these cases the contractor needs to make some adjustment for the compensation he believes will be paid for increasing prices. Since most index-based methods fit no case exactly, it is a matter of fine judgement as to the value of this compensation.

Even if the calculation of the compensation resulted in an accurate payment for price rises, in times of relatively high inflation even this recompense may be eroded as is shown in the following example.

Example 9.2

A contractor estimates the cost of a project to be £100 000 at current prices.

He adds a profit margin of 15% on which he can expect to pay tax at a rate of 50%. If inflation is running at 2% per month and the owner fully compensates the contractor for price increases due to inflation but delays payment until $1\frac{1}{2}$ months after certification, what is the contractor's true profit margin after tax?

Let m = contractor's profit margin = $\dfrac{P}{C}$ or $\dfrac{P_c}{C}$

where P = profit (£)

$\quad\quad C$ = estimated cost (£)

$\quad\quad t$ = tax rate

$\quad\quad P_c$ = profit after compensation for inflation

Profit, less tax, from construction = $P = (1 + m)C - C - mtC$
$$= mC(1 - t)$$

If the owner fully compensates the contractor, for an inflation rate of $d\%$ for only specified periods of time, the contractor's profit after tax becomes

$$P_c = C \left[\frac{1 + m}{1 + d}\right] - C - mtC$$

or the effective profit margin would be

$$\frac{P_c}{C} = \left[\frac{1 + m}{1 + d} - 1 - mt\right]$$

Therefore the contractor's true profit margin after tax is

$$\frac{P_c}{C} = \left[\frac{1 + 0.15}{1 + 0.03} - 1 - 0.15 \times 0.50\right] = 0.04 \text{ or } \underline{4\%}$$

If inflation were 0% then

$$\frac{P_c}{C} = (1 + 0.15 - 1 - 0.15 \times 0.50) = \underline{7.5\%}$$

9.8 The nature and classification of costs

For a manager to be provided with the necessary relevant information in order to be able to control the cost of the operations for which he is responsible, and to make the appropriate decisions, there is a need for the cost data which are collected to be analysed and classified. Before a cost accounting and control system can be established, it is necessary to set out the classification of costs which will be used. The total cost of carrying out work is not difficult to establish on a historical basis. From this can be obtained average unit costs, for example the average output per head of labour in terms of money, or the cost per square foot of floor area for a building. However, something more than average figures over large units of work is required if close and objective control of work is to be exercised.

In classifying costs, the objective is to identify each cost with some activity, department, product, contract or other unit. Within a company, or smaller production unit, costs are normally classified within certain functions. These functions may be more readily identified by departments or sections of the firm and will include production, administration, selling, distribution, services and perhaps research and development. The extent to which each function is represented depends largely on the precise nature of the business which is being undertaken. In addition, certain industries use particular terminology which may or may not represent exactly the equivalent division of the functions within another industry. For the purposes of cost control, costs within these functions are first allocated to *cost centres* which may for example be chosen to represent some or more items of plant or mechanical equipment grouped and working together, a group or gang of labourers or craftsmen, or perhaps a physical or geographical location on a construction site, such as a particular building which can readily be isolated and identified. A cost centre represents some activity or operation which contributes to production and to which the accumulated costs can be readily allocated. Having been identified and classified under a cost centre, costs are then allocated to a *cost unit* which may be a particular product group or contract or unit of production.

Costs allocated to cost centres and/or cost units can be further broken down into those for materials, labour, sub-contractors, plant and other expenses. This further breakdown is into what are then known as *cost elements*. Each cost element can be further classified into either *direct* or *indirect* costs. A direct cost is defined as one which may be readily identified with a specific activity or operation in the form of a cost centre or a cost unit. An indirect cost is one which cannot readily be allocated because of its general nature and it must be apportioned on some other suitable basis. Indirect costs are such that they are rarely related, other than in a general way, to the actual quantity of productive work carried out. They are costs which are often incurred, whether production is in progress or not, and are frequently divorced from the rate of production as well as the quantitative progress. The *prime cost* of production consists of the direct material cost plus the direct labour cost plus other direct expenses. Direct material costs are the costs of all those materials which can be firmly allocated to the product. They can be accurately measured and allocated. In construction, the amount of cement used per cubic metre of concrete can be measured and costed with some reasonable degree of accuracy and then allocated to a cost centre. This is, therefore, a direct material cost. It must always be borne in mind that it is rarely worth the expense of measuring and allocating absolutely all direct costs in detail to cost centres, but rather to treat them as overall production expenses. The cost of screws and nails in the preparation and construction of timber formwork falls into this category.

Direct labour costs are, like direct material costs, those costs of labour that can be associated directly, clearly and undeniably with a cost centre or a cost unit. These will normally be the wages of labour associated with specific cost accounts. In some instances it may be possible to allocate the cost of supervision to specific cost centres, even though such costs would normally be identified and allocated as indirect costs. Such an allocation will then qualify unit costs of production and this must be borne in mind when comparison with other unit costs (which may be on a different basis) is made. There are likely to be instances where the allocation of labour costs may appear to be possible on a direct basis but where, in fact, the cost of doing so and the accuracy with which it can be done mitigate against it. An example is that of a driver of a derrick crane which may be involved in excavation, concreting, structural steel erection, and, say, piling operations, on an intermittent basis throughout the day. In such circumstances the labour cost may better be treated as an indirect cost and as a fixed rather than a variable cost.

Direct expenses are those costs which can be directly allocated to a cost centre or cost unit, but which do not fall under either the heading of direct labour or direct materials. In some instances it may be possible to make specific allocations of cost for services. For example, an extraordinary requirement of cost for a supply of water or fuel might be put under this heading. The requirement for shutter oil may not be capable of allocation if formwork is being used over a wide range of cost centres, in which case it will be more convenient and sensible to treat it as an expense. The use of fuel for plant may well fall into the same category.

In construction, at least two other direct costs are usually identified and separated where they occur – those for plant or mechanical equipment and those for sub-contractors. Plant and mechanical equipment direct costs are self-explanatory and are usually kept as a separate item in the same way that direct materials and labour have been described above. Sub-contractors may be highlighted as a separate cost account so that some control is exercised over this area. Alternatively, they may be incorporated under the other direct costs according to the requirements of the costing exercise.

All other expenses, over and above the prime cost of production which consists of direct materials, labour, plant, sub-contractors and expenses, are known collectively as overheads. The bulk of overhead costs consists of administrative costs and, in production and other similar industries, a considerable proportion may arise out of distribution, selling, research and development expenditure. Overheads are therefore the indirect material, indirect labour and indirect expenses which cannot be allocated to a cost centre or cost unit. In the case of a construction company it may be convenient or necessary to divide overheads proportionally between contracts and the head office. In this case, they are usually known as site oncosts and head office administration charges, respectively. Amongst a contractor's site oncosts, one would expect to find the cost of management and supervisory salaries, office staffs and equipment costs, office rentals or rates if levied, insurance, consumable stores, canteens, welfare facilities, etc.

There is a tendency for specialist services to be centralized in many organizations. The installation of a computer is an example of a high capital investment which is required to provide a service throughout a company, not necessarily within the immediate geographical environment of the building in which it is housed. This must be absorbed within the cost units and rapid technological advances in this and allied equipment, overheads are becoming an increasingly important feature of cost over which strict control must be exercised.

9.9 Construction resources in cost control

Some further explanation is needed of the problems which face a construction contractor and areas which will require particular attention. Like projects in a general sense, a contractor faces the one-off situation wherein he rarely executes two identical projects under identical environmental conditions. Construction projects offer great scope for ingenuity, but they are more difficult than work carried out in a factory or a similar establishment, largely because of the temporary nature of the project organization. Work is not carried out, for example, within the confines of four walls and a roof, but is subject to the vagaries of the weather. Labour is not of a regular nature, but to a large extent is casual and, therefore, its quality takes some time to establish. Experience of local conditions in trade and other respects has to be built up over a period of time and all too often the construction contractor has to break new ground which is unfamiliar to him. The risks involved in this aspect of work, therefore, tend to be greater than in many other industries and the estimation of the cost of work, prior to the work commencing, is a difficult and skilled operation. The need for an accurate system of cost control is paramount and from such a system there is also the need for feedback to the estimating organization of a contractor.

The standards of cost by which a construction contract must be controlled are set at the time that an estimate of cost for the work is prepared. This assumes that the type of contract favoured by an owner is one in which a contractor is required to agree the basis on which payment will be made for the completed work before either party enters into a contract. Methods of estimating, however, rarely result in the cost estimate appearing in the form that is required for subsequent comparison and control. It is almost invariably necessary for the form of the estimate to be rearranged before the cost standards can be set up for use in the control function or conversely for it to fit the tendering documents. As far as control is concerned, it is desirable to have cost under the headings of labour, plant, materials, sub-contracts, and overheads which might include profit, finance, insurance as well as the many services provided on a centralized basis from the Head Office of the company. Profit is usually identified as a separate item in its own right. The method of dealing with each of these divisions of cost can vary widely because of the different nature of each. The extent to which control over each can be exercised will vary, as will the ability to collect useful data and undertake the control from that point onwards. Some influences on control, and ultimately profitability, will be outside the jurisdiction of a site manager. Increases in basic wage rates on a nationally agreed basis, the weather and changing site conditions are examples.

In construction work, there is little doubt that the most difficult of these divisions to control is that of labour. The difficulty starts at the estimating stage because the productivity of labour is variable and therefore average productivity figures based on historic information must be used. A prediction of how labour will perform in the future on a particular project must be incorporated into an estimate of cost. The cost of labour to a contractor falls into two subdivisions – the first is that of the hourly cost of the man arising out of his basic rate plus any extras which arise from skills or bonus payments, in whatever way these may be calculated. The second includes the cost of all the remaining items paid either to the man or on his behalf. The former are the direct costs of an activity on which the man is employed, but the latter are not necessarily related to the length of time that the man works on any particular operation. In the second category fall such payments as travelling time, fares, guaranteed time for inclement weather, National Insurance contributions, holiday pay stamps, tool money, etc. The labour cost of an operation, therefore, can be calculated by multiplying the hourly direct cost of a man (including his overtime premiums) by the time in hours that he spends on a particular operation by the number of men which is allocated to that operation. To that sum must be added a percentage which is calculated in the knowledge of the payments which are made and belongs to the second category of costs. Obtaining these costs will enable the unit labour cost of production to be calculated, a unit cost being the total direct cost of carrying out one unit of the work being undertaken. Examples of unit costs are the cost of excavation of one cubic metre of soil, placing one cubic metre of concrete or fixing one square metre of formwork. These unit costs can then be compared with the estimated rates for doing the work and comparisons can be made between performances and costs over different periods of time.

The establishment of labour costs begins with the collection of the hours spent by each man in the field. These are recorded on a form with a brief description of the activity on which the men have been engaged. The responsibility for recording these data lies with the foreman, the chargehands or the gangers responsible for the supervision of the men. The hours are then allocated to various code numbers in accordance with the breakdown of costs that is required. This function is usually the responsibility of a cost clerk who also has the job of reconciling the hours recorded on the allocation sheets with those recorded on clock cards (if these are in use) and ultimately the hours which are paid for through wages sheets. If costs which are collected are to be assessed in a meaningful fashion, it is necessary to establish a value of the work carried out for a comparison to be made between actual cost and payment. This will normally be made by an engineer responsible for the work or a quantity surveyor performing a similar function. At this latter stage, a check should be made to see that the allocation of hours to code numbers is accurate and within the requirements of the cost control system. A detailed statement of the labour cost performance of the contract can then be established in accordance with the requirements of the manager responsible for the work.

Plant is the second major division already referred to above. The control of expenditure on plant, especially of the major items of mechanical equipment found on a contract site, is much more readily achieved than for labour. It is usually easier to associate the direct cost of plant with a particular activity, and major items of plant are rarely switched from one activity to another with the ease and frequency of labour resources. Neither are they flexible in their application. In preparing an estimate of cost for work, it is usual to separate the cost of transporting plant from depot to site or from site to site, so that this can be dealt with as a lump sum item. It is usual to treat site repairs and maintenance as overhead costs. When recording operating plant hours, it is necessary to differentiate between plant hired outside the user's company and company owned plant, particularly to control the use of hired equipment properly. Plant costs can therefore be controlled, not on an hour-to-hour basis, but rather in the light of the total length of time the equipment is actually on the site compared with the duration that was estimated for it to be needed. The productivity of machines over short periods of time can only be a relatively meaningless way of controlling their cost.

The control of the cost of materials provides a different problem from either labour or plant. An important factor is to ensure that materials are not being wasted or stolen and that they are delivered at the right time. Where centralized buying functions are being used, the total cost of materials delivered and incorporated into the work, or used for temporary works, does not come forward immediately and very often not until the invoice is paid by the purchaser. There is, therefore, always some considerable delay in arriving at material cost statements and it is not realistic to control the cost on a day-to-day basis, but rather to use a unit such as a month as the minimum feasible unit of time.

Sub-contractors provide no problem as far as cost control is concerned, so long as proper consideration is given to the placing of a sub-contract in the first instance. This is a problem which is best tackled by ensuring that a thorough examination of the sub-contractor's business is carried out beforehand, rather than leaving it until it is too late. Sub-contracts are generally let under the same conditions as the main contract and on these will depend the necessity for examining such items as increased costs and variations to work as it proceeds.

Overhead costs lend themselves less than labour and plant costs to dynamic control on a short-term basis and probably are also best carried out on a monthly basis. Comparison of actual costs with estimated value of work done can be carried out on a simple and ready basis. The nature of overhead costs is such that immediate effective action is rarely possible since the response time between action and its results is somewhat longer than with hourly paid labour for example.

9.10 Design of a contractor's project cost control system

When designing a cost control system for a construction contractor, there is a wide variety of aspects which need to be considered. One of the first and of major importance is the size of the company itself and type of work which it undertakes. The type of work has two important influences. If the work tends to be dominated by civil engineering, or a specialized activity such as large earthworks or piling, then the cost system will need to be different from that which would be designed if the company's business were directed largely at the construction of domestic and small multi-storey buildings. This difference stems largely from the different range of operations that are undertaken in each type and the different contractual arrangements that relate to each. The type of contract which is being undertaken will clearly be a strong influence. An admeasurement contract will have quite different implications for the control of cost from those, for example, of a lump sum contract.

There can be differences of opinion as to whether the scheduling process and the cost control process should be controlled by an integrated reporting system. If cost control systems are planning and schedule orientated, they usually have the advantage that the one system can be used for control at the same time. However, integrated systems are at a disadvantage in the sense that either simplicity or attention to cost detail is sacrificed. It is the experience of many practitioners that separate schedule and cost control systems provide a cheaper means of good control and the output from the two separate systems is more easily understood. The processing of the data for separate systems does, of course, need to be integrated.

In order for a cost control system to be used effectively and easily it is necessary to be able to identify with accuracy, and to record the resources clearly identified against the activities or operations for which they have been used. If this is to be done, then an appropriate coding is required in order to identify the various attributes of a particular section of the work. Further, a cost control system must exhibit the quality of relevance. All the information that is presented to a manager must be entirely relevant to his responsibility and his position in an organizational hierarchy. It must be sufficient for him to make managerial decisions readily and to apply his judgement in the process. It must have the appropriate material content and the timing of its arrival in the manager's hands must be exact. The cost control system must also have the quality of reliability. There must be neither errors in the input and the output nor as a result of the processing of data before presenting it as information. If a cost control system is to be applied successfully, it must be economic and the cost of its installation and application must be less than the increase in efficiency which it is believed will accrue to its use. These quantities of allocation, relevance, economy, and accuracy required of a cost control system militate against the use of a standard package which has been prepared on the basis that it will be used to control costs in all types of construction companies that are undertaking all sorts of different work and with many patterns of organization structure. There is little doubt that a system specifically designed on the basis of flexibility so as to allow ease of updating and fitting to the organization structure of the company concerned (and adapting to the changes in organization structure as time passes by) is the most efficient and worthwhile.

In setting up a cost control system, the first action must be to identify elements of cost. These must reflect the way in which the work is broken down into operations of activities and care must be taken to see that the level of breakdown provides sufficient detail for good control. On the other hand, care must be taken to see that the breakdown of work (and hence cost) is not into such small elements that control becomes unjustified and leads to erratic results. The budget upon which the cost control system will be based will have, in most instances, to spring from the estimate of cost for the work whether it be prepared for a tender or in some other way. If the estimate is closely followed and it is in the usual form of such estimates, then control in the field will provide feedback which will be recognized as being useful to the estimating process.

There is little point in breaking work down for cost control purposes into elements that cannot be recognized by constructors in the field. The breakdown into elements will preferably follow a pattern which is reasonably standard throughout a particular company so that comparisons can be made between the unit cost of the various items which are placed in similar situations on a number of different jobs. This also has the advantage that usage of the system will create familiarity with this type of breakdown.

Variance from budget costs will normally be the result of one or more of four principal causes:

(a) construction performance;
(b) poor technical and administrative performance, such as in design of the facility or in the purchase of materials;

(c) errors in the preparation of the estimate or budget;
(d) special circumstances which will have affected a site in particular, such as strikes, poor foundation conditions, poor weather, etc.

All extraordinary items must be noted in feedback so that the estimating process can be updated and be as reliable a possible. When work is in progress, costs must be collected in the field and apportioned to the particular activities which form the elements of the work and against which comparisons will be made. The critical element is to assign these costs to the correct cost element. As well as allocating labour costs to cost elements, it is necessary to undertake the same process with regard to materials so that productivity data can be maintained. It is particularly important that scrap and wastage is noted where the use of materials is concerned.

It is essential in any cost control system that accurate measurement of progress to date is made, not only from the point of view of controlling the actual cost of work but also to give an indication of the amount of income that can be expected at the end of each valuation period. Every attempt should be made to forecast or estimate the cost of work to completion. Progress estimates are necessary on a weekly basis since problems in construction can occur very quickly and can have an extreme effect on profitability.

Most companies must necessarily consider the employment of a computer-based system; at the other extreme of the spectrum a small builder may well consider the use of a straightforward, simple, manually operated process as being the best suited to his own needs.

When assessing the desirability of having a system installed in a construction company, there are a number of facets that need careful investigation. Firstly, there will be the cost of installing the system so that it can cover and be appropriate to the majority of the construction company's projects. This is not a matter to be taken lightly since, not only has the system, if it is computerized, to be installed on the company's computer system (or machines to which the company has access) but there is the question of instituting the necessary processes through from the operational level at each site to various levels of management and into the associated administrative areas such a accounts, pay-roll, invoicing, etc. If structural design operations are part of the company's activity, then these too will need to be incorporated into the system as well as the company's estimating process so that it has the benefit of the feedback from the operational level. Finally, as part of the installation process, it is necessary to train all the staff in the system's operation, not only at the sites but also within the head office and the centralized departments.

Secondly, the cost of collecting the data and information and ensuring that they are accurate for the purpose and are input to the cost control system without error needs to be considered. In this recording process, there is a need to record the cost of all resources such as labour and equipment together with the cost of materials which are associated or incorporated in the work. This can be a complex and costly process and once the system is installed and running, these costs accrue on a day-to-day basis for the total duration of the project.

Thirdly, there is the question of the preparation of plans, schedules and budgets prior to the commencement of work and that set the standards against which the performance of the construction activities will be measured. Much of this work, except perhaps in terms of cost, will have to be undertaken whether a cost control system is installed or not. However, the imposition of costs on these other activities is again an expensive business. The costs for this preliminary work are usually incurred in the pre-contract phase and the early days of the contract itself.

Fourthly, and finally, is the question of the preparation of reports, being the result of the output from the system and as a result of the manipulation and processing and calculation of the data which have been input to the system. Any problems with the processing and calculation will need to be resolved at the point in time at which reports are prepared.

In order to have a complete picture of the cost control on a construction site, it is necessary to have a wide range of reports and not that which is just a manipulation of the various input data that are collected, processed and printed on a report form in quantitative terms. The cost control system will need to be seen in the context of a written report by the project or site manager who will be making his own personal judgements about the site activities that can be associated with the current cost report. There may also be such reports about unusual or extraordinary incidents from managers at a lower level in the hierarchy than that of the project manager. Cost reports need to be considered in association with reports concerning the general progress of work and particularly in relation to the preconceived plan and the schedule. In some instances there will be checks on construction methods by work study engineers who will be checking that production, as a result of study they have previously made, is as was anticipated and if not, why this is so. All of these sundry reports help to build up a general picture of the cost and progress situation on a construction site.

A construction cost report has three main functions. These are:

(a) to compare the actual cost of construction as built with the budget for the equivalent work and to calculate the difference between the two so that reasons for any variance can be established. This clearly necessitates the work to be broken down into comparatively detailed activities of short duration and low cost if a thorough investigation is to be made of productivity;

(b) to report certain statistics of the construction costs. These will show the actual cost of work constructed and total sums of money incurred such as the cost of labour and of mechanical equipment. Some of these statistics will be those that are required by a project manager out of his experience in managing similar activities. They may include the amounts of bonuses that are earned by individuals, break-down of labour into trades, the progress of sub-contractors on the site, the total quantities of construction items such as concrete and earthworks in major elements of subdivisions of the project and so on;

(c) a forecast of what is likely to be the end result of the project in terms of cost in the light of the experience of work already carried out. This can then be compared with the cost for the total works that was estimated at the outset of the work and, in particular, it will be necessary to establish the contribution that the particular project will make to the overall activities of the contracting firm by way of margin and profit together with that made towards the overhead costs of the company as a whole.

Managerial use of cost control systems extends to the following:

(a) valuation of profitability of the project under construction;
(b) calculation of the variances between the budgeted resources for a particular activity and the actual consumption in construction, the purpose being to take control action as necessary;
(c) measurement and examination of the efficiency of the various subdivisions, subsections, departments of the organization and within the company;
(d) the preparation of records concerning those resources which are used in production together with their costs and their productivity in use for feedback purposes. Such information is then available for future use in preparing plans, budgets, or methods analysis.

9.11 Budgetary control for contract costs

Budgetary control and reporting has been illustrated in the previous chapter in the form that it might be applied by an owner to the control of a project, wherein most or all of the work is carried out by contractors rather than by direct labour. A different form of budgetary control and standard costing should, however, be used by contractors as a means of controlling their own cost and production performance. Figure 9.11 shows a typical layout for a budgetary control report in the early stages of a construction contract.

The breakdown into budget centres and elements of cost, and then the collection of the cost data, need to be carried out in such a way that they represent the form in which specific costs can be readily identified. Initially, each element of cost must refer to a portion or an element of the work for which an estimated or budgeted cost can be extracted from the contract

Budget cost report for contract CB185 – October 1994. Total contract value: £273 000

Cost code no.	Element of cost / Description	Total value of work (£)	Cumulative					Month			
			Budgeted value of work done (£)	Value of work done (£)	Actual cost of work done (£)	Variance (£)	Variance (%)	Value of work done (£)	Actual cost of work done (£)	Variance (£)	Variance (%)
	Direct Labour										
003	Excavation to foundations	18 734	7 328	6 243	6 947	(704)	(11.28)	2 731	2 842	(111)	(4.06)
005	Excavation to drains	9 376	1 260	1 360	1 240	120	8.82	563	530	33	5.86
017	Concrete blinding	1 020	90	98	102	(4)	(4.08)	60	80	(20)	(33.33)
	Direct Materials										
117	Concrete aggregates	3 700	80	90	80	10	11.11	50	60	(10)	(20)
	Plant										
203	Excavation to foundations	31 670	10 346	9 207	9 500	(293)	(3.18)	5 346	5 200	146	2.73
205	Excavation to drains	25 500	2 341	2 460	2 370	90	3.66	1 026	900	126	12.28
	Totals – Labour, materials and plant	90 000	21 445	19 458	20 239	(781)	(4.01)	9 776	9 612	164	1.68
997	Site oncosts	9 500	2 450	2 500	2 400	100	4	1 050	1 250	(200)	(19.04)
998	H.O. overheads	4 500	1 100	1 200	1 060	140	11.67	570	550	20	3.51
999	Profit	9 000	2 400	2 500	2 390	110	4.40	960	957	3	0.31
	Totals	113 000	27 395	25 658	26 089	(431)	(1.68)	12 356	12 369	(13)	(0.11)

Fig. 9.11   Variance analysis for contract.

documents or from the pre-tender estimate. This will normally be related in construction work to the operational method of carrying out the work. It is often the case that a single excavation is represented by many different items in a bill of quantities, and in different sections of the bill, classified by characteristics some of which may be depth, width, type of material, in cofferdam, in open cut. Even though this may be the case, the excavation process as qualified by these characteristics, may vary from time to time, though it may still be carried out by the same equipment operating in a broadly similar fashion.

The element of work used in a budget must be such that it is possible to collect data for it. Using an example involving concrete, a concrete batching and mixing plant may be producing a wide variety of different mixes during a single day's operation in order to satisfy the demands of a number of separate concreting operations. It will be almost impossible, under normal circumstances, to divide the labour cost of the mixing plant operating crew precisely between each cubic metre of concrete produced. Some rationalization has therefore to take place in the form of an averaging process.

Having identified those elements of the work over which control needs to be exercised, and can be exercised – and this will rarely be every single item of work involved in a project – a breakdown into at least labour, materials, plant and sub-contractor costs for each element of work is desirable. This is because of the different nature of the control process involved in each of these four or more areas.

Labour, for example, can be switched from job to job on a short-term basis, has a very variable productive quality and needs a high degree of skilled supervision. There are many views as to the extent of materials control required. As far as material cost is concerned, much of the control is taken out of the hands of a purchaser so long as a genuine attempt is made to get the best price available. There is certainly a need, however, to control materials from the point of view of wastage, misuse and theft and as a result it is necessary to have some check on the costs involved. Plant and mechanical equipment present a different situation – one of utilization and therefore organization and planning rather than cost control. A high degree of utilization of mechanical equipment will automatically ensure the most economic use if the plant has been selected properly in the first instance. Major items of plant tend to be kept in one location for as long as possible as it is not economic to move them frequently from location to location without some regard to the high cost of doing so. Each of these points is reflected in the structure of the contract variance analysis of Fig. 9.11.

In addition to the direct costs of labour, materials and plant, control needs to be exercised on overheads generally, whether these be site oncosts which are specifically incurred in the management, supervision and administration of the work at the contract site, or the head office overhead and administration charges resulting from those centralized activities necessary to both the operation of the contract site and the company as a whole. The desirability, necessity or practicality of including head office overheads and profit as items in a site control statement may well be questionable.

In Fig. 9.11 it will be seen that the budget has been broken down into easily recognizable elements of cost and each has been coded. Direct costs are totalled separately, to which are then added oncosts, head office overheads and profits.

The budget statement is divided vertically into two main parts. An updated statement for the month just completed is set out on the right-hand side and a cumulative statement is produced on the left. These two statements enable assessments to be made of current and total performance and also of trends in performance. Total values of work to be carried out are included so as to give a measure of the size of the total element against which completion so far achieved can be measured. At this level of control, a straightforward variance as between the value of the work done and the actual cost of the work done is established, and unfavourable or adverse variances are indicated by the use of brackets. Where the variance, either favourable or unfavourable, is considered to be significant, an analysis of a more detailed nature can be made in order to establish the possible cause. The variance as a percentage of the value of the work so far completed is included so as to give a better idea of the trends which may have been established.

It will be seen from the report for the last month that an unfavourable variance occurred for direct labour costs in *excavation to foundations*. By comparing the value and actual cost of work done during the month with the cumulative totals it will be seen that the trend of the unfavourable variance is down by proportion. Nearly one half of the total item 003 so far carried out was completed during the past month, but the unfavourable variance for the month is considerably less than that for the total *excavation to foundations* so far carried out. Small unfavourable variances are highlighted for *concrete blinding labour* and *concrete/materials*. Since these elements of the work are clearly in their early stages, the variances are probably not worth further investigation at the present time as a result of this report. An unfavourable cumulative variance for plant costs on *excavation to foundations* has been helped by a favourable variance on a large amount of work during the month.

Taking direct costs of work overall, there is still a cumulative adverse variance of £781 though the operations in the last month reported produced a favourable variance of £164. The trend is therefore favourable. Of the oncosts, overheads and profit, site oncosts look as though they are rising too steeply, resulting in the conversion of a previous cumulative favourable variance of £100 into an unfavourable one of £200 following last month's work. Clearly, further investigation is required. Profit is being recovered at a slightly faster rate than was anticipated by the margin allowed in the estimate.

Not only can the report be analysed as far as cost is concerned, but also the performance against programme can be measured. Looking at the final line of the direct cost report (although this comment applies in principle to each line of the report) it will be seen that a total of £21 445 of work was budgeted to be carried out up until this date, but only £19 458 of value was in fact achieved. There is a shortfall arising out of the *excavation to foundations*. Some extra effort is needed in this area in order to get back on programme.

The budget cost report set out in Fig. 9.11 is designed for use by the contract site management and the level of detail has been set accordingly. On a large project it would be necessary to have a hierarchy of reporting levels since the level of detail of this cost report would prove overwhelming for one manager to use if the work involved was extensive. In a similar fashion, the organization structure of a contractor will be such that a contracts manager/director will have responsibility for several contracts and he will require summarized information in the first instance. Figure 9.12 shows a typical monthly budget control statement for four projects – one of which is Contract CB/185 which is the subject of the detailed cost report of Fig. 9.11.

The variance analysis summary sheet collects together the bottom lines of the detailed variance analysis for each project. It enables the contracts manager to look for the salient features of the financial performance of the projects and to seek further information where this is thought to be necessary. For example, Contract CB/210 has clearly had a bad month, though the project is only in its early stages. A former favourable variance has been converted to an unfavourable one and the work is now behind programme since only £13 746 worth of work has been carried out, whereas £15 634 worth was budgeted. Contract CB/201 has moved from an adverse to a favourable variance situation, it is well up to programme, though in that respect it fell behind last month in spite of favourable variance. This may need some investigation.

9.12 Variances and standard costs

It is one thing to make arrangements to collect project data concerning cost achievement in the field, but it is quite another to use these data or to process them into information and then interpret them in a useful way. In order to make a meaningful interpretation of such information, it is necessary to have some form of *norm* or *standard* against which the actual performance can be measured. The standards for the one-off project will almost certainly be generated from past experience of like or near-like situations where possible, and they will then be used in establishing the initial budget. Alternatively, standards may be derived from the use, for example, of work measurement techniques where historic records do not exist. Any subsequent departure from the set standard in practice is known as a *variance*. An *unfavourable* or

Monthly Budget Control Statement for all contracts – October, 1994

Contact	Description	Month			Cumulative				
		Value of work done (£)	Actual cost of work done (£)	Variance (£)	Budgeted value of work done (£)	Value of work done (£)	Actual cost of work done (£)	Variance (£)	Contract value (£)
CB/185	Single storey factory – J. Soap	12 356	12 369	(13)	27 395	25 658	26 089	(431)	123 346
CB/201	Block of flats – Robbers Ltd	25 634	23 216	2 418	103 246	107 421	105 241	2 180	183 246
CB/210	Office block – Stouts Ltd	10 346	12 431	(2 085)	15 634	13 746	14 843	(1 097)	73 281
CB/215	Brewery – Hale Ltd	20 261	18 340	1 921	100 246	102 368	99 874	2 494	175 674
Total for all contracts		68 597	66 356	2 241	246 521	249 193	246 047	3 146	555 547

Fig. 9.12 Variance analysis summary sheet.

adverse variance is one in which the planned or budgeted standard has been exceeded in the cost and/or use of resources; a *favourable* variance represents a situation in which less cost or resource has been incurred in order to complete an operation or activity than was budgeted. The use of a standard, therefore, enables a comparison to be made, whereas the actual costs or performance data alone are relatively meaningless. In some instances it may be both useful and desirable to set a standard in the form of a range of values of cost within which an operation might be carried out.

Because of the extensive use of standards in budgetary control, it is virtually indistinguishable from *standard costing*. In fact, before standard costs can be derived, some form of budgeting must be implemented. The two will be treated as being inseparable. Standard costs may be defined as those predetermined costs of material, labour and overheads which are budgeted to be spent in the production of a unit or units of a product for some defined period of time in the immediate future. The standard cost must be what is regarded as the true or real cost of carrying out the work. Views vary as to how a standard cost should be set. Some believe that standard costs should be based on absolutely ideal conditions and therefore they represent an extremely high standard of efficiency which can be achieved under only the very best conditions and with the very best resources. Such a basis for standards cannot be used in construction since this type of standard is rarely, if ever, achieved. If used, it results in poor morale and a poor attitude on the part of the personnel involved due to a lack of satisfaction because of their failure to achieve the standard. It is better to set a realistic standard in the light of known conditions – in effect what a construction estimator is attempting to do at the time an estimate of cost is being prepared. Standard costing employs a standard cost and an actual cost for calculation of the variance.

A variance between the standard and the actual cost can only occur because of one or both of the following reasons:

(a) *more money, or a greater price, has to be paid for the resources that are used in production than was anticipated at the time of the preparation of the standard costs and the budget; or*
(b) *either more or less resources are used in the production process to achieve the required budgeted output.*

A process or method which is used to control the cost of work carried out must therefore concentrate attention directly on to these two factors – the quantity of resources in use and the prices paid for them.

There are several advantages to be gained from using standard costs for control. Firstly, it enables the variances to be calculated and analysed in monetary terms thus leading to the exposure of inefficiencies, both favourable and unfavourable. This leads to a second advantage, that of the application of management by exception implied in the first. Action can be concentrated in a limited area. Thirdly, a distinction can be drawn between fluctuations in the price of resources as opposed to inefficiency in their use. In other words, the reason why a variance occurs is revealed as a result of the analysis. Standard costing enables distinctions to be made between those deviations from the budget that are controllable and those that are not.

A continuous monitoring system is provided to study the various effects of changes in prices and the use of resources on the overall project status.

If it is required to relate what actually happens in terms of the costs and the quantities of resources used for any particular operation, the following generalizations for standard costing are true:

$$\text{Actual cost (AC)} = \text{actual quantity (AQ)} \times \text{actual price (AP)}$$
$$\text{Standard cost (SC)} = \text{standard quantity (SQ)} \times \text{standard price (SP)}$$
$$\text{Total cost variance} = \text{standard cost (SC)} - \text{actual cost (AC)}$$
$$= \text{from above (SQ} \times \text{SP)} - (\text{AQ} \times \text{AP})$$

This general basis for the calculation of variances will be extended to cover the detail of monitoring and controlling the various elements of cost for a construction project.

9.13 Material variances

An effective system for controlling and monitoring costs must be based on placing the responsibility for cost achievement on a designated individual. Even so, there will be some areas of activity where an individual, though he is closely associated with a production operation, has practically no control over some aspects of that operation. The purchase of materials is often one area where the user of the materials has little option but to accept delivery at one price, because that is the best that can be negotiated, knowing full well that the standard or budgeted cost is different. There is a need, therefore, to identify in detail those items which are the responsibility of specific individuals. In the case of materials, a distinction can be made between the variances of price and the changes in the level of their use due to the efficiency or otherwise of the activity. The *material price variance* may be defined as the difference between standard price (SP) and actual price (AP) for the quantity of materials used. The *material usage variance* may be defined as the difference between standard quantity (SQ) and actual quantity (AQ) of materials used. It will be seen therefore that

(a) Material price variance $= AQ(SP - AP)$
(b) Material usage variance $= SP(SQ - AQ)$
(c) Material cost variance $= SC - AC$
$\qquad\qquad\qquad\qquad\quad = (SQ \times SP) - (AQ \times AP)$

The calculation of material usage variance follows a convention adopted in the United Kingdom, that variations in *quantities actually used* against *standard quantities* are priced at the *standard price*, since this buying price is very frequently outside the control of individuals responsible for using the materials.

An adverse *material price variance* may well occur because market prices have increased generally in the area of the materials being purchased. Alternatively, it may be because the purchasing function has not purchased in the best market. Favourable material price variances may result from the purchase of materials which are not to the specification as perceived at the stage when the standards were set. In such cases greater wastage may ensue when used in the construction and an unfavourable material usage variance may also result.

The *material usage variance* is expressed in monetary terms and results from comparing the actual quantity used in the work with the standard quantity that should be used. Unfavourable variances often arise out of careless materials handling, insecure storage of materials leading to pilferage and an abnormal quantity failing to pass inspection during the construction process.

The *material cost variance* may be regarded as a total overall variance consisting of the arithmetic sum of the *material price* and *usage variances*.

Example 9.3

At the time of tendering for the construction of the first floor of a reinforced concrete framed building, the material take-off showed a total of 13.29 tonnes of mild steel reinforcing bar were required. The weighted average cost of the various sizes of bar, by quotation, came to £440 per tonne delivered to site. The weight included the company's usual allowances for wastage.

When the reinforcement had been cut, bent and fixed in position, a check of the stock showed that 13.97 tonnes had been used for this work and that the actual weighted average cost of the material came to £432 per tonne.

$$\text{Standard price} = £440/\text{tonne}$$
$$\text{Standard quantity} = 13.29 \text{ tonnes}$$
$$\text{Standard cost} = 13.29 \times 440 = £5847.60$$
$$\text{Actual price} = £432/\text{tonne}$$
$$\text{Actual quantity} = 13.97 \text{ tonnes}$$
$$\text{Actual cost} = 13.97 \times 432 = £6035.04$$

Therefore,

$$\text{Material cost variance} = 5847.60 - 6035.04$$
$$= -£187.44 \text{ (unfavourable)}$$

$$\text{Material price variance} = AQ(SP - AP) = 1397(440 - 432)$$
$$= £111.76 \text{ (favourable)}$$

$$\text{Material usage variance} = SP(SQ - AQ) = 440(13.29 - 13.97)$$
$$= -£299.20 \text{ (unfavourable)}$$

$$\text{Material price variance} + \text{material usage variance}$$
$$= 111.76 - 299.20$$
$$= \underline{-£187.44} \text{ (unfavourable)}$$

9.14 Labour variances

The *labour cost variance* is a total variance made up of two subdivisions – the *labour rate* (or *wages rate*) and the *labour efficiency variances*.

The labour rate variance is the difference between the standard (SR) and the actual labour (AR), or wage rate of the operatives employed multiplied by the actual hours work (AH). Therefore,

$$\text{Labour rate variance} = AH(SR - AR)$$

A labour rate variance can occur because of changes taking place in basic wage rates, large amounts of overtime being worked, and operatives on grades that have different wage rates from those originally envisaged being employed on the work. For example, an apprentice may be called upon to do work originally envisaged as being carried out by a tradesman.

The *labour efficiency variance* is the difference between the *standard* (SH) and the *actual hours* (AH) measured at the *standard rate* (SR). Therefore,

$$\text{Labour efficiency variance} = SR(SH - AH)$$

The efficiency of labour may be affected by a poor working environment, inefficient organization, the use of badly chosen and unreliable equipment and incompetent management and supervision.

$$\text{Labour cost variance} = \text{labour rate variance} + \text{labour efficiency variance}$$
$$= AH(SR - AR) + SR(SH - AH)$$
$$= (SH \times SR) - (AH \times AR)$$
$$= \text{standard cost} - \text{actual cost}$$

Example 9.4

Carpenters are employed on making formwork for a large complex junction of cooling water ducts formed in reinforced concrete. The all-in rate for the carpenters is calculated as £6.75/hour. A work study establishes that the carpenters should take 520 hours to complete the formwork. In fact they take 502 hours and their actual all-in-rate is calculated at £6.85/hour.

$$\text{Labour rate variance} = AH(SR - AR) = 502(6.75 - 6.85)$$
$$= -£50.20 \text{ (unfavourable)}$$

$$\text{Labour efficiency variance} = SR(SH - AH) = 6.75(520 - 502)$$
$$= £121.5 \text{ (favourable)}$$

$$\text{Labour cost variance} = -50.20 + 121.50 = \underline{£71.30} \text{ (favourable)}$$

9.15 Overhead variances

Overheads can be classified into a number of categories, two of which are *variable* and *fixed*. *Variable overheads* is normally that category wherein the costs vary directly with the rate of production; *fixed overheads* do not vary with the level of production unless there is a conscious management decision made to increase/decrease them. Examples of variable costs are those for electric power, equipment repairs and maintenance, operatives' fares, lodging allowances and equipment and driver hire where this is fixed plant and cannot be attributable to specific items of work. Fixed overheads do not vary with the level of production but remain unchanged and incur the same cost, whatever the level of production. At a construction site the provision of site offices is an example of a fixed overhead – the cost of their supply, installation and maintenance will remain constant whether work is progressing on the site or not. Temporary roads, administrative staff, toilets and so on are other examples of fixed overheads and are frequently priced in the *preliminaries* section of a bill of quantities.

An overhead budget, therefore, may consist of both fixed and variable components and, as such, is known as a *flexible budget*. Whilst the fixed part of a flexible budget may be fairly spread over a period of time, such as a contract duration, in an even fashion, the flexible component needs to change in relation to varying levels of activity.

The computation of overhead variances in general construction is best undertaken in units of hours and costs rather than in units of production as may be more suitable in production engineering situations.

For overhead costs an *overhead variance* is the standard overhead for the period or production in question minus the overhead which is actually incurred. The standard overhead for the production is the standard hours of the production multiplied by the standard cost per hour (or other unit of time where appropriate). This overhead variance may be analysed into three sub-variances in normal use – the *volume*, the *efficiency* and the *overhead cost variances*, each associated with a specific cause of variation.

The *overhead volume variance* represents the difference between the budgeted and actual output multiplied by the standard overhead rate per unit. Therefore,

Overhead volume variance = standard cost of actual units
— flexible budget for actual units

An unfavourable volume variance may be interpreted as indicating overhead facilities which have been provided for the production being underutilized.

The *overhead efficiency variance* is the difference between the standard overhead in terms of cost/time units for the actual production achieved and the standard cost of the actual time units taken. Therefore,

Overhead efficiency variance = standard cost of standard units
— standard cost of actual units

This variance may be interpreted in a similar way to that of the *labour efficiency variance*.

The *overhead cost variance* is the difference between the budgeted cost for the time units actually worked and the actual overhead incurred. Therefore,

Overhead cost variance = budgeted overhead for actual units worked
— actual overhead incurred

Example 9.5

At a construction site the overhead budget for the project contains £120 000 for fixed costs and £48 000 for variable costs. The contract has been based on a duration of 280 working days. In reviewing the work at the outset it was assessed that one self-contained section of the project should be allowed 170 working days for completion. In fact it took 181 days and an overhead of £115 000 was incurred.

Calculate the overhead variances for the section of the project completed.

Section budget for fixed costs		
= (120 000)170/280	=	72 857
Section budget for variable costs		
= (48 000)170/280	=	29 143
Total overhead budget cost for section		£102 000
Standard days allowed for section	=	170
Standard fixed overhead/day	=	428.57
Standard variable overhead/day	=	171.43
Standard total overhead/day		£600.00

Actual duration taken $\qquad = $ 181 working days

Actual overhead incurred $\qquad = £115\,000$

Flexible overhead for actual duration taken $\quad = \quad 72\,857 + (181 \times 171.43)$
$$= £103\,886$$

Standard overhead for section $= 170 \times 600 \quad = £102\,000$

Overhead cost variance $= 103\,886 - 115\,000 \quad = \underline{£(11\,114)}$

Overhead volume variance
$$= (181 \times 600) - 103\,886 \qquad\qquad = \underline{£4714}$$

Overhead efficiency variance
$$= (170 - 181)600 \qquad\qquad = \underline{£(6600)}$$

Overhead variance $= 102\,000 - 115\,000$
$$= £(13\,000) \text{ or } (11\,114) + 4714 + (6600) = \underline{£(13\,000)}$$

The *overhead variance* is the overall variance covering the general situation and it is unfavourable as it shows a result of £(13 000), indicating that £13 000 more than was budgeted was spent on overheads for this section. The *cost variance* was unfavourable and resulted from the cost incurred being greater than it should have been for the extent of work completed. The *volume variance* is favourable since the flexible budget for the actual duration of the work is less than the standard cost for the actual duration. This variance identifies that part of the total *overhead variance* which is due to actual production being different from budgeted production. The variance arises out of the fixed cost element of the standard cost being for only 170 days whereas the variable element is for 181 days. The calculation of the overhead variances is shown diagrammatically in Fig. 9.13.

It should be noted that there is no commonly agreed method of classifying overhead variances and there is no terminology which is standardized in describing such variances. A great many variations in overhead variances, both in their descriptions and in the methods of their calculation, appear in the literature and in practice.

9.16 Random influences on cost variance

In construction projects, the budget of standards which is set as a level of achievement is usually determined by the estimated cost of the work established beforehand. It is sometimes questionable as to the degree of detail into which such an estimate can be broken, and at the same time provide a realistic standard against which performance can be measured and control can be exercised. Estimates of cost are produced in the best of faith, relying

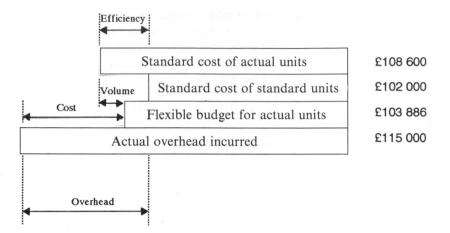

Fig. 9.13 Overhead variances for Example 9.5.

on historic data and previous experience, though frequently they are produced in the hope and expectation that the 'swings' will take care of the 'roundabouts'. There are two different attitudes from which cost estimating is carried out, depending on the position of the estimator. In the case of an owner who will be employing a contractor or some other outside agency to carry out the work, the estimate of project cost is rarely based on a detailed knowledge of the co-ordination and control of the actual resources that will be used in the construction process. More often it takes the form of a distillation of historic prices for completed elements of work of a similar nature, factored by an index to take account of current conditions. Frequently, the object of such an estimate is to allow a check to be made on the economic viability of the project by facilitating the calculation of a rate of return which should result from the capital investment. Approval to spend the money is often sought on the basis of such an estimate and, depending on the sensitivity of the rate of return to changes in the variables concerned, some leeway will probably be given to the limit within which expenditure can be made.

On the other hand, when a contractor prepares an estimate of cost for carrying out work he does so knowing that if he is successful in getting the work from the owner, his profit on that project depends, to a major degree, on how successfully he is able to predict his performance beforehand. His estimate of cost, therefore, needs to be based on considerably more detailed knowledge of the management of the resources involved.

In the former case, the owner retains control of the project cost by reviewing each order for work before it is placed and comparing its cost with the budgeted standard. Because there are many alternative ways of carrying out most work of this nature, the owner has the facility of delaying, withdrawing or changing the work at many stages throughout the programme, depending on the contractual process by which he has elected to carry out the work. In such cases, the contractor, once he is awarded the contract, usually has little option but to produce the required facility to the specification and in accordance with the required programme.

This consideration of different points of view about project control leads to the consideration of the level of achievement that is required of a controller who is managing the work. Much discussion can arise out of the setting of such standards. Should, for example, a level of achievement known to be incorrect or difficult, perhaps impossible, to achieve be retained as a standard in a budget? As has already been seen, the extent of the variance which results from activity will depend on the standard set. Often it is difficult to instil the necessary motivation in managers who are working to a budget known to be generous for the extent of the work involved; often it is even more difficult to motivate managers who know their budget to be so low that it is impossible to achieve. If the variances in a budget turn out to be zero, what does this in fact mean? It may be coincidence; it may mean that an estimator can accurately predict the output of every machine and man used in the work. It is unlikely to be the latter, if for no other reason than that human beings are involved, though clearly an estimator with experience of the work for which he is estimating will be more likely to be right than someone with little or no experience of this type of process.

One of the ways in which an attempt can be made to overcome the setting of an arbitary budget standard is by using the concept of expected value which was described in Chapter 5. An expected value is the weighted average of a range of values that might be provided by a number of those skilled estimators associated with, and having experience of, the work. Therefore, if a budget standard for carrying out an element of work is estimated to have a profitability of 0.10 that it will be £1000, 0.25 that it will be £1200, 0.30, £1300, 0.25, £1400 and 0.10 that it will be £1500, the expected value of the standard will be equal to £100 + 300 + 390 + 350 + 150 = £1290. In calculating a budget standard in such a fashion, the implication is that this expected value is a central tendency, or mean, of a distribution of values and that random variances arising out of the work being carried out will tend to fall on either one or other side of the mean with equal probability. In other words, the distribution of budgeted cost is assumed to be normal about the mean of the expected cost. In such a case there is an equal probability that the variances (in the budget sense) will either be favourable or unfavourable.

The expected or mean value of the distribution alone is not enough to define its characteristics. In addition, the value of the standard deviation as a measure of the range of values involved must be known. This can be found in either of two ways. Firstly, historic data concerning variances which have arisen from non-controllable random causes can be plotted and analysed by regression. Alternatively, the standard deviation can be calculated by obtaining an estimate of the range of values within which the actual value of the budgeted cost has an even chance of falling, due again to random non-controllable influences. For simplicity, the assumption needs to be made that the distribution is either normal or near-normal. If the expected value for a budget standard calculated above is referred to, that is £1290, it may be decided that there is an even chance (a 50/50 chance or a 0.50 probability) that the actual costs incurred will fall within the range £1190 to £1390. In other words, 50 per cent of the area under the normal curve of the probability that one or other actual value in the complete range will be achieved must lie between these two values. This is the area cross-hatched in Fig. 9.14. By reference to statistical tables, for a normal probability distribution factor, 50 per cent or the area of such a curve will be contained between ±0.675 standard deviations from the mean. 0.675 of a standard deviation has a value of £100, in this case (1390 − 1290), and a standard deviation equals 100/0.675 = £148.

If the actual cost of this element of work is now recorded as being, say, £1500, the probability that the adverse variance arose from random non-controllable causes can be examined. For £1500 there is a variation of £1500 − 1290 = £210 from the mean. In Fig. 9.14 it will be seen that this is 210/148 = 1.42 standard deviations to the right of the mean. The probability of this value arising from non-controllable or random factors is 0.08 (from statistical tables of the normal probability distribution function). Since this is very low, management's attention needs to be drawn to the fact that it is unlikely to have arisen out of random considerations.

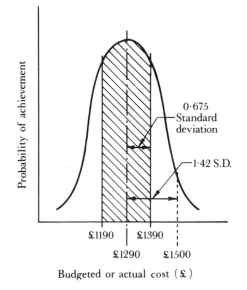

Fig. 9.14   Variance probabilities.

In calculating these probabilities, therefore, an assessment in numerical terms is being made of the degree to which variances will arise out of non-controllable considerations, and different elements of work can then be investigated in different degrees to suit each situation. In conjunction with the probability of randomness affecting the outcome, the size of the variation in absolute terms must be taken into account, since clearly a variation of £10 000 is of greater significance than one of £5.

Such a method of analysis can be complex and can involve considerable expense in both the analysis and collection of data. Before embarking upon the use of such methods, management needs to be reasonably certain that value will be obtained for the effort and cost involved and that the return can be measured in terms of not only money, but also the accuracy of the information obtained. Such methods frequently lend themselves to production or process operations which are repetitive or continuous rather than to the activities which form a significant part of a one-off project.

Example 9.6

A project manager divides a small section of this project into three parts for the purpose of cost control. The first part is estimated to have a total cost of £14 000, the second £5000 and the third £10 000. From his previous experience the project manager gauges that he has a 50/50 chance of achieving a cost of between £13 000 and £15 000 for the first section, between £4000 and £6000 for the second and between £9500 and £10 500 for the third. Actual costs for each part are established as £13 500, £5300 and £10 400. Which part is unlikely to have varied in cost out of random considerations and therefore warrants further consideration before using it as feedback?

First part:

$$\text{Standard deviation} = \frac{(15\,000 - 14\,000)}{0.675} = £1481$$

$$\text{Variation of actual from mean} = (14\,000 - 13\,500)$$
$$= £500$$

$$£500 = \frac{500}{1481} = 0.338 \text{ standard deviation from mean.}$$

Probability of this value arising from non-controllable factors = 0.27

Second part:

$$\text{Standard deviation} = \frac{(5000 - 4000)}{0.675} = £1481$$

$$\text{Variation of actual from mean} = (5300 - 5000)$$
$$= £300$$

$$£300 = \frac{300}{1481} = 0.203 \text{ standard deviation from mean.}$$

Probability of this value arising from non-controllable factors = 0.42

Third part:

$$\text{Standard deviation} = \frac{(10\,500 - 10\,000)}{0.675} = £741$$

$$\text{Variation of actual from mean} = (10\,400 - 10\,000)$$
$$= £400$$

$$£400 = \frac{400}{741} = 0.541 \text{ standard deviation from mean.}$$

Probability of this value arising from non-controllable or random factors = 0.295

Therefore, in ascending order of attention required by management, the three parts will be 1, 3 and 2. This is because there is apparently less chance of the variance in part 1 being due to a random factor than for either part 3 or part 2.

Problems

9.1 A contracting firm is invited to tender for a small hydroelectric scheme in a developing country overseas. The firm calculates its actual costs to be £5 000 000 and this figure is accepted by the client subject to a condition regarding payment of the money for completed work. The condition is that £1 500 000 will be paid at the end of each of the third, fourth, fifth and sixth years after construction is commenced, to cover the actual costs of the contractor plus his requirement for profit.

The contractor programmes the work to cover a three-year period and estimates that the costs he will incur will amount to £1 000 000 by the end of the first year, £3 000 000 in total by the end of the second year and £5 000 000 in total by completion of the work at the end of the third year. The contractor estimates that he will have to find all the money to finance the work from sources outside the firm and this will have to be borrowed at an interest rate of 10 per cent per year. Produce a budget statement for the contractor, year by year, assuming that all payments and receipts are on a year-end basis. If inflation is assumed to be at the rate of 3 per cent per year establish the true profit/loss for the contract reckoned in terms of the value of money at the commencement of the contract.

9.2 During the preparation of an estimate of cost for a small length of sea wall, a contractor needs to assess the cost of financing the work.

Table 9.7

End of month no.	Cumulative value of work completed (£)
2	20 000
4	58 000
6	120 000
8	240 000
10	260 000
12	275 000

His total tender price, linked to a duration of 12 months, is £275 000. Table 9.7 lists the cumulative value of the work the contractor anticipates that he will have completed at the end of each two-month period after commencement of the work.

The first valuation of work completed for payment is to be submitted at the end of two months and further valuations at intervals of two months thereafter. Payments are made by the client up to the limit of 90 per cent of each of the first five cumulative valuations. Ninety-five per cent of the sixth cumulative valuation will then be paid and, if there are no claims outstanding against the contractor, the balance of 5 per cent will be paid six months later.

The contractor assumes that he will not receive the client's cheque in payment for work carried out until one month after the date of each valuation. He estimates that capital costs him 8 per cent per year. What interest charges will he have to pay on his working capital if his cumulative expenditure is estimated to be as in Table 9.8? If the contractor can invest his surplus income over expenditure for this contract at 5 per cent interest per year, can he balance his outgoings against receipts as far as interest charges are concerned? What is the difference between the two amounts?

Table 9.8

End of month no.	Cumulative expenditure (£)
2	5 000
4	12 000
6	85 000
8	130 000
10	200 000
12	220 000
14	245 000
16	247 000

9.3 Having prepared a tender for six reinforced concrete bunkers, there only remains the calculation of the cost of providing the necessary working capital. It is estimated that the total value (except the cost of finance) of the work will amount to £300 000 and that it will be completed during a contract duration of 16 months. A budget is drawn up based upon the anticipated programme of work and this is set down in Table 9.9.

Table 9.9

End of month no.	Cumulative value of work completed (£)
2	30 000
4	60 000
6	85 000
8	140 000
10	200 000
12	260 000
14	280 000
16	300 000

The contract documents provide for the contractor to be paid at intervals of two months from the commencement of the work. Payment will be made on the basis of 90 per cent of the cumulative value of the work completed by the end of each payment period. At the end of the contract, half of the 10 per cent retention will be paid and the remaining 5 per cent will be paid promptly six months after the end of the contract.

If the contractor estimates that his actual costs will be 90 per cent of the work completed and that he will receive his money for work completed $1\frac{1}{2}$ months after it is due to him, i.e. $1\frac{1}{2}$ months after the end of each period of two months, what will his working capital cost him over the duration of the contract only at an interest rate of 10 per cent?

9.4 In the example used in Section 9.1, Table 9.1, examine the effect on working capital requirements for the three turnovers given, that is £100 000, £140 000 and £200 000, if the credit period allowed to trade debtors is reduced to six weeks. Assume that 75 per cent of them will pay within this period and that the remainder will still require eight weeks to pay.

Examine also the effect of giving $2\frac{1}{2}$ per cent discount for payment within 28 days on the basis of 25 per cent, 50 per cent, and then 75 per cent of the debtors taking advantage of this offer, the remainder requiring eight weeks as before.

9.5 The standard costs for the materials involved in mixing a compound are £0.60 per kg of Material X, £0.75 per kg of Material Y and £0.53 per kg of Material Z. The prices paid for each material in practice are £0.65 per kg, £0.73 per kg and £0.55 per kg, respectively. The specification of the compound calls for a standard mix of 100 kg of X, 270 kg of Y and 135 kg of Z. A check on the material stocks shows that 105 kg, 267 kg and 140 kg respectively are being used. Calculate the material variances for the actual operation.

9.6 The standard cost details for a precast concrete component are as follows:

Standards for one
   100 kg cement at £0.007 per kg
   200 kg sand at £0.002 per kg
   350 kg ballast at £0.0025 per kg
   30 kg reinforcement at £0.08 per kg.

The actual data concerning materials collected at the time of making 200 components are as follows:

Price paid for cement £0.0075 per kg; 20 100 kg used

Price paid for sand £0.0019 per kg; 39 000 kg used

Price paid for ballast £0.0026 per kg; 71 000 kg used

Price paid for reinforcement £0.085 per kg; 600 kg used.

Calculate the material cost, price and usage variances.

9.7 The following data relate to the overhead costs concerned with the production of some standard formwork for a retaining wall in a joiners' shop:
   Calculate the following overhead variances:

(a) Budget   (c) Volume
(b) Efficiency   (d) Calendar

	Budget	Actual
Overhead cost	£1000	£950
Production	300 m^2	280 m^2
Number of days	15	16
Number of hours	750	800

9.8 The following data are provided for a small building modification:

	Budget	Actual
Material A, use	2 000 kg	2 200 kg
Material B, use	20 000 kg	19 800 kg
Material C, use	9 000 kg	9 100 kg
Material A, price	£0·06/kg	£0·065/kg
Material B, price	£0·13/kg	£0·125/kg
Material C, price	£0·09/kg	£0·092/kg
Labour, rate	£0·39/hour	£0·40/hour
Labour, output	600 m³	620 m³
Labour, hours	800	860
Overhead	£250	£240
Number of days	10	12

Produce a statement of actual cost against budgeted cost and a full variance analysis for the data collected.

9.9 In examining the cost variances for a series of overhead charges the following statement is revealed:

	Expected budget	Likely range	Actual cost
Salaries	£12 000	£10 000 — 14 000	£13 000
Administration	10 000	9 000 — 11 000	10 700
Computer	£15 000	12 500 — 17 500	13 750
Accounts	£12 000	11 500 — 12 500	12 300

Complete the item which is most likely and that which is least likely to have been influenced by random factors.

Selected Further Reading

General Management and Control

Anthony, R.N. (1965) *Planning and Control Systems – a Framework for Analysis*. Harvard University Press.

Barrie, D.S. & Paulson, B.C. (1984) *Professional Construction Management*, 2nd edn. McGraw-Hill, New York.

Cleland, D.I. & King, W.R. (1982) *Systems Analysis and Project Management*, 2nd edn. McGraw-Hill, New York.

Cyert, R.M. & March, J.G. (1963) *A Behavioral Theory of the Firm*. Prentice-Hall, Englewood Cliffs, NJ.

Davies, D. & McCarthy, C. (1967) *Introduction to Technological Economics*. Wiley, London.

Hendrickson, C. & Au, T. (1989) *Project Management for Construction*. Prentice-Hall, Englewood Cliffs, NJ.

Mintzberg, H. (1973) *The Nature of Managerial Work*. Harper and Row, New York.

Oglesby, C.H., Parker, H.W. & Howell, G.A. (1989) *Productivity Improvement in Construction*. McGraw-Hill, New York.

Pilcher, R. (1993) *Principles of Construction Management*, 3rd edn. McGraw-Hill, London.

Taylor, F.W. (1967) *The Principles of Scientific Management*. Norton, New York.

Financial Control

Allen, M.W. & Myddelton, D.R. (1987) *Essential Management Accounting*. Prentice-Hall, London.

Barnes, M. (ed.) (1990) *Financial Control*. Thomas Telford, London.

Companies Act 1985, HMSO, London.

Foster, G. (1986) *Financial Statement Analysis*, 2nd edn. Prentice-Hall, London.

Horngren, C.T. & Sundem, G.L. (1987) *Introduction to Financial Accounting*, 3rd edn. Prentice-Hall, London.

Institute of Chartered Accountants in England and Wales, *Statements of Standard Accounting Practice* (SSAPs) and *Statements of Recommended Practice*.

Millichamp, A.H. (1992) *Finance for Non-financial Managers*, DP Publications, London.

Padet, P. (1991) *Accounting in the Construction Industry*. CIMA.

Reid, W. & Myddelton, D.R. (1982) *The Meaning of Company Accounts*, 3rd edn. Gower, London.

Rockley, L.E. (1984) *Finance for the Non-accountant*, 4th edn. London.

Thomas, C. (1990) *Company Law*, 3rd edn. Teach Yourself Books, Hodder and Stoughton, Sevenoaks.

Engineering Economic Analysis

Couper, J.R. & Radar, W.H. (1986) *Applied Finance and Economic Analysis for Scientists and Engineers*. Van Nostrand Reinhold, New York.

Grant, E.L., Ireson, W.G. & Leavenworth, R.S. (1990) *Principles of Engineering Economy*, 8th edn. Wiley, New York.

Institution of Civil Engineers (1969) *An Introduction to Engineering Economics*. ICE, London.

Lindley, D.V. (1971) *Making Decisions*. Wiley, London.

Merrett, A.J. & Sykes, A. (1973) *Capital Budgeting and Company Finance*, 2nd edn. Longmans, London.

Merrett, A.J. & Sykes, A. (1973) *The Finance and Analysis of Capital Projects*, 2nd edn. Longmans, London.

Pilcher, R. (1993) *Principles of Construction Management*, 3rd edn. McGraw-Hill, London.

Riggs, J.L. & West, T.M. (1986) *Engineering Economics*, 3rd edn. McGraw-Hill, New York.

White, J.A., Agee, M.H. & Case, K.E. (1989) *Principles of Engineering Economic Analysis*, 3rd edn. Wiley, New York.

Risk and Uncertainty

Ansell, J. & Wharton, F. (1992) *Risk Analysis Assessment and Management*. Wiley, Chichester.

Benjamin, I.R. & Cornell, C.A. (1970) *Probability, Statistics and Decision for Civil Engineers*. McGraw-Hill, New York.

Bunn, D. (1982) *Analysis for Optimal Decisions*. Wiley, Chichester.

Cooper, D.F. & Chapman, C.B. (1987) *Risk Analysis for Large Projects, Model Methods and Cases*. Wiley, Chichester.

Hertz, D.B. & Thomas, H. (1983) *Risk Analysis and its Application*. Wiley.

Lindley, D.V. (1971) *Making Decisions*. Wiley-Interscience, London.

Moore, P.G. (1983) *The Business of Risk*. Cambridge University Press, Cambridge.

Pilcher, R. (1993) *Principles of Construction Management*, 3rd edn. McGraw-Hill, London.

Reichmann, W.J. (1970) *Use and Abuse of Statistics*. Methuen, London.

Savage, L.J. (1972) *The Foundations of Statistics*. Wiley, New York.

Schlaiffer, R. (1969) *Analysis of Decisions Under Uncertainty*. McGraw-Hill, New York.

Ward, S.C. & Chapman, C.B. (1991) *Extending the Use of Risk Analysis in Project Management*, International Journal of Project Management, Vol. 9, 1991, pp. 117–123.

Cost and Budgetary Control

American Society of Civil Engineers (1985) *Construction Cost Control*. The Society, New York.

Antill, J.M. & Woodhead, R.W. (1982) *Critical Path Methods in Construction Practice*, 3rd edn. Wiley, New York.

Ballard, E.H. (1972) *The Control of Resources Required for the Construction of a Civil Engineering Project*, Proceedings Institution of Civil Engineers, Part 1, 52, pp. 291–304.

Batty, J. (1970) *Standard Costing*, 3rd edn. McDonald and Evans, London.

Beeston, D.T. (1983) *Statistical Methods for Building Price Data*. Spon, London.

Bierman, H. & Smidt, S. (1975) *The Capital Budgeting Decision*. Macmillan, New York.

Carr, R.I. (1977) *Paying the Price for Construction Risk*, Journal of the Construction Division, ASCE, 103, no. C01, pp. 152–161.

Carsberg, B. (1975) *Economics of Business Decisions*, Penguin Books, Harmondsworth.

Department of Energy (1976) *North Sea Costs Escalation Study*. HMSO, London.

Douglas, J. (1975) *Construction Equipment Policy*. McGraw-Hill, New York.

Drake, B.E. (1978) *A Mathematical Model for Expenditure Forecasting Post Contract*. Procs. CIB W–65 Second Sumposium on Organisation and Management of Construction, Haifa, 31 Oct. – 2 Nov, pp. II–163 to II–183.

EDC for Building (1969) *Formulae Methods of Price Adjustment in Building Contracts*. NEDC, London.

Erickson, C.A., Boyer, L.T. & O'Connor, M.J. (1978) *Risk Assignment in Construction Contracts*. Proceedings CIB W–65 Second Symposium on Organisation and Management of Construction, Vol. 3, Haifa, pp. 137–155.

Gobourne, J. (1982) *Site Costing in the Construction Industry*. Butterworth, London.

Hague, D.C. (1971) *Managerial Economics. Analysis for Business Decisions*. Longmans, London.

Kharbanda, O.P., Stallworthy, E.A. & Williams, L.F. (1980) *Project Cost Control in Action*. Gower, Farnborough.

Knight, W.D. & Weinwurm, E.H. (1964) *Managerial Budgeting*. Macmillan, New York.

Likierman, J.A. (1977–8) *Analysing Project Cost Escalation: the Case Study of North Sea Oil*. Accounting and Business Research, Vol. 8, pp. 51– 57.

McDowell, I. (1960) The economical planning period for engineering works. *Operational Research*, July–August.

Middleton, K.A. (1968) Standard costing overhead variances, *Management Accounting*.

Miller, R.W. (1963) *Schedule, Cost and Profit Control with PERT*. McGraw-Hill, New York.

Ministry of Public Building and Works (1967) *Network Analysis in Construction Design*. HMSO, London.

Moder, J.J., Philips, C.R. & Davis, E.W. (1983) *Project Management with CPM, PERT and Precedence Diagramming*, 3rd edn. Van Nostrand Reinhold, New York.

O'Brien, J.J. (1993) *CPM in Construction Management*, 4th edn. McGraw-Hill, New York.

Park, W.R. (1966) *The Strategy for Contracting for Profit*. Prentice-Hall, Englewood Cliffs, NJ.

Parker, E.J. (1969) The planning of project finance, *Proceedings of the Institution of Civil Engineers, London*, **43**, June, pp. 261–271, and discussion in **44**, December, pp. 360–362.

Peer, S. (1982) *Application of Cost-flow Forecasting Models*, Journal of the Construction Division, ASCE, Vol. 108, No. C02, pp. 226–232.

Perry, W.W. (1970) *Automation in Estimation Contractor Earnings*, The Military Engineer, No. 410, Nov.–Dec., pp. 393–395.

Pilcher, R. (1993) *Principles of Construction Management*, 3rd edn. McGraw-Hill, London.

Reynold, A.J. (1992) *The Finances of Engineering Companies. An Introduction for Students and Practising Engineers*. Arnold, London.

Risk, J.M. (1956) *The Classification and Coding of Accounts*. The Institute of Cost and Works Accountants, London.

Savage, C.I. & Small, J.R. (1975) *Introduction to Managerial Economics*. Hutchinson, London.

Schrader, C.R. (1971) Optimum replacement life of large scrapers, *Journal of the Construction Division, Proceedings of the American Society of Civil Engineers*, Vol. 97, No. C01, March, pp. 37–52.

Shaw, L.W. (1967) *Management Information and Statistical Method*. The

General Educational Trust of the Institute of Chartered Accountants in England and Wales, London.

Sizer, J. (1969) *An Insight into Management Accounting*. Penguin Books, London.

Swaim, R.O. (1966) *Utility Theory – Insights into Risk Taking*, Harvard Business Review, **44**, No. 6, Nov. – Dec., pp. 123–136.

Tysoe, B.A. (1981) *Construction Cost and Price Indices: Description and Use*. Spon, London.

Walker, T.M. (1980) *Understanding Standard Costing*. Gee, London.

Warszawski, A. (1982) *Financial Analysis under Inflation in Construction*, Journal of the Construction Division, ASCE 108, No. C02, pp. 341–354.

Warszawski, A. (1982) *Managerial Decisions under Inflation*, Journal of the Construction Division, ASCE 108, No. C01, March, pp. 147–157.

Welsch, H. (1971) *Budgeting: Profit Planning and Control*, 3rd edn. Prentice-Hall, Englewood Cliffs.

Willenbrock, J.M. (1973) *Utility Function Determination for Bidding Models*, Journal of the Construction Division, ASCE 99, C01, July, pp. 133–153.

Management and Information Control Systems

Anthony, R.N. (1965) *Planning and Control Systems – a Framework for Analysis*. Harvard University Press.

Cleland, D.I. & King, W.R. (1982) *Systems Analysis and Project Management*, 2nd edn. McGraw-Hill, New York.

Edwards, C. (1980) *Identifying Determinants of Successful Information Systems Maintenance*, unpublished PhD thesis, University of Strathclyde.

Fisher, N. & Yen, S.L. (1992) *Information Management within a Contractor*. Thomas Telford, London.

Galbraith, J. (1980) *Design of Complex Organizations*. Addison Wesley, Reading, Mass.

Guevara, J.M. & Boyer, L.T. (1981) Communication problems within construction, *Journal of the Construction Division*, ASCE, Vol. 107, No. C04, December, pp. 551–557.

Lucas, H.C. (1986) *Information Systems Concepts for Management*, 3rd edn. McGraw-Hill, New York.

O'Brien, J.J. (1970) *Management Information Systems*. Van Nostrand Reinhold, New York.

Rasdorf, W.J. & Herbert, M.J. (1990) *Automated identification systems – focus on bar coding*, Journal of Computing in Civil Engineering, ASCE, Vol. 4, No. 3, July, pp. 279–296.

Russell, A.D. & Triassi, E. (1982) *General contractor project control practices and MIS*, Journal of the Construction Division, ASCE, Vol. 108, No. C03, September, pp. 419–437.

Scott, G.M. (1986) *Principles of Management Information Systems.* McGraw-Hill, New York.

Songer, A.D., Considine, C. & Ibbs, C.W. (1989) *Integrating voice recognition systems with the collection of project control data*, Proceedings of 1st Construction Congress, ASCE, pp. 409–415.

Tricker, R.I. & Boland, R.J. (1982) *Management Information and Control Systems*, 2nd edn. Wiley, Chichester.

Solutions to Problems Given in Chapter 2

2.1

Authorized	1 000 000 shares of £1 each	£1 000 000
Issued	870 000 shares of £1 each	870 000
Called-up	870 000 shares at 75p each	652 500
Calls in arrears	70 000 shares at 75p each	52 500
Paid up	£625 500 less 52 500	600 000

2.2

	£	£	£
Net profit			100 000
Net profit brought forward			17 500
Total available for appropriation			117 500
Corporation tax			33 000
			84 500
Appropriations:			
To general reserve		14 500	
Ordinary dividend, interim	22 500		
final	45 000	67 500	82 000
Balance carried forward			2 500

2.4

PARTNERS CONSTRUCTION

Trading and profit and loss account

	£	£
Sales		345 000
Cost of sales:		
Labour	83 400	
Materials	163 000	
Sub-contracts	47 000	293 400

	£	£
Gross profit		51 600
Operating expenses:		
Office expenses	22 000	
Insurance	5 500	
Legal, accountancy and audit fees	11 760	
Depreciation	2 600	
Administrative expenses	22 690	
Taxation	7 000	
Dividends	2 500	
Repairs and maintenance	7 900	81 950
Operating profit/(loss)		(30 350)

2.5

JASMIN CONSTRUCTION PLC

Trading and profit and loss account for year ending 31 December 19x5

	£000	£000
Sales		1 500
Cost of sales:		
Labour	250	
Materials	320	
Sub-contracts	210	780
Gross profit		720
Operating expenses:		
Office expenses	82	
Depreciation	72	
Administrative expenses	250	404
Operating profit/(loss)		316

	£000	£000
Net profit		316
Net profit brought forward from last year		4
Total available for appropriation		320
Corporation tax		34
		286
Appropriations:		
To general reserve	100	
Ordinary dividend	165	265
Balance carried forward		21

JASMIN CONSTRUCTION PLC

Balance sheet for year ending 31 December 19x5

	£000	£000	£000
Assets employed			
Fixed assets			
Buildings		300	
less depreciation		123	177
Plant and equipment		175	
less depreciation		101	74
Motor vehicles		110	
less depreciation		54	56
Total fixed assets			307
Current assets			
Stock	134		
Debtors	320		
Cash at bank and in-hand	125		
Total current assets		579	
Less current liabilities			
Creditors, amounts, falling due within one year:			
Trade creditors	224		
Taxation on profits	34		
Proposed dividends	165		
Total current liabilities		423	
Net current assets (working capital)			156
Total assets less total liabilities (capital employed)			463
Less **creditors, amounts falling due after more than one year:**		nil	
Net worth of company			463
The above net worth of the company was financed by:			
Called-up share capital			
330 000 ordinary shares at £1 each		330	330
Reserves			
General reserve		112	
Profit and loss account transferred this year		21	133
			463

2.6 The main features between the two years, before ratio analysis, are:

(a) turnover has increased by 7.2%;
(b) office expenses, wages and salaries, administrative cost and directors' emoluments have all increased;
(c) profit retained reduced from £41 000 to £4000;
(d) fixed asset investment static;
(e) creditors increased by 75% over the year 19x6;
(f) short-term bank loans up to 80%;
(g) dividends cut during the year;
(h) 12% long-term bank loan not being repaid.

Ratio analysis as follows:

Gross profit ratio	429/1250 = 34.3%	337/1340 = 28.1%
Return on capital employed	209/474 = 44.1%	102/487 = 20.9%
Current ratio	232/228 = 1.02	236/217 = 1.09
Acid test	(232–121)/228 = 0.49	(236–117)/217 = 1.09
Stock turnover	(121/821)365 = 32.4	(117/963)365 = 44.3
Debt turnover	(111/1250)365 = 32.4	(119/1340)365 = 32.4
Asset utilization	1250/474 = 2.64	1340/487 = 2.75
Gearing ratio	220/150 = 1.47	229/150 = 1.52

All the key ratios appear to have deteriorated except debt ratio which has remained the same.

The key ratio of gross profit has fallen which may indicate that increased costs may not be truly reflected in the prices asked of customers or that possibly lower margin products are being sold at the expense of higher margin ones. The increased sales volume may be at the expense of these margins and has been enabled by lower prices than those of the competition.

Management appears not to be making too good a job of working the assets effectively – the ratio has fallen badly over the two years.

The current ratio appears to be satisfactory; the acid test shows a shortage of working capital. These figures may or may not be significant and it is difficult to interpret them in the environment in which they have been given.

The gearing ratio is important in the context of the question. The company has a high gearing ratio and may need to decrease the long-term debt soon in the context of a falling off of profitability, etc. otherwise it may become difficult to service. Certainly it would be unwise to increase the long-term loans. Introducing more equity may be the answer though there are other aspects, such as the general overheads, the creditors and bank loans, which also need immediate attention.

It would help to have the results of an interfirm comparison against which the company's ratios could be compared on an industry-wide basis.

TACKS CONSTRUCTION PLC

Balance Sheet as at 31 December 1993

	£000	£000	£000
Assets employed			
Fixed assets			
Land and buildings		740.0	
less depreciation		120.0	620.0
Vehicles, plant and machinery		425.0	
less depreciation		114.5	310.5
Fixtures and fittings		205.0	
less depreciation		52.9	152.1
Investments			26.4
Total fixed assets			1109.0
Current assets			
Land and developments	1207.4		
Raw materials and consumables	22.4		
Work in progress	332.3		
Debtors	942.1		
Short-term investments	20.0		
Cash at bank and in-hand	521.7		
Total current assets		3045.9	
Less current liabilities			
Creditors, amounts falling due within one year:			
Bank loans and overdrafts	423.1		
Obligations under financial leases	54.3		
Payments received on account	120.3		
Trade creditors	750.2		
Taxation on profits	94.0		
Proposed dividends	100.0		
Other creditors	52.6		
Total current liabilities		1594.5	
Net current assets (working capital)			1451.4
Total assets less total liabilities (capital employed)			2560.4
Less **creditors, amounts falling due after more than one year:**			
Bank loan repayable June 1994		75.0	
10% unsecured loan stock 1992–97		100.0	
Obligations under financial leases		37.2	212.2
Net worth of company			2348.2

The above net worth of the company was financed by:

Called-up share capital

7% cumulative preference shares of £1 each	200.0	
1 750 000 ordinary shares at £1 each	1750.0	1950.0

Reserves

Share premium account	273.0	
Revaluation reserve	(70.0)	
Other reserves	27.0	
Profit and loss account at 1 January 1993	23.7	
Profit and loss account transferred this year	144.5	398.2
		2348.2

Interest Tables

1% DISCRETE COMPOUND INTEREST FACTORS

	Single payment		Uniform series			
n	Compound amount factor	Present worth factor	Sinking fund factor	Compound amount factor	Capital recovery factor	Present worth factor
	F/P	*P/F*	*A/F*	*F/A*	*A/P*	*P/A*
1	1.0100	0.9901	1.0000	1.0000	1.0100	0.9901
2	1.0201	0.9803	0.4975	2.0100	0.5075	1.9704
3	1.0303	0.9706	0.3300	3.0301	0.3400	2.9410
4	1.0406	0.9610	0.2463	4.0604	0.2563	3.9020
5	1.0510	0.9515	0.1960	5.1010	0.2060	4.8534
6	1.0615	0.9420	0.1625	6.1520	0.1725	5.7955
7	1.0721	0.9327	0.1386	7.2135	0.1486	6.7282
8	1.0829	0.9235	0.1207	8.2857	0.1307	7.6517
9	1.0937	0.9143	0.1067	9.3685	0.1167	8.5660
10	1.1046	0.9053	0.0956	10.4622	0.1056	9.4713
11	1.1157	0.8963	0.0865	11.5668	0.0965	10.3676
12	1.1268	0.8874	0.0788	12.6825	0.0888	11.2551
13	1.1381	0.8787	0.0724	13.8093	0.0824	12.1337
14	1.1495	0.8700	0.0669	14.9474	0.0769	13.0037
15	1.1610	0.8613	0.0621	16.0969	0.0721	13.8651
16	1.1726	0.8528	0.0579	17.2579	0.0679	14.7179
17	1.1843	0.8444	0.0543	18.4304	0.0643	15.5623
18	1.1961	0.8360	0.0510	19.6147	0.0610	16.3983
19	1.2081	0.8277	0.0481	20.8109	0.0581	17.2260
20	1.2202	0.8195	0.0454	22.0190	0.0554	18.0456
21	1.2324	0.8114	0.0430	23.2392	0.0530	18.8570
22	1.2447	0.8034	0.0409	24.4716	0.0509	19.6604
23	1.2572	0.7954	0.0389	25.7163	0.0489	20.4558
24	1.2697	0.7876	0.0371	26.9735	0.0471	21.2434
25	1.2824	0.7798	0.0354	28.2432	0.0454	22.0232
26	1.2953	0.7720	0.0339	29.5256	0.0439	22.7952
27	1.3082	0.7644	0.0324	30.8209	0.0424	23.5596
28	1.3213	0.7568	0.0311	32.1291	0.0411	24.3164
29	1.3345	0.7493	0.0299	33.4504	0.0399	25.0658
30	1.3478	0.7419	0.0287	34.7849	0.0387	25.8077
31	1.3613	0.7346	0.0277	36.1327	0.0377	26.5423
32	1.3749	0.7273	0.0267	37.4941	0.0367	27.2696
33	1.3887	0.7201	0.0257	38.8690	0.0357	27.9897
34	1.4026	0.7130	0.0248	40.2577	0.0348	28.7027
35	1.4166	0.7059	0.0240	41.6603	0.0340	29.4086
40	1.4889	0.6717	0.0205	48.8864	0.0305	32.8347
45	1.5648	0.6391	0.0177	56.4811	0.0277	36.0945
50	1.6446	0.6080	0.0155	64.4632	0.0255	39.1961
55	1.7285	0.5785	0.0137	72.8525	0.0237	42.1472
60	1.8167	0.5504	0.0122	81.6697	0.0222	44.9550
70	2.0068	0.4983	0.0099	100.6763	0.0199	50.1685
80	2.2167	0.4511	0.0082	121.6715	0.0182	54.8882
90	2.4486	0.4084	0.0069	144.8633	0.0169	59.1609
100	2.7048	0.3697	0.0059	170.4814	0.0159	63.0289

Project Cost Control in Construction

5% DISCRETE COMPOUND INTEREST FACTORS

	Single payment		Uniform series			
n	Compound amount factor	Present worth factor	Sinking fund factor	Compound amount factor	Capital recovery factor	Present worth factor
	F/P	P/F	A/F	F/A	A/P	P/A
1	1.0500	0.9524	1.0000	1.0000	1.0500	0.9524
2	1.1025	0.9070	0.4878	2.0500	0.5378	1.8594
3	1.1576	0.8638	0.3172	3.1525	0.3672	2.7232
4	1.2155	0.8227	0.2320	4.3101	0.2820	3.5460
5	1.2763	0.7835	0.1810	5.5256	0.2310	4.3295
6	1.3401	0.7462	0.1470	6.8019	0.1970	5.0757
7	1.4071	0.7107	0.1228	8.1420	0.1728	5.7864
8	1.4775	0.6768	0.1047	9.5491	0.1547	6.4632
9	1.5513	0.6446	0.0907	11.0266	0.1407	7.1078
10	1.6289	0.6139	0.0795	12.5779	0.1295	7.7217
11	1.7103	0.5847	0.0704	14.2068	0.1204	8.3064
12	1.7959	0.5568	0.0628	15.9171	0.1128	8.8633
13	1.8856	0.5303	0.0565	17.7130	0.1065	9.3936
14	1.9799	0.5051	0.0510	19.5986	0.1010	9.8986
15	2.0789	0.4810	0.0463	21.5786	0.0963	10.3797
16	2.1829	0.4581	0.0423	23.6575	0.0923	10.8378
17	2.2920	0.4363	0.0387	25.8404	0.0887	11.2741
18	2.4066	0.4155	0.0355	28.1324	0.0855	11.6896
19	2.5270	0.3957	0.0327	30.5390	0.0827	12.0853
20	2.6533	0.3769	0.0302	33.0660	0.0802	12.4622
21	2.7860	0.3589	0.0280	35.7193	0.0780	12.8212
22	2.9253	0.3418	0.0260	38.5052	0.0760	13.1630
23	3.0715	0.3256	0.0241	41.4305	0.0741	13.4886
24	3.2251	0.3101	0.0225	44.5020	0.0725	13.7986
25	3.3864	0.2953	0.0210	47.7271	0.0710	14.0939
26	3.5557	0.2812	0.0196	51.1135	0.0696	14.3752
27	3.7335	0.2678	0.0183	54.6691	0.0683	14.6430
28	3.9201	0.2551	0.0171	58.4026	0.0671	14.8981
29	4.1161	0.2429	0.0160	62.3227	0.0660	15.1411
30	4.3219	0.2314	0.0151	66.4388	0.0651	15.3725
31	4.5380	0.2204	0.0141	70.7608	0.0641	15.5928
32	4.7649	0.2099	0.0133	75.2988	0.0633	15.8027
33	5.0032	0.1999	0.0125	80.0638	0.0625	16.0025
34	5.2533	0.1904	0.0118	85.0670	0.0618	16.1929
35	5.5160	0.1813	0.0111	90.3203	0.0611	16.3742
40	7.0400	0.1420	0.0083	120.7998	0.0583	17.1591
45	8.9850	0.1113	0.0063	159.7002	0.0563	17.7741
50	11.4674	0.0872	0.0048	209.3480	0.0548	18.2559
55	14.6356	0.0683	0.0037	272.7126	0.0537	18.6335
60	18.6792	0.0535	0.0028	353.5837	0.0528	18.9293
70	30.4264	0.0329	0.0017	588.5285	0.0517	19.3427
80	49.5614	0.0202	0.0010	971.2288	0.0510	19.5965
90	80.7304	0.0124	0.0006	1594.6073	0.0506	19.7523
100	131.5013	0.0076	0.0004	2610.0252	0.0504	19.8479

10% DISCRETE COMPOUND INTEREST FACTORS

	Single payment		Uniform series			
n	Compound amount factor	Present worth factor	Sinking fund factor	Compound amount factor	Capital recovery factor	Present worth factor
	F/P	*P/F*	*A/F*	*F/A*	*A/P*	*P/A*
1	1.1000	0.9091	1.0000	1.0000	1.1000	0.9091
2	1.2100	0.8264	0.4762	2.1000	0.5762	1.7355
3	1.3310	0.7513	0.3021	3.3100	0.4021	2.4869
4	1.4641	0.6830	0.2155	4.6410	0.3155	3.1699
5	1.6105	0.6209	0.1638	6.1051	0.2638	3.7908
6	1.7716	0.5645	0.1296	7.7156	0.2296	4.3553
7	1.9487	0.5132	0.1054	9.4872	0.2054	4.8684
8	2.1436	0.4665	0.0874	11.4359	0.1874	5.3349
9	2.3579	0.4241	0.0736	13.5795	0.1736	5.7590
10	2.5937	0.3855	0.0627	15.9374	0.1627	6.1446
11	2.8531	0.3505	0.0540	18.5312	0.1540	6.4951
12	3.1384	0.3186	0.0468	21.3843	0.1468	6.8137
13	3.4523	0.2897	0.0408	24.5227	0.1408	7.1034
14	3.7975	0.2633	0.0357	27.9750	0.1357	7.3667
15	4.1772	0.2394	0.0315	31.7725	0.1315	7.6061
16	4.5950	0.2176	0.0278	35.9497	0.1278	7.8237
17	5.0545	0.1978	0.0247	40.5447	0.1247	8.0216
18	5.5599	0.1799	0.0219	45.5992	0.1219	8.2014
19	6.1159	0.1635	0.0195	51.1591	0.1195	8.3649
20	6.7275	0.1486	0.0175	57.2750	0.1175	8.5136
21	7.4002	0.1351	0.0156	64.0025	0.1156	8.6487
22	8.1403	0.1228	0.0140	71.4027	0.1140	8.7715
23	8.9543	0.1117	0.0126	79.5430	0.1126	8.8832
24	9.8497	0.1015	0.0113	88.4973	0.1113	8.9847
25	10.8347	0.0923	0.0102	98.3471	0.1102	9.0770
26	11.9182	0.0839	0.0092	109.1818	0.1092	9.1609
27	13.1100	0.0763	0.0083	121.0999	0.1083	9.2372
28	14.4210	0.0693	0.0075	134.2099	0.1075	9.3066
29	15.8631	0.0630	0.0067	148.6309	0.1067	9.3696
30	17.4494	0.0573	0.0061	164.4940	0.1061	9.4269
31	19.1943	0.0521	0.0055	181.9434	0.1055	9.4790
32	21.1138	0.0474	0.0050	201.1378	0.1050	9.5264
33	23.2252	0.0431	0.0045	222.2515	0.1045	9.5694
34	25.5477	0.0391	0.0041	245.4767	0.1041	9.6086
35	28.1024	0.0356	0.0037	271.0244	0.1037	9.6442
40	45.2593	0.0221	0.0023	442.5926	0.1023	9.7791
45	72.8905	0.0137	0.0014	718.9048	0.1014	9.8628
50	117.3909	0.0085	0.0009	1163.9085	0.1009	9.9148
55	189.0591	0.0053	0.0005	1880.5914	0.1005	9.9471
60	304.4816	0.0033	0.0003	3034.8164	0.1003	9.9672
70	789.7470	0.0013	0.0001	7887.4696	0.1001	9.9873
80	2048.4002	0.0005	0.0000	20474.0021	0.1000	9.9951
90	5313.0226	0.0002	0.0000	53120.2261	0.1000	9.9981
100	13780.6123	0.0001	0.0000	137796.1234	0.1000	9.9993

15% DISCRETE COMPOUND INTEREST FACTORS

	Single payment		Uniform series			
n	Compound amount factor	Present worth factor	Sinking fund factor	Compound amount factor	Capital recovery factor	Present worth factor
	F/P	P/F	A/F	F/A	A/P	P/A
1	1.1500	0.8696	1.0000	1.0000	1.1500	0.8696
2	1.3225	0.7561	0.4651	2.1500	0.6151	1.6257
3	1.5209	0.6575	0.2880	3.4725	0.4380	2.2832
4	1.7490	0.5718	0.2003	4.9934	0.3503	2.8550
5	2.0114	0.4972	0.1483	6.7424	0.2983	3.3522
6	2.3131	0.4323	0.1142	8.7537	0.2642	3.7845
7	2.6600	0.3759	0.0904	11.0668	0.2404	4.1604
8	3.0590	0.3269	0.0729	13.7268	0.2229	4.4873
9	3.5179	0.2843	0.0596	16.7858	0.2096	4.7716
10	4.0456	0.2472	0.0493	20.3037	0.1993	5.0188
11	4.6524	0.2149	0.0411	24.3493	0.1911	5.2337
12	5.3503	0.1869	0.0345	29.0017	0.1845	5.4206
13	6.1528	0.1625	0.0291	34.3519	0.1791	5.5831
14	7.0757	0.1413	0.0247	40.5047	0.1747	5.7245
15	8.1371	0.1229	0.0210	47.5804	0.1710	5.8474
16	9.3576	0.1069	0.0179	55.7175	0.1679	5.9542
17	10.7613	0.0929	0.0154	65.0751	0.1654	6.0472
18	12.3755	0.0808	0.0132	75.8364	0.1632	6.1280
19	14.2318	0.0703	0.0113	88.2118	0.1613	6.1982
20	16.3665	0.0611	0.0098	102.4436	0.1598	6.2593
21	18.8215	0.0531	0.0084	118.8101	0.1584	6.3125
22	21.6447	0.0462	0.0073	137.6316	0.1573	6.3587
23	24.8915	0.0402	0.0063	159.2764	0.1563	6.3988
24	28.6252	0.0349	0.0054	184.1678	0.1554	6.4338
25	32.9190	0.0304	0.0047	212.7930	0.1547	6.4641
26	37.8568	0.0264	0.0041	245.7120	0.1541	6.4906
27	43.5353	0.0230	0.0035	283.5688	0.1535	6.5135
28	50.0656	0.0200	0.0031	327.1041	0.1531	6.5335
29	57.5755	0.0174	0.0027	377.1697	0.1527	6.5509
30	66.2118	0.0151	0.0023	434.7451	0.1523	6.5660
31	76.1435	0.0131	0.0020	500.9569	0.1520	6.5791
32	87.5651	0.0114	0.0017	577.1005	0.1517	6.5905
33	100.6998	0.0099	0.0015	664.6655	0.1515	6.6005
34	115.8048	0.0086	0.0013	765.3654	0.1513	6.6091
35	133.1755	0.0075	0.0011	881.1702	0.1511	6.6166
40	267.8635	0.0037	0.0006	1779.0903	0.1506	6.6418
45	538.7693	0.0019	0.0003	3585.1285	0.1503	6.6543
50	1083.6574	0.0009	0.0001	7217.7163	0.1501	6.6605
55	2179.6222	0.0005	0.0001	14524.1479	0.1501	6.6636
60	4383.9987	0.0002	0.0000	29219.9916	0.1500	6.6651

20% DISCRETE COMPOUND INTEREST FACTORS

	Single payment		Uniform series			
n	Compound amount factor	Present worth factor	Sinking fund factor	Compound amount factor	Capital recovery factor	Present worth factor
	F/P	*P/F*	*A/F*	*F/A*	*A/P*	*P/A*
1	1.2000	0.8333	1.0000	1.0000	1.2000	0.8333
2	1.4400	0.6944	0.4545	2.2000	0.6545	1.5278
3	1.7280	0.5787	0.2747	3.6400	0.4747	2.1065
4	2.0736	0.4823	0.1863	5.3680	0.3863	2.5887
5	2.4883	0.4019	0.1344	7.4416	0.3344	2.9906
6	2.9860	0.3349	0.1007	9.9299	0.3007	3.3255
7	3.5832	0.2791	0.0774	12.9159	0.2774	3.6046
8	4.2998	0.2326	0.0606	16.4991	0.2606	3.8372
9	5.1598	0.1938	0.0481	20.7989	0.2481	4.0310
10	6.1917	0.1615	0.0385	25.9587	0.2385	4.1925
11	7.4301	0.1346	0.0311	32.1504	0.2311	4.3271
12	8.9161	0.1122	0.0253	39.5805	0.2253	4.4392
13	10.6993	0.0935	0.0206	48.4966	0.2206	4.5327
14	12.8392	0.0779	0.0169	59.1959	0.2169	4.6106
15	15.4070	0.0649	0.0139	72.0351	0.2139	4.6755
16	18.4884	0.0541	0.0114	87.4421	0.2114	4.7296
17	22.1861	0.0451	0.0094	105.9306	0.2094	4.7746
18	26.6233	0.0376	0.0078	128.1167	0.2078	4.8122
19	31.9480	0.0313	0.0065	154.7400	0.2065	4.8435
20	38.3376	0.0261	0.0054	186.6880	0.2054	4.8696
21	46.0051	0.0217	0.0044	225.0256	0.2044	4.8913
22	55.2061	0.0181	0.0037	271.0307	0.2037	4.9094
23	66.2474	0.0151	0.0031	326.2369	0.2031	4.9245
24	79.4968	0.0126	0.0025	392.4842	0.2025	4.9371
25	95.3962	0.0105	0.0021	471.9811	0.2021	4.9476
26	114.4755	0.0087	0.0018	567.3773	0.2018	4.9563
27	137.3706	0.0073	0.0015	681.8528	0.2015	4.9636
28	164.8447	0.0061	0.0012	819.2233	0.2012	4.9697
29	197.8136	0.0051	0.0010	984.0680	0.2010	4.9747
30	237.3763	0.0042	0.0008	1181.8816	0.2008	4.9789
31	284.8516	0.0035	0.0007	1419.2579	0.2007	4.9824
32	341.8219	0.0029	0.0006	1704.1095	0.2006	4.9854
33	410.1863	0.0024	0.0005	2045.9314	0.2005	4.9878
34	492.2235	0.0020	0.0004	2456.1176	0.2004	4.9898
35	590.6682	0.0017	0.0003	2948.3411	0.2003	4.9915
40	1469.7716	0.0007	0.0001	7343.8578	0.2001	4.9966
45	3657.2620	0.0003	0.0001	18281.3099	0.2001	4.9986
50	9100.4382	0.0001	0.0000	45497.1908	0.2000	4.9995
55	22644.8023	0.0000	0.0000	113219.0113	0.2000	4.9998
60	56347.5144	0.0000	0.0000	281732.5718	0.2000	4.9999

Index